ANALOG ELECTRONIC CIRCUITS

Theory and Practical

ANALOG ELECTRONIC CIRCUITS
Theory and Practical

AUTHORS

Dr. Venu Sonti
Assistant Professor,
Department of Electrical Engineering,
National Institute of Technology,
Raipur, Chhattisgarh.

Dr. Sachin Jain
Professor,
Department of Electrical Engineering,
National Institute of Technology,
Raipur, Chhattisgarh.

REVIEWER

Dr. Dipankar Saha
Professor,
Department of Electrical Engineering,
Indian Institute of Technology, Bombay, Maharashtra.

KHANNA BOOK PUBLISHING CO. (P) Ltd.
4C/4344, Ansari Road, Darya Ganj,
New Delhi, India

All India Council for Technical Education
Nelson Mandela Marg, Vasant Kunj,
New Delhi, India

Dear Readers,

To prevent the piracy, this book is secured with HIGH SECURITY HOLOGRAM on the front title cover. In case you don't find the hologram on the front cover title, please write us to at contact@khannabooks.com or whatsapp us at +91-99109 09320 and avail special gift voucher for yourself.

Specimen of Hologram on front Cover title:

Moreover, there is a SPECIAL DISCOUNT COUPON for you with EVERY HOLOGRAM.

How to avail this SPECIAL DISCOUNT:

Step 1: Scratch the hologram
Step 2: Under the scratch area, your "coupon code" is available
Step 3: Logon to www.khannabooks.com
Step 4: Use your "coupon code" in the shopping cart and get your copy at a special discount
Step 5: Enjoy your reading!

Thanks,
Khanna Publishing House

ISBN: 978-93-5538-213-9

Book Code: UG417EN

ANALOG ELECTRONIC CIRCUITS
Theory and Practical
by **Dr. Venu Sonti, Dr. Sachin Jain**

First Edition: 2026

To view complete list of books, please scan the QR Code:

Published by:
Khanna Book Publishing Co. (P) Ltd.
CIN: U22110DL1998PTC095547

Visit us at: www.khannabooks.com
Write us at: contact@khannabooks.com

Printed in Bharat

प्रो. टी. जी. सीताराम
अध्यक्ष
Prof. T. G. Sitharam
Chairman

सत्यमेव जयते

आजादी का अमृत महोत्सव

अखिल भारतीय तकनीकी शिक्षा परिषद्
(भारत सरकार का एक सांविधिक निकाय)
(शिक्षा मंत्रालय, भारत सरकार)
नेलसन मंडेला मार्ग, वसंत कुंज, नई दिल्ली-110070
दूरभाष : 011-26131498
ई-मेल : chairman@aicte-india.org

ALL INDIA COUNCIL FOR TECHNICAL EDUCATION
(A STATUTORY BODY OF THE GOVT. OF INDIA)
(Ministry of Education, Govt. of India)
Nelson Mandela Marg, Vasant Kunj, New Delhi-110070
Phone : 011-26131498
E-mail : chairman@aicte-india.org

FOREWORD

Engineers are the backbone of any modern society. They are the ones responsible for the marvels as well as the improved quality of life across the world. Engineers have driven humanity towards greater heights in a more evolved and unprecedented manner.

The All India Council for Technical Education (AICTE), have spared no efforts towards the strengthening of the technical education in the country. AICTE is always committed towards promoting quality Technical Education to make India a modern developed nation emphasizing on the overall welfare of mankind.

An array of initiatives has been taken by AICTE in last decade which have been accelerated now by the National Education Policy (NEP) 2020. The implementation of NEP under the visionary leadership of Hon'ble Prime Minister of India envisages the provision for education in regional languages to all, thereby ensuring that every graduate becomes competent enough and is in a position to contribute towards the national growth and development through innovation & entrepreneurship.

One of the spheres where AICTE had been relentlessly working since past couple of years is providing high quality original technical contents at Under Graduate & Diploma level prepared and translated by eminent educators in various Indian languages to its aspirants. For students pursuing 2nd year of their Engineering education, AICTE has identified 88 books, which shall be translated into 12 Indian languages - Hindi, Tamil, Gujarati, Odia, Bengali, Kannada, Urdu, Punjabi, Telugu, Marathi, Assamese & Malayalam. In addition to the English medium, books in different Indian Languages are going to support the students to understand the concepts in their respective mother tongue.

On behalf of AICTE, I express sincere gratitude to all distinguished authors, reviewers and translators from the renowned institutions of high repute for their admirable contribution in a record span of time.

AICTE is confident that these outcomes based original contents shall help aspirants to master the subject with comprehension and greater ease.

(Prof. T. G. Sitharam)

Preface

The book titled "Analog Electronic Circuits: Theory and Practical" is an outcome of the rich experience of our teaching of analog circuit courses. The initiation of writing this book is to expose the basic of the analog electronic circuit to the engineering student. The book covers all the basic electronic devices typically employed in electronic circuits starting from a simple diode to complex op-amps devices. The book also presents the fundamental aspects of solving analog circuits using simple circuit analysis techniques. Keeping in mind the purpose of wide coverage as well as to provide essential supplementary information, we have included the topics recommended by AICTE, in a very systematic and orderly manner throughout the book. Efforts have been made to explain the fundamental concepts of the subject in the simplest possible way.

During the process of preparation of the book, we have considered simple circuitry approach for solving various analog circuits. The books have the complex derivation of the analog circuitry using the basic equivalent circuit of the devices. The book starts with the elementary diode, and its frequently employed application in rectifier, clipper and clamper circuit. The chapter presents the detailed steps in solving the analog circuits employing the diodes using basic circuitry analysis which can be easily grasped by the students. Complex analysis using the exact equivalent model for the diode was avoided. While preparing the different sections emphasis has also been laid on making the analysis simple and accurate. Students will be able to solve any complex analog circuitry using diode based on the steps given in the chapter. The same is applicable to the other chapters of BJTs, MOSFETs and op-amps. Similarly, book also covers different applications of the electronic devices in various signal conditioning circuits. The chapters on the linear and non-linear applications of the op-amps cover complete design details and illustrations. The analysis given is using equivalent electrical circuit model of the op-amp. The usage of equivalent electrical circuit model of the op-amp makes the complex op-amp circuitry very simple.

Thus, the book covers all types of medium and advanced level derivations for the given analog circuitry. Using the given book, the engineering students will be able to design various analog circuits using semiconductor devices like diodes, BJTs, MOSFETs, op-amps etc. Using this book and knowledge of analog electronics students will be able to tackle 21st century and onward engineering challenges and address the related aroused questions. The subject matters are presented in a constructive manner so that an engineering degree prepares students to work in different sectors or in national laboratories at the very forefront of technology.

We sincerely hope that the book will inspire the students to learn and discuss the ideas behind basic principles analog electronics and will surely contribute to the development of a solid foundation of the subject. We would be thankful to all beneficial comments and suggestions which will contribute to the improvement of the future editions of the book. It gives us immense pleasure to place this book in the hands of the teachers and students. It was indeed a big pleasure to work on different aspects covering in the book.

Dr. Venu Sonti
Dr. Sachin Jain

Acknowledgement

The authors are grateful to the authorities of AICTE, particularly Prof. T. G. Sitharam, Chairman; Dr. Abhay Jere, Vice-Chairman; Prof. Rajive Kumar, Member-Secretary; and Dr. Sunil Luthra, Director, Training and Learning Bureau for their planning to publish the books on Analog Electronic Circuits: Theory and Practical. We sincerely acknowledge the valuabl contributions of the reviewer of the book Prof. Dipankar Saha, Professor, Department of Electrical Engineering, Indian Institute of Technology Bombay for his constant encouragement, suggestions, support and cooperation.

I would like to extend my heartfelt gratitude to all those who played a direct and indirect role in helping me complete this book. My deepest and most sincere thanks go to my parents S.S. Lakshmi Narayana and M. Ratna Manikyam, whose unwavering prayers, constant encouragement, and blessings have been my guiding light throughout this journey. I consider myself truly fortunate to have the unwavering support of my wife, Navya Sree. Her unwavering support not only lightened my load but also allowed me to dedicate more time for completing this book. I would like to express my profound gratitude and thanks to my supervisor and co-author of this book, Dr. Sachin Jain, for affording me the invaluable opportunity to work alongside him. Without his support, this journey would have remained an unattainable dream. Through his insightful discussions and resourceful ideas, I was able to delve into various facets of analog electronics and seamlessly incorporate these topics into the book. With the utmost sincerity, I wish to convey my heartfelt gratitude to Dr. Narendra D. Londhe, Professor in the Department of Electrical Engineering, for believing in me and for suggesting me to write this book. I extend my heartfelt thanks to Dr. Anamika Yadav, Professor and HOD, Electrical Engineering Department and administration of the National Institute of Technology, Raipur, for their invaluable support and for creating an environment that facilitated the writing of this book.

This book is an outcome of various suggestions of AICTE members, experts and authors who shared their opinion and thought to further develop the engineering education in our country. Acknowledgements are due to the contributors and different workers in this field whose published books, review articles, papers, photographs, footnotes, references and other valuable information enriched us at the time of writing the book.

Dr. Venu Sonti
Dr. Sachin Jain

Outcome Based Education

For the implementation of an outcome based education the first requirement is to develop an outcome based curriculum and incorporate an outcome based assessment in the education system. By going through outcome based assessments evaluators will be able to evaluate whether the students have achieved the outlined standard, specific and measurable outcomes. With the proper incorporation of outcome based education there will be a definite commitment to achieve a minimum standard for all learners without giving up at any level. At the end of the programme running with the aid of outcome based education, a student will be able to arrive at the following outcomes:

PO1. **Engineering knowledge**: Apply the knowledge of mathematics, science, engineering fundamentals, and an engineering specialization to the solution of complex engineering problems.

PO2. **Problem analysis**: Identify, formulate, review research literature, and analyze complex engineering problems reaching substantiated conclusions using first principles of mathematics, natural sciences, and engineering sciences.

PO3. **Design / development of solutions**: Design solutions for complex engineering problems and design system components or processes that meet the specified needs with appropriate consideration for the public health and safety, and the cultural, societal, and environmental considerations.

PO4. **Conduct investigations of complex problems**: Use research-based knowledge and research methods including design of experiments, analysis and interpretation of data, and synthesis of the information to provide valid conclusions.

PO5. **Modern tool usage**: Create, select, and apply appropriate techniques, resources, and modern engineering and IT tools including prediction and modeling to complex engineering activities with an understanding of the limitations.

PO6. **The engineer and society**: Apply reasoning informed by the contextual knowledge to assess societal, health, safety, legal and cultural issues and the consequent responsibilities relevant to the professional engineering practice.

PO7. **Environment and sustainability**: Understand the impact of the professional engineering solutions in societal and environmental contexts, and demonstrate the knowledge of, and need for sustainable development.

PO8. **Ethics**: Apply ethical principles and commit to professional ethics and responsibilities and norms of the engineering practice.

PO9. **Individual and team work**: Function effectively as an individual, and as a member or leader in diverse teams, and in multidisciplinary settings.

PO10. **Communication**: Communicate effectively on complex engineering activities with the engineering community and with society at large, such as, being able to comprehend and write effective reports and design documentation, make effective presentations, and give and receive clear instructions.

PO11. Project management and finance: Demonstrate knowledge and understanding of the engineering and management principles and apply these to one's own work, as a member and leader in a team, to manage projects and in multidisciplinary environments.

PO12. Life-long learning: Recognize the need for, and have the preparation and ability to engage in independent and life-long learning in the broadest context of technological change.

Course Outcomes

After completion of the course, the students will be able to:

CO-1: Analyze and evaluate the analog electronic circuits with diodes, BJTs, MOSFETs or op-amps.

CO-2: Analyze and design the analog electronic circuits for various signal conditioning like clipping, clamping and regulation and waveform generation like square wave, sine wave triangular wave etc.

CO-3: Analyze and design the analog electronic circuits for given gain and attenuations of input signal.

CO-4: Analyze and design various amplifier circuits like differential amplifiers and power amplifiers.

CO-5: Analyze and design the analog electronic circuits having op-amps for their linear and non-linear application

Course Outcomes	Expected Mapping with Programme Outcomes (1- Weak Correlation; 2- Medium correlation; 3- Strong Correlation)											
	PO-1	PO-2	PO-3	PO-4	PO-5	PO-6	PO-7	PO-8	PO-9	PO-10	PO-11	PO-12
CO-1	3	3	3	3	1	-	-	-	-	-	-	-
CO-2	3	3	3	2	1	-	-	-	-	-	-	-
CO-3	3	3	3	1	1	-	-	-	-	-	-	-
CO-4	3	3	3	1	1	-	-	-	-	-	-	-
CO-5	3	3	3	1	1	-	-	-	-	-	-	-

Symbols and Abbreviations

LIST OF ABBREVIATIONS

General Terms			
Abbreviations	**Full form**	**Abbreviations**	**Full form**
ADC	Analog to Digital Converter	MOSFET	Metal Oxide Semiconductor Field Effect Transistor
BJT	Bipolar Junction Transistor	Op-amp	Operational Amplifier
CMRR	Common Mode Rejection Ratio	PSSR	Power Supply Rejection Ratio

List of Symbols

Symbols	Description	Symbols	Description
q	Charge of an electron	h_{oE}	Output admittance of the transistor in common emitter configuration
α_d	Ideality factor the diode	h_{iC}	Input impedance of the transistor in common collector configuration
K_B	Boltzmann constant	h_{rC}	Reverse voltage gain of the transistor in common collector configuration
I_O	Reverse saturation current of diode	h_{rC}	Reverse voltage gain of the transistor in common emitter configuration
V_T	Thermal voltage drop of the diode	h_{fC}	Forward current gain of the transistor in common emitter configuration
α	Current gain in common base configuration	R_I	Input impedance of the amplifier
β	Current gain in common emitter configuration	A_V	Voltage gain of the amplifier
γ	Current gain in common collectorconfiguration	Y_O	Output admittance of the amplifier
h_{iB}	Input impedance of the transistor in common base configuration	A_I	Current gain of the amplifier

h_{rB}	Reverse voltage gain of the transistor in common base configuration	g_m	Transconductance
h_{fB}	Forward current gain of the transistor in common base configuration	A_{OL}	Open loop gain of the op-amp
h_{oB}	Output admittance of the transistor in common base configuration	R_{in}	Input resistance of the op-amp
h_{iE}	Input impedance of the transistor in common emitter configuration	R_o	Output resistance of the op-amp
h_{rE}	Reverse voltage gain of the transistor in common emitter configuration	A_d	Differential gain
h_{fE}	Forward current gain of the transistor in common emitter configuration	A_c	commonmode gain

Guidelines for Teachers

To implement Outcome Based Education (OBE) knowledge level and skill set of the students should be enhanced. Teachers should take a major responsibility for the proper implementation of OBE. Some of the responsibilities (not limited to) for the teachers in OBE system may be as follows:

- Within reasonable constraint, they should manoeuvre time to the best advantage of all students.
- They should assess the students only upon certain defined criterion without considering any other potential ineligibility to discriminate them.
- They should try to grow the learning abilities of the students to a certain level before they leave the institute.
- They should try to ensure that all the students are equipped with the quality knowledge as well as competence after they finish their education.
- They should always encourage the students to develop their ultimate performance capabilities.
- They should facilitate and encourage group work and team work to consolidate newer approach.
- They should follow Blooms taxonomy in every part of the assessment.

Bloom's Taxonomy

Level	Teacher should Check	Student should be able to	Possible Mode of Assessment
Create	Students ability to create	Design or Create	Mini project
Evaluate	Students ability to justify	Argue or Defend	Assignment
Analyse	Students ability to distinguish	Differentiate or Distinguish	Project/Lab Methodology
Apply	Students ability to use information	Operate or Demonstrate	Technical Presentation/ Demonstration
Understand	Students ability to explain the ideas	Explain or Classify	Presentation/ Seminar
Remember	Students ability to recall (or remember)	Define or Recall	Quiz

Guidelines for Students

Students should take equal responsibility for implementing the OBE. Some of the responsibilities (not limited to) for the students in OBE system are as follows:

- Students should be well aware of each UO before the start of a unit in each and every course.
- Students should be well aware of each CO before the start of the course.
- Students should be well aware of each PO before the start of the programme.
- Students should think critically and reasonably with proper reflection and action.
- Learning of the students should be connected and integrated with practical and real life consequences.
- Students should be well aware of their competency at every level of OBE.

List of Figures

Unit 1: PN Junction Diode

Fig. 1.1.1	PN junction diode.	2
Fig. 1.1.2	Symbol of PN junction diode.	2
Fig. 1.2.1	Diffusion in PN junction diode.	3
Fig. 1.2.2	Operation of unbiased PN junction diode.	3
Fig. 1.3.1	Forward-biased and reverse-biased PN junction diode.	4
Fig. 1.3.1.1	Operation of PN junction diode in the forward-biased condition.	5
Fig. 1.3.2.1	Operation of PN junction diode in the reverse-biased condition.	5
Fig. 1.4.1	Operation of the ideal PN junction diode in the forward and reverse biased conditions.	6
Fig. 1.4.2	I-V characteristics of the ideal PN junction diode.	6
Fig. 1.6.1	PN junction diode circuit with a series resistive load and input source of V_{in}.	10
Fig. 1.6.2	Load line and I-V characteristics of the PN junction diode.	11
Fig. 1.6.1.1	Piecewise linearization of the PN junction diode characteristics.	11
Fig. 1.6.1.2	Electrical equivalent circuit model of the PN junction diode.	12
Fig. 1.6.1.3	Operating characteristics of the PN junction diode with a time-varying voltage source	13
Fig. 1.8.1	Operation of the PN junction diode for an AC input signal.	14
Fig. 1.8.1.1	Circuit diagram of a half wave rectifier connected to a resistive load R_L.	15
Fig. 1.8.1.2	Operation of a half wave rectifier connected to a resistive load R_L.	16
Fig. 1.8.1.3	Waveforms of a half wave rectifier connected to a resistive load R_L.	16
Fig. 1.8.2.1	Circuit diagram of a full wave rectifier connected to a resistive load R_L.	18
Fig. 1.8.2.2	Operation of a full wave rectifier connected to a resistive load RL.	18
Fig. 1.8.2.3	Waveforms of a full wave rectifier connected to a resistive load RL.	19
Fig. 1.8.2.4	Circuit diagram of a full wave bridge rectifier connected to a resistive load RL.	21
Fig. 1.8.2.5	Operation of a full wave bridge rectifier connected to a resistive load RL.	21
Fig. 1.9.1	Symbol of zener diode.	22
Fig. 1.9.2	I-V characteristics of the zener diode.	23
Fig. 1.9.3	Circuit schematic for the typical zener diode used for voltage protection.	24
Fig. 1.9.4	Circuit schematic for the typical zener diode used for voltage protection with variable input voltage.	25

Fig. 1.9.5	Circuit schematic for the typical zener diode used for voltage protection with variable load resistance.	26
Fig. 1.10.1.1	Circuit of a series type clipper.	29
Fig. 1.10.1.2	Circuit of a series type clipper by replacing the PN junction diode by open circuit (Step-I).	29
Fig. 1.10.1.3	Circuit of a series type clipper by applying Thevenin voltage across the PN junction diode (Step-II).	29
Fig. 1.10.1.4	Series type clipper in which the PN junction diode is replaced the closed switch (Step-IV).	30
Fig. 1.10.1.5	Series type clipper in which the PN junction diode is replaced open switch (Step-V).	30
Fig. 1.10.1.6	Waveforms of the series type clipper.	31
Fig. 1.10.2.1	Circuit of a parallel type clipper.	31
Fig. 1.10.2.2	Circuit of a parallel type clipper used in Step-I.	32
Fig. 1.10.2.3	Circuit of a parallel type clipper used in Step-II.	32
Fig. 1.10.2.4	Circuit of a parallel type clipper used in Step-III.	33
Fig. 1.10.2.5	Circuit of a parallel type clipper used in Step-IV.	33
Fig. 1.10.2.6	Waveforms of the parallel type clipper.	34
Fig. 1.11.1.1	Circuit of a positive clamper circuit.	34
Fig. 1.11.1.2	Waveforms of the positive clamper circuit.	35
Fig. 1.11.2.1	Circuit of a negative clamper circuit.	35
Fig. 1.11.2.2	Waveforms of the negative clamper circuit.	35

Unit 2: Bipolar Junction Transistor

Fig. 2.1.1	Construction of BJT.	41
Fig. 2.1.2	Construction of N-P-N type and P-N-P type transistors.	41
Fig. 2.1.3	Symbol of N-P-N type and P-N-P type transistors.	41
Fig. 2.2.1	N-P-N type and P-N-P type transistors at equilibrium condition.	43
Fig. 2.2.1.1	Biasing for the transistor to operate in the active region.	43
Fig. 2.2.1.2	Biasing for the transistor to operate in the saturation region.	44
Fig. 2.2.1.3	Biasing for the transistor to operate in the cut-off region.	44
Fig. 2.2.2.1.1	Operation of N-P-N transistor in the active region.	45
Fig. 2.3.1.1	Common base, common emitter and common collector configurations of the transistor.	46
Fig. 2.3.1.1.1	Common base configuration of the transistor.	46
Fig. 2.3.1.1.2	Input characteristics of common base configuration.	47
Fig. 2.3.1.1.3	Output characteristics of common base configuration.	48
Fig. 2.3.1.2.1	Common emitter configuration of the transistor.	48

Fig. 2.3.1.2.2	Input characteristics of common emitter configuration.	50
Fig. 2.3.1.2.3	Output characteristics of common emitter configuration.	51
Fig. 2.3.1.3.1	Common collector configuration of the transistor.	51
Fig. 2.3.1.3.2	Input characteristics of common collector configuration.	52
Fig. 2.3.1.3.3	Output characteristics of common collector configuration.	53
Fig. 2.5.1	Circuit considered for explaining the Q-point of transistor.	54
Fig. 2.5.2	DC load line and output characteristics of the common emitter configuration.	54
Fig. 2.6.1	Q-point is near to the saturation region.	55
Fig. 2.6.2	Q-point is at the center of the DC load line.	56
Fig. 2.6.3	Q-point is near to the cut-off region.	56
Fig. 2.8.1.1	Fixed bias circuit.	58
Fig. 2.8.2.1	Collector-to-base bias circuit.	59
Fig. 2.8.3.1	Voltage divider bias circuit.	61
Fig. 2.8.3.2	Thevenin's equivalent circuit for the voltage divider bias circuit.	61
Fig. 2.9.1.1	BJT as an open switch.	63
Fig. 2.9.2.1	BJT as a closed switch.	64
Fig. 2.10.1.1	Common base amplifier	64
Fig. 2.10.2.1	Common emitter amplifier	65
Fig. 2.11.1	h-parameter equivalent circuit of the transistor	67
Fig. 2.11.2	h-parameter equivalent circuit of the transistor in the common base configuration.	68
Fig. 2.11.3	h-parameter equivalent circuit of the transistor in the common emitter configuration.	68
Fig. 2.11.4	h-parameter equivalent circuit of the transistor in the common collector configuration.	69
Fig. 2.12.1	CE amplifier.	70
Fig. 2.12.2	CE amplifier replacing all the coupling capacitors and the bypass capacitors with a short circuit.	70
Fig. 2.12.3	Replacing the external dc power supplies used for the biasing with a short circuit.	71
Fig. 2.12.4	Replacing the transistor with its h-parameter model.	71
Fig. 2.13.1	Simplified h-parameter equivalent circuit of the transistor.	74
Fig. 2.14.1	Current mirror circuit schematic for the using transistor.	74
Fig. 2.15.1	Frequency response for (a) RC coupled; (b) Transformer coupled; (c) Direct coupled amplifier	77
Fig. 2.15.2	Parasitic capacitance within the transistor device.	78

Fig. 2.15.3	Circuit schematic showing low frequency response analysis at input side for resistor divider transistor circuit.	79
Fig. 2.15.4	Circuit schematic for Miller capacitance derivation.	81
Fig. 2.14.5	Circuit schematic showing high frequency response analysis at input side for resistor divider transistor circuit.	83

Unit 3: MOSFET

Fig. 3.1.1	Construction of N-channel MOSFET's.	88
Fig. 3.1.2	Symbol of (a) N-P-N depletion type; (b) P-N-P depletion type; (c) N-P-N enhancement type, and (d) P-N- enhancement type MOSFET's.	89
Fig. 3.2.1	Depletion mode MOSFET's circuit with (a) no biasing, (b) biasing with variable drain to source voltage V_{DD} and shorted gate to source terminal, and (c) biasing with variable drain to source voltage V_{DD} and gate to source voltage V_{GS} .	89
Fig. 3.2.2	Transfer and drain characteristics of the MOSFET.	91
Fig. 3.2.3	MOSFET's resistance between drain and source terminal for, (a) ohmic region operating; (b) cut-OFF region operating, and (c) saturation region operating.	92
Fig. 3.3.1	Common gate, common source and common drain configurations of the MOSFET.	92
Fig. 3.4.1	Graphical computation of transconductance 'g_m' for the MOSFET	94
Fig. 3.4.2	Analytical computation of transconductance 'g_m' for the MOSFET and its variation with gate to source voltage V_{GS}	95
Fig. 3.4.3	Drain characteristics for the for ideal and practical MOSFET operating in saturation region.	97
Fig. 3.4.4	AC equivalent circuit for the three terminal MOSFET device operating in ohmic region.	97
Fig. 3.6.1	The circuit diagram for the fixed biased circuit using depletion MOSFET.	99
Fig. 3.6.2	The equivalent circuit diagram for DC analysis of the fixed biased circuit using DEPLETION MOSFET.	99
Fig. 3.6.3	The equivalent circuit diagram for AC analysis of the fixed biased circuit using depletion MOSFET.	100
Fig. 3.7.1	The circuit diagram for the self-biased circuit using depletion MOSFET.	101
Fig. 3.7.2	The equivalent circuit diagram for DC analysis of the self biased circuit using depletion MOSFET.	101
Fig. 3.7.3	The equivalent circuit diagram for AC analysis of the self biased circuit using depletion MOSFET.	103
Fig. 3.8.1	The circuit diagram for the voltage divider biased circuit using depletion MOSFET.	104
Fig. 3.8.2	The equivalent circuit diagram for DC analysis of the voltage divider biased circuit using depletion MOSFET.	104

Fig. 3.8.3	The equivalent circuit diagram for AC analysis of the voltage divider biased circuit using depletion MOSFET.	105
Fig. 3.9.1	The circuit diagram for the common gate circuit using depletion MOSFET.	106
Fig. 3.9.2	The equivalent circuit diagram for DC analysis of the common gate circuit using depletion MOSFET.	107
Fig. 3.9.3	The equivalent circuit diagram for AC analysis of the common gate circuit using depletion MOSFET.	108
Fig. 3.9.4	The AC equivalent circuit diagram for determining input impedance zi for the common gate circuit using depletion MOSFET.	108
Fig. 3.9.5	The AC equivalent circuit diagram for determining output impedance zo for the common gate circuit using depletion MOSFET.	110
Fig. 3.10.1	The circuit diagram for the common drain circuit using depletion MOSFET.	111
Fig. 3.10.2	The equivalent circuit diagram for DC analysis of the common drain circuit using depletion MOSFET.	112
Fig. 3.10.3	The equivalent circuit diagram for AC analysis of the common drain circuit using depletion MOSFET.	113
Fig. 3.10.4	The AC equivalent circuit diagram for determining input impedance z_i for the common drain circuit using depletion MOSFET.	113
Fig. 3.10.5	The AC equivalent circuit diagram for determining output impedance z_o for the common drain circuit using depletion MOSFET.	115

Unit 4: Power Amplifier and Differential Amplifier

Fig. 4.2.1.1	Q-point for class A amplifier.	122
Fig. 4.2.2.1	Q-point for class B amplifier.	122
Fig. 4.2.3.1	Q-point for class C amplifier.	123
Fig. 4.2.4.1	Q-point for class AB amplifier.	124
Fig. 4.3.1	Circuit schematic of differential amplifier using CE transistor.	125
Fig. 4.3.2	Equivalent circuit schematic of differential amplifier for DC analysis.	126
Fig. 4.3.3	Equivalent circuit schematic of differential amplifier for AC analysis.	126

Unit 5: Operational Amplifiers

Fig. 5.2.1	Symbol of op-amp.	131
Fig. 5.2.2.1	Various packages of the op-amp.	132
Fig. 5.2.3.1	Power supply connections of op-amp.	133
Fig. 5.2.3.2	Balanced and unbalanced power supply connections of the op-amp.	133
Fig. 5.2.3.3	Circuits for obtaining the balanced power supply for the op-amp from the single power supply.	134
Fig. 5.3.1	Pin configuration of IC 741 op-amp.	134
Fig. 5.4.1	Block diagram of op-amp.	135

Fig. 5.5.1	Equivalent circuit of the ideal op-amp.	136
Fig. 5.6.1	Transfer curve of the ideal op-amp.	137
Fig. 5.7.1	Transfer curve of the practical op-amp.	138
Fig. 5.8.1	Equivalent circuit of the practical op-amp.	139
Fig. 5.8.2	Op-amp input bias currents.	139
Fig. 5.9.1	Op-amp with negative feedback.	141

Unit 6: Linear Applications of Operational Amplifiers

Fig. 6.1.1	Op-amp (a) symbol, (b) practical equivalent circuit schematic, and (c) ideal equivalent circuit schematic	147
Fig. 6.1.2	Inverting op-amp (a) circuit schematic, (b) practical equivalent circuit schematic for inverting op-amp, and (c) ideal equivalent circuit schematic for inverting op-amp.	147
Fig. 6.2.1	Inverting op-amp (a) circuit schematic, (b) ideal equivalent circuit schematic for inverting op-amp.	148
Fig. 6.3.1	Non-inverting op-amp (a) circuit schematic, (b) ideal equivalent circuit schematic for non-inverting op-amp.	149
Fig. 6.4.1	Voltage follower using op-amp (a) circuit schematic, (b) ideal equivalent circuit schematic.	150
Fig. 6.5.1	Summer circuits using op-amp (a) circuit schematic, (b) ideal equivalent circuit schematic.	151
Fig. 6.6.1	Integrator circuit using op-amp (a) circuit schematic, (b) ideal equivalent circuit schematic.	152
Fig. 6.7.1	Differentiator circuit using op-amp (a) circuit schematic, (b) ideal equivalent circuit schematic.	153
Fig. 6.8.1	Instrumentation amplifier circuit using op-amp (a) circuit schematic, (b) ideal equivalent circuit schematic.	154
Fig. 6.9.1	Active low pass filter circuit using op-amp (a) circuit schematic, (b) ideal equivalent circuit schematic.	156
Fig. 6.9.2	Plot of frequency versus gain for low pass filter circuit.	156
Fig. 6.10.1	Active high pass filter circuit using op-amp (a) circuit schematic, (b) ideal equivalent circuit schematic.	158
Fig. 6.10.2	Plot of frequency versus gain for high pass filter circuit.	159
Fig. 6.11.1	Plot of frequency versus gain for band pass filter circuit.	160
Fig. 6.11.2	Active band pass filter circuit using op-amp (a) circuit schematic, (b) ideal equivalent circuit schematic.	161

Fig. 6.12.1	Proportional controller circuit using op-amp (a) circuit schematic, (b) ideal equivalent circuit schematic.	162
Fig. 6.13.1	Proportional integral controller circuit using op-amp (a) circuit schematic, (b) ideal equivalent circuit schematic.	163
Fig. 6.15.1	Proportional integral derivative controller circuit using op-amp (a) circuit schematic, (b) ideal equivalent circuit schematic.	165
Fig. 6.15.2	Ladder network using resistor network for converting digital value to its analog equivalent.	166
Fig. 6.15.3	Typical ADC IC using op-amp.	166
Fig. 6.15.4	Ladder output voltage corresponding to digital input from counter.	167
Fig. 6.16.1	Voltage regulator circuit using op-amp.	167
Fig. 6.17.1	Phase shift oscillator circuit using op-amp (a) circuit schematic, (b) ideal equivalent circuit schematic.	168
Fig. 6.18.1	Wien bridge oscillator circuit using op-amp.	169
Fig. 6.19.1	Lag lead compensator circuit using op-amp (a) circuit schematic, (b) ideal equivalent circuit schematic.	170

Unit 7: Nonlinear Applications of Operational Amplifiers

Fig. 7.2.1.1	Circuit of the noninverting hysteresis comparator.	176
Fig. 7.2.1.2	Input and output waveforms of noninverting hysteresis comparator.	177
Fig. 7.2.1.3	Transfer characteristics of noninverting hysteresis comparator.	177
Fig. 7.2.2.1	Circuit of the inverting hysteresis comparator.	178
Fig. 7.2.2.2	Input and output waveforms of inverting hysteresis comparator.	179
Fig. 7.2.1.3	Transfer characteristics of inverting hysteresis comparator.	179
Fig. 7.3.1.1	Circuit of the noninverting zero crossing detector.	180
Fig. 7.3.1.2	Input and output waveforms of noninverting zero crossing detector.	180
Fig. 7.3.2.1	Circuit of the noninverting zero crossing detector.	181
Fig. 7.4.1.1	Circuit of the square-wave generator.	182
Fig. 7.4.1.2	Operation of square wave generator when $V_O = +V_{sat}$.	182
Fig. 7.4.1.3	Operation of square wave generator when $V_O = -V_{sat}$.	183
Fig. 7.4.1.4	Waveforms of the output voltage V_O and waveform of the voltage across the capacitor V_C in the square-wave generator.	183
Fig. 7.4.2.1	Circuit of the triangular-wave generator.	184

Fig. 7.4.2.2	Waveforms of triangular-wave generator.	184
Fig. 7.5.1.1	Circuit schematic of the half wave precision rectifier.	185
Fig. 7.5.1.2	Circuit schematic of the half wave precision rectifier during (a) positive half cycle of ac input voltage; (b) negative half cycle of ac input voltage.	186
Fig. 7.5.1.3	Circuit schematic of the half wave precision rectifier with zero output impedance.	186
Fig. 7.5.2.1	Circuit schematic of the full wave precision rectifier.	187
Fig. 7.5.2.2	Circuit schematic of the full wave precision rectifier during (a) positive half cycle of ac input voltage; (b) negative half cycle of ac input voltage.	187
Fig. 7.6.1	Basic positive peak detector.	188
Fig. 7.6.2	Waveforms of basic positive peak detector.	188
Fig. 7.7.1	Monostable multivibrator.	189
Fig. 7.7.2	Waveforms of monostable multivibrator.	190

Contents

Foreword v
Preface vii
Acknowledgement ix
Outcome Based Education xi
Course Outcomes xiii
Abbreviations and Symbols xv
Guidelines for Teachers xvii
Guidelines for Students xix
List of Figures xxi

Unit 1: PN Junction Diode* *1-38

Unit specifics 1
Rationale 1
Pre-requisites 1
Unit outcomes 2
1.1 PN junction diode 2
1.2 Operation of an unbiased PN junction diode 3
1.3 Operation of a biased PN junction diode 4
1.3.1 Operation of PN junction diode in the forward-biased condition 4
1.3.2 Operation of PN junction diode in the reverse-biased condition 5
1.4 I-V characteristics of a PN junction diode 6
1.5 I-V characteristics equation of a PN junction diode 7
1.6 Operating point calculation in PN junction diode 9
1.6.1 Piecewise linearization of the PN junction diode characteristics 11
1.7 Application of the PN junction diodes 14
1.8 Rectifier 14
1.8.1 Half wave rectifier 14
1.8.2 Full wave rectifier 17
1.9 Zener diode 22
1.10 Clipping Circuit 28
1.10.1 Series type clipper circuit 28
1.10.2 Parallel type clipper circuit 31

1.11 Clamping circuit 34
1.11.1 Positive clamper circuit 34
1.11.2 Negative clamper circuit 35
Unitg Summary 36
Exercises 36
Know More 38
References 38

Unit 2: Bipolar Junction Transistor 39-85

Unit specifics 39
Rationale 39
Pre-requisites 40
Unit outcomes 40
2.1 Introduction to the BJT 40
2.2 Operation of the transistor 42
2.2.1 Operation of the biased transistor 42
2.2.1.1 Operation of the transistor in the active region 44
2.3 Various configurations of the transistor 45
2.3.1 Common base configuration 45
2.3.2 Common emitter configuration 48
2.3.3 Common collector configuration 53
2.4 Biasing of the transistor 53
2.5 Quiescent point (Q-Point) of the transistor 53
2.6 Selection of quiescent point (Q-Point) of the transistor 55
2.7 Biasing stabilization 57
2.8 Biasing circuits for the transistors 57
2.8.1 Fixed bias circuit 57
2.8.2 Collector-to-base bias circuit 59
2.8.3 Voltage divider bias circuit 60
2.9 BJT as a switch 63
2.9.1 BJT as an open switch 63
2.9.2 BJT as a closed switch 63
2.10 BJT as an amplifier 64
2.10.1 Common base amplifier 64
2.10.2 Common emitter amplifier 65

2.10.3 Common collector amplifier ... 66
2.11 Hybrid parameter model of the transistor ... 66
2.12 Small signal analysis of the amplifier using the h-parameters... 69
2.13 Small signal analysis of the amplifier using the simplified h-parameters... 73
2.14 Current Mirror ... 74
2.15 High-frequency response of the amplifiers... 75
Unit Summary... 83
Exercises... 84
Know More... 85
References... 85

Unit 3: MOSFET 86-119

Unit specifics ... 86
Rationale ... 86
Pre-requisites... 87
Unit outcomes ... 87
3.1 Introduction to the MOSFET's ... 87
3.2 Depletion mode MOSFET transfer and drain characteristics... 89
3.3 Various configurations of the MOSFET ... 92
3.4 Analysis of MOSFET's in ohmic region... 92
3.5 Biasing of the MOSFET's... 97
3.6 Fixed biased circuit... 98
3.7 Self biased circuit... 100
3.8 Voltage divider circuit ... 103
3.9 Common gate circuit ... 106
3.10 Common drain circuit ... 111
Unit Summary... 116
Exercises... 117
Know More... 119
References... 119

Unit 4: Power Amplifier and Differential Amplifier... 120-128

Unit specifics ... 120
Rationale ... 120
Pre-requisites... 120
Unit outcomes ... 120

4.1 Introduction to power amplifiers 121
4.2 Classification of power amplifiers 121
4.2.1 Class A power amplifiers 121
4.2.2 Class B power amplifiers 121
4.2.3 Class C power amplifiers 123
4.2.4 Class AB power amplifiers 123
4.3 Differential amplifiers 124
Unit Summary 127
Exercises 127
Know More 128
References 128

Unit 5: Operational Amplifiers 129-145

Unit specifics 129
Rationale 129
Pre-requisites 130
Unit outcomes 130
5.1 Introduction 130
5.2 Basics of the op-amp 130
5.2.1 Symbol of op-amp 130
5.2.2 Popular packages of op-amp 131
5.2.3 Power supply requirement for op-amp 132
5.3 IC 741 op-amp 134
5.4 Internal structure of an op-amp 135
5.4.1 Input stage 135
5.4.2 Intermediate stage 135
5.4.3 Level-shifting stage 135
5.4.4 Output stage 136
5.5 Ideal op-amp 136
5.6 Characteristics of ideal op-amp 137
5.7 Practical op-amp 138
5.8 Characteristics of practical op-amp 138
5.9 Negative feedback in op-amp 140
5.10 Procedure for analysing the op-amp circuits 143
Unit Summary 143

Exercises 143
Know More 144
References 144

Unit 6: Linear Applications of Operational Amplifiers 145-173

Unit specifics 145
Rationale 146
Pre-requisites 146
Unit outcomes 146
6.1 Assumptions for op-amp analysis 146
6.2 Op-amp as inverting amplifier 148
6.3 Non-inverting amplifier 149
6.4 Voltage follower (unity follower) 149
6.5 Summing amplifier circuit 150
6.6 Integrator circuit 151
6.7 Differentiator circuit 152
6.8 Instrumentation amplifier 153
6.9 Op-amp as low pass filter 155
6.10 Op-amp as high pass filter 157
6.11 Op-amp as band pass filter 160
6.12 Proportional controller 161
6.13 Proportional integral controller 162
6.14 Proportional integral derivative controller 164
6.15 Analog to digital converter 164
6.16 Voltage regulator 167
6.17 Phase shift oscillator 168
6.18 Wien bridge oscillator 169
6.19 Lag lead compensator 169
Unit Summary 171
Exercises 171
Know More 173
References 173

Unit 7: Nonlinear Applications of Operational Amplifiers 174- 191

Unit specifics 174
Rationale 174

Pre-requisites 174

Unit outcomes 174

7.1 Introduction 175

7.2 Hysteresis comparators 175

7.2.1 Noninverting hysteresis comparators 175

7.2.2 Inverting hysteresis comparators 177

7.3 Zero crossing detector 179

7.3.1 Noninverting zero crossing detector 180

7.3.2 Inverting zero crossing detector 180

7.4 Square-wave and triangular-wave generators 181

7.4.1 Square-wave generator 181

7.4.2 Triangular-wave generator 182

7.5 Precision rectifier 185

7.5.1 Half wave rectifier 185

7.5.2 Full wave rectifier 186

7.6 Peak detector 187

7.7 Monostable multivibrator 189

Unit Summary 190

Exercises 190

Know More 191

References 191

Appendices* *192

Index* *250

01 PN JUNCTION DIODE

UNIT SPECIFICS

Through this unit we have discussed the following aspects:

- *Introduction to PN junction diode;*
- *Operation of an unbiased PN junction diode;*
- *Operation of a biased PN junction diode;*
- *I-V characteristics of PN junction diode;*
- *I-V characteristics equation of a PN Junction diode;*
- *Operating point calculation in PN Junction diode;*
- *Application of the PN junction diodes;*
- *Rectifier;*
- *Zener diode;*
- *Clipping circuit;*
- *Clamping circuit;*

RATIONALE

This fundamental unit on the PN junction diode helps students to get a primary idea about the operation of the PN junction diode under both unbiased and biased operating conditions. Once the operation of the PN junction diode is clear, then the I-V characteristics of the PN junction diode are discussed to understand the electrical behaviour of the PN junction diode. Then the I-V characteristic equation of the PN junction diode is explained to determine the relationship between the current and voltage in the PN junction diode. This characteristics equation is necessary to identify the behaviour of the various operating conditions. All these concepts are required for obtaining the operating point of the PN junction diode. The operating point will always ensure that the diode always operates in stable and desired operating conditions. Next, the practical applications of the PN junction diode are discussed to identify its real-time applications like rectifiers, clipping circuits and clamping circuits. The unit also explains the construction, working and characteristics of the zener diode All these concepts are essential to in understanding the operation of various electronic circuits. The laboratory experiments related to the I-V characteristics of the PN junction diode, zener diode, clipping and clamping circuits, and design of the rectifiers were presented in the appendix of the book.

PRE-REQUISITES

Physics: Semiconductor (Class XII)

UNIT OUTCOMES

List of outcomes of this unit is as follows:

U1-O1: Understand the basics of PN junction diode

U1-O2: Identify the I-V characteristics of the PN junction diode

U1-O3: Identify the operating point of the PN junction diode

U1-O4: Realize the applications of the PN junction diode

U1-O5: Understand the basics of zener diode along with the I-V characteristics

Unit-1 Outcomes	EXPECTED MAPPING WITH COURSE OUTCOMES *(1- Weak Correlation; 2- Medium correlation; 3- Strong Correlation)*				
	CO-1	CO-2	CO-3	CO-4	CO-5
U1-O1	3	3	-	-	1
U1-O2	3	3	-	-	1
U1-O3	3	3	-	-	1
U1-O4	3	3	-	-	1
U1-O5	3	3	-	-	1

1.1 PN JUNCTION DIODE

A PN junction diode is a semiconductor device that is formed by a combination of P-type and N-type semiconductor materials. In P-type semiconductor material, the holes are the majority charge carriers whereas the electrons are the minority charge carriers. Similarly, in the N-type semiconductor material, the electrons are the majority charge carriers whereas the holes are the minority charge carriers. Fig. 1.1.1 shows the PN junction diode. The PN junction diode generally has two terminals: anode and cathode. The anode is always connected with a metallic contact on the P-type semiconductor material while the cathode terminal is always connected with a metallic contact on the N-type semiconductor material. Here, the anode is denoted by the term A and the cathode is denoted by the term K. Fig. 1.1.2 shows the symbol of the PN junction diode. As the diode is one of the fundamental elements that is used in a large number of electronic circuits, it is always essential to study the operation of the diodes.

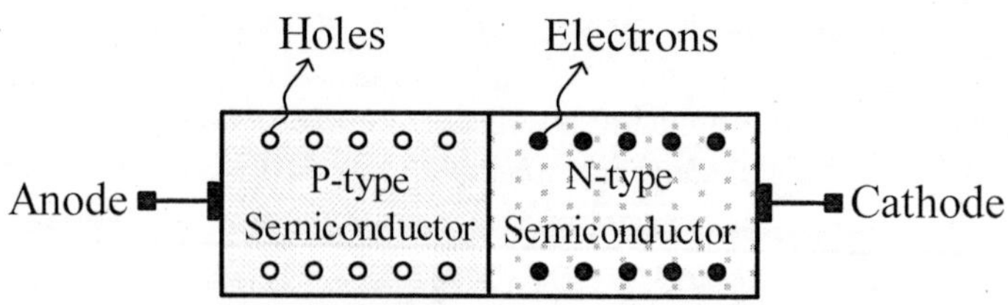

Fig. 1.1.1 PN junction diode.

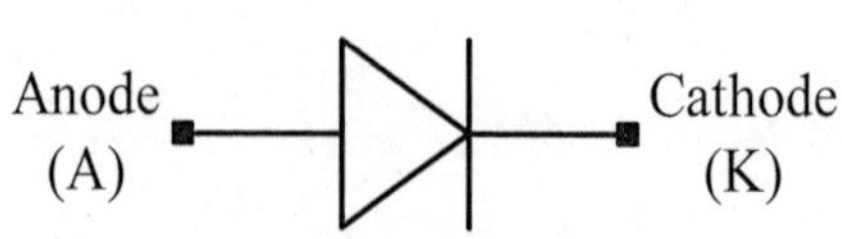

Fig. 1.1.2 Symbol of PN junction diode.

1.2 OPERATION OF AN UNBIASED PN JUNCTION DIODE

A PN junction diode without any biasing or a battery is called a PN junction diode under equilibrium. In an unbiased PN junction diode, due to the difference in the charge carrier concentrations, the charge carriers will transfer from the higher concentration side to the lower concentration side. The process of transfer of charge carriers from the higher concentration side to the lower concentration side is called diffusion. In the PN junction diode because of the diffusion, the holes which are the majority charge carriers in the P-type semiconductor will transfer from the P-type semiconductor (higher concentration side) to the N-type semiconductor (lower concentration side). Similarly, the electrons which are the majority charge carriers in the N-type semiconductor will transfer from the N-type semiconductor (higher concentration side) to the P-type semiconductor (lower concentration side). Fig. 1.2.1 shows the process of diffusion in the PN junction diode.

The holes that are crossing the junction and entering the N-type semiconductor due to diffusion combine with the electrons in the N-type semiconductor. Similarly, the electrons that are crossing the junction and entering the P-type semiconductor due to diffusion combine with the holes in the P-type semiconductor. Due to this combination of electrons and holes, a layer is formed in the PN junction diode which is depleted of free charges. This layer is called the depletion layer or depletion region or

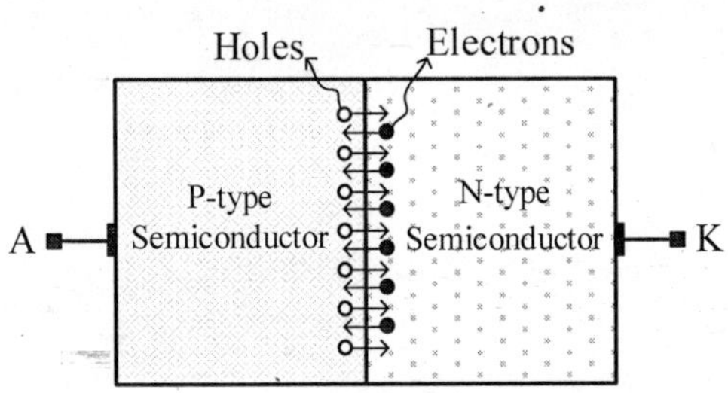

Fig. 1.2.1 Diffusion in PN junction diode.

space charge region. The width of the depletion layer will increase and this widening of the depletion region will stop at a point beyond which there is no diffusion of holes and electrons. Therefore, this depletion region will act as a barrier for the crossing of electrons and holes in the equilibrium condition. Fig. 1.2.2 shows the operation of an unbiased PN junction diode. If the P-type and N-type semiconductors have the same charge concentration, then the width of the depletion layer will be equal in both P-type and N-type semiconductors. If the charge concentration of the P-type semiconductor is more compared to the N-type semiconductor, then the width of the depletion layer is more in the N-type semiconductor compared to the P-type semiconductor.

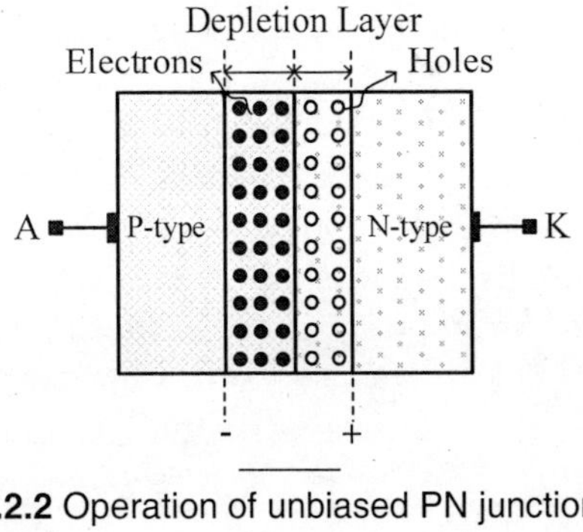

Fig. 1.2.2 Operation of unbiased PN junction diode.

Due to the presence of holes in the N-type semiconductor and electrons in the P-type semiconductor, there exists an electric field (E) at the junction of the PN junction diode. This electric field at the junction leads to the formation of barrier potential. The barrier potential is approximately 0.7 V for Si made PN junction diode and 0.3 V for the Germanium-made PN junction diode.

1.3 OPERATION OF A BIASED PN JUNCTION DIODE

A PN junction diode for which the biasing or the battery is connected is called the biased PN junction diode. Based on the direction of the biasing, the PN junction diode operates in two regions: forward-biased and reverse-biased. In the forward-biased condition, the positive of the power supply or battery is connected to the anode terminal while the negative of the power supply or battery is connected to the cathode terminal. In the reverse-biased condition, the negative of the power supply or battery is connected to the anode terminal while the positive of the power supply or battery is connected to the cathode terminal. Fig. 1.3.1 shows the forward-biased and reverse-biased PN junction diode. A current limiting resistor is always connected in series with the diode to limit the starting current.

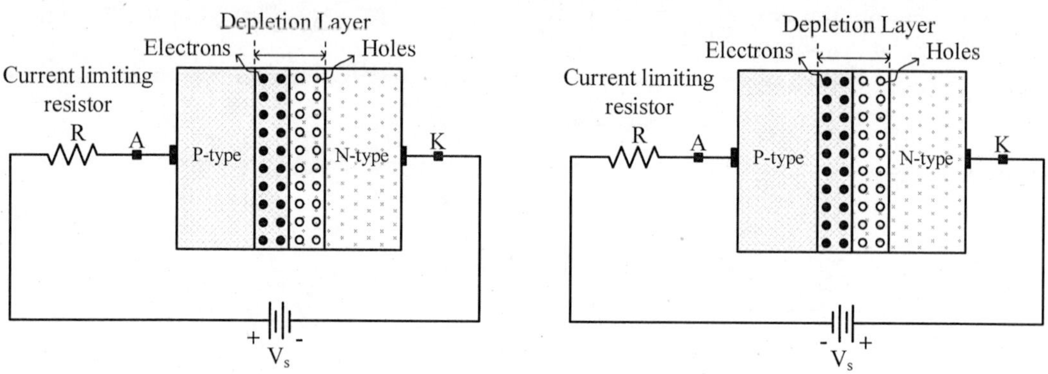

(a) Forward biased condition **(b) Reverse biased condition**

Fig. 1.3.1 Forward-biased and reverse-biased PN junction diode.

1.3.1 Operation of PN Junction Diode in the Forward-Biased Condition

In the forward-biased condition, if the applied voltage to the diode is less than or equal to the barrier potential voltage, the diode current I_D will be zero. Whenever the applied voltage is greater than the barrier potential voltage, the holes present in P-type semiconductor will move into the depletion region due to the repulsion from the positive of the power supply. Similarly, the electrons present in N-type semiconductor will move into the depletion region due to the repulsion from the negative of the power supply. As a result, the width of the depletion region will decrease in the forward-biased condition. The holes from the P-type semiconductor cross the junction and enters the N-type semiconductor and move towards the negative of the voltage source or battery. Similarly, the electrons from the N-type semiconductor cross the junction and enters the P-type semiconductor and move towards the positive of the voltage source or battery. This results in flow of diode current I_D. The direction of this conventional positive current is always in the direction of positive charges (holes). Therefore, the direction of forward current I_D will be in the direction of holes. Fig. 1.3.1.1 shows the operation of PN junction diode in the forward-biased condition. Majority charge carriers are responsible for the flow of current in forward bias condition.

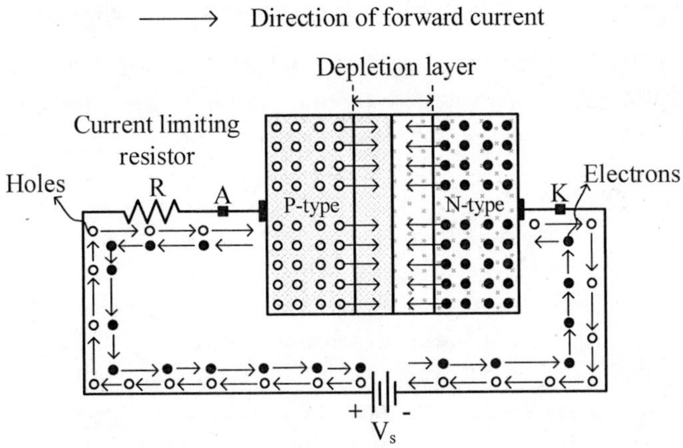

Fig. 1.3.1.1 Operation of PN junction diode in the forward-biased condition.

1.3.2 Operation of PN Junction Diode in the Reverse-Biased Condition

In the reverse-biased condition of the PN junction diode, the holes in the P-type semiconductor are attracted toward the positive terminal of the power supply. Similarly, the electrons in the N-type semiconductor are attracted toward the negative terminal of the power supply. This results in an increase in the width of the depletion region. The polarity of the applied voltage and the barrier potential voltage is the same so that the effective barrier potential voltage increases. This increase in effective barrier potential voltage restricts the transfer of holes from P-type semiconductor to N-type semiconductor and the electrons from N-type semiconductor to P-type semiconductor. The minority charge carriers in both P-type and N-type semiconductors will cross the junction due to the applied voltage. These minority charge carriers will

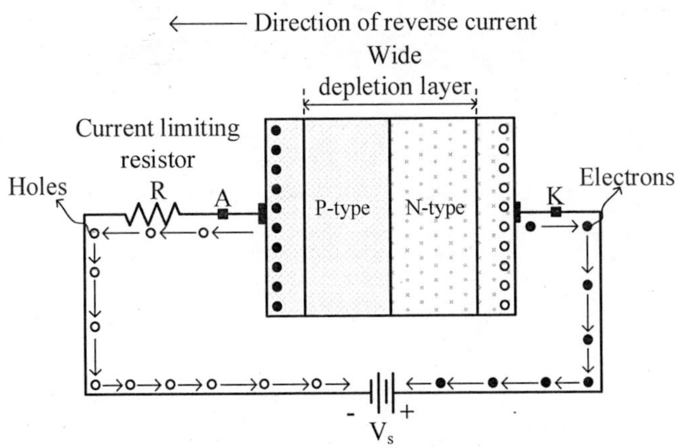

Fig. 1.3.2.1 Operation of PN junction diode in the reverse-biased condition.

constitute the anode current I_D. As this current is due to the minority charge carriers, the magnitude of anode I_D is in very small. In the reverse-biased condition, the diode acts as an insulator. Fig. 1.3.2.1 shows the operation of PN junction diode in the reverse-biased condition.

1.4 I-V CHARACTERISTICS OF A PN JUNCTION DIODE

Before studying the I-V characteristics of a PN junction diode, it is important to identify the difference between the ideal and the practical PN junction diodes. An ideal PN junction diode, when forward-biased, acts as a short circuit (offers zero resistance) and can flow infinite current through it. Similarly, when reversed-biased, an ideal PN junction diode acts as an open circuit (offers infinite resistance) and can block infinite voltage across it. Fig.1.4.1 shows the operation of the ideal PN junction diode in the forward and reverse biased conditions without the current limiting resistors. The terms V_D, I_D and R_D denotes the voltage across the diode, the current flowing through the diode and the internal resistance of the diode. The forward-biased ideal PN junction diode acts as a closed switch so that the magnitude of V_D and R_D are zero and the current I_D flowing through the diode is infinite. Similarly, the reverse-biased ideal PN junction diode acts as an open switch so that the magnitude of V_D and R_D are zero and the current I_D flowing through the diode is infinite. The I-V characteristic of the ideal diode is shown in Fig. 1.4.2. Since for the reverse-biased ideal PN junction diode, the voltage applied is negative, the I-V characteristics will be in observed in the second quadrant.

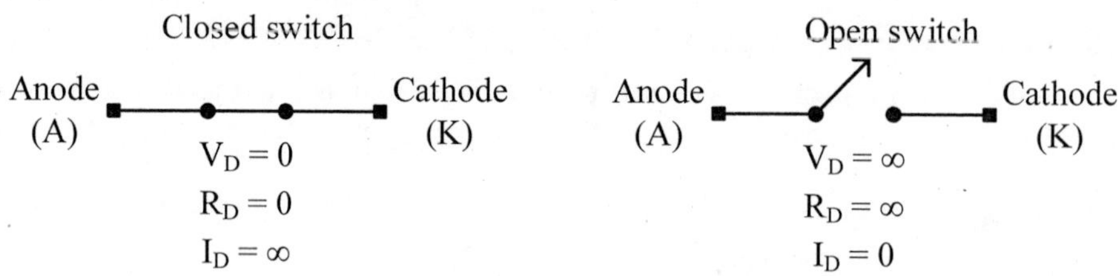

(a) Forward-biased ideal PN junction diode **(b) Reverse-biased ideal PN junction diode**

Fig. 1.4.1 Operation of the ideal PN junction diode in the forward and reverse biased conditions.

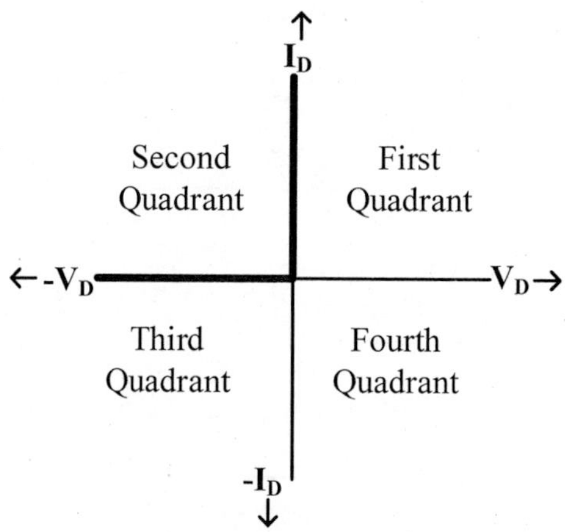

Fig. 1.4.2 I-V characteristics of the ideal PN junction diode.

However, for the practical PN junction diode, there exist a voltage drop exists across the diode when forward-biased (offer non-zero resistance) and can flow finite current through it. Similarly, a practical PN junction diode when reversed biased a very small current (leakage or reversed saturation current) flows through it (offers very high impedance) and can block finite voltage across it. Fig. 1.4.3 shows the I-V characteristics of a practical PN junction diode. In the practical PN junction diode, during the forward-biased condition, as the forward voltage VD is increasing up to the knee voltage or cut in voltage, the current ID flowing through the diode is zero. The knee voltage is the voltage at which the forward-biased PN junction diode starts conducting. This happens because the applied voltage is less than or equal to the barrier potential voltage and the charge carriers cannot cross the junction. Once the applied voltage is greater than the barrier potential voltage, the current flowing through the diode increases exponentially. The I-V characteristics of the practical PN junction diode during the forward-biased condition will exist in the first quadrant. During the reverse-biased condition of the practical PN junction diode, the reverse current ID will flow due to the minority charge carriers. The direction of the reverse current will be in the direction opposite to the forward current. As the applied voltage to the PN junction diode is negative and the current ID flowing through the diode is negative, the I-V characteristics of the practical PN junction diode will exist in the third quadrant. During the reverse-biased condition of the practical PN junction diode, as the applied voltage increases, the current ID will increase initially and remains constant to the value IO after a certain voltage. This current IO is called reverse saturation current. If the applied voltage increases further, then the avalanche breakdown will happen, leading to the puncture of the depletion junction and the large reverse current will be flowing through the diode. The large applied voltage and large reverse current will deliver a large power. The large power will be dissipated in the form of heat. Due to this heat, the temperature of the device will get increased beyond the safe limits. This voltage at which this breakdown happens is generally referred as the breakdown voltage.

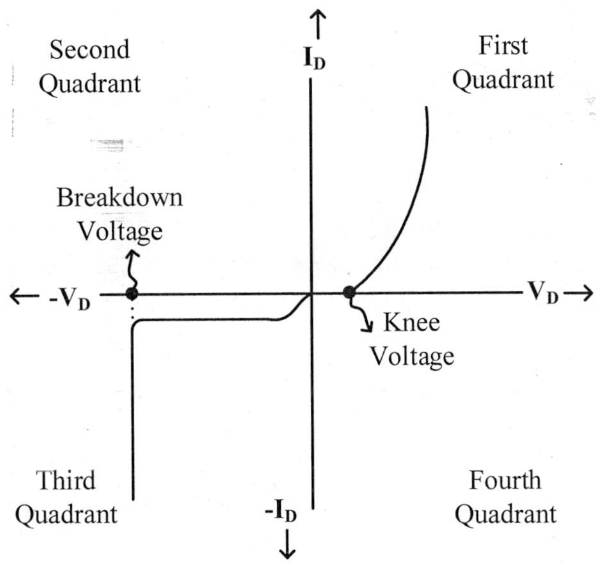

Fig. 1.4.3 I-V characteristics of the practical PN junction diode.

1.5 I-V CHARACTERISTICS EQUATION OF A PN JUNCTION DIODE

The I-V characteristic equation of the PN junction diode in the forward-biased and reverse-biased conditions is given by the Shockley equation as given below:

$$I_D = I_O\left(e^{\frac{qV_D}{\alpha_d K_B T}} - 1\right) \quad (1.5.1)$$

where I_D is the PN junction diode current; V_D is the voltage drop across the PN junction diode; q is the electroniç charge having a value of value 1.6×10^{-19} Coulombs; α_d is the ideality factor having a value between 1 and 2; K_B is the Boltzmann constant having a value $1.3\times10^{-23}m^2kgs^{-2}K^{-1}$; T is the temperature in Kelvin and I_O is the reverse saturation current.

Further, the value

$$V_T = \frac{KT}{q} \tag{1.5.2}$$

where V_T, is the thermal voltage drop. And its value increases with the increase of the temperature. Thus, the Shockley equation can be re-written as,

$$I_D = I_O\left(e^{\frac{V_D}{\alpha V_T}} - 1\right) \tag{1.5.3}$$

In forward-biased condition, the exponential term will be much higher than 1. Thus, the voltage drop can be approximated as,

$$V_D = \ln\frac{I_d}{I_o} - \propto V_T \tag{1.5.4}$$

Thus, as temperature increases, the thermal voltage drop V_T increases, which decreases the PN junction diode voltage drop for the same value of the diode current or diode has a negative temperature coefficient. As the temperature increases, typically, the PN junction diode voltage drop decreases. Another critical thing to be noted is PN junction diode starts conducting when the voltage across the diode is higher than the thermal voltage drop V_T. The same can be concluded from the Shockley equation where the PN junction diode current I_D will rise exponentially when the power term in the exponential has a value greater than 1. Or in other words, the current through the PN junction diode is nearly zero when the forward-biased voltage across it is less than the thermal voltage drop V_T. This current would increase exponentially if the applied forward-biased voltage becomes more than thermal voltage drop V_T.

Similarly, the PN junction diode voltage VD becomes negative in the reverse-biased condition. Thus, the power term in the exponent becomes negative and leads to a very small current flow. The PN junction diode will keep blocking the externally applied reverse-biased voltage until the voltage reaches the value of the avalanche break down voltage or maximum blocking voltage capability of the PN junction diode. Further, as the applied reverse-biased voltage is increased, PN junction length of the depletion region increases to block the applied voltage. Further, the electric field strength remains nearly constant till the depletion region length keeps on increasing. However, depletion region length becomes fixed when the voltage reaches near break down voltage of the diode. Thus, a further increase in the applied reverse voltage will increase the electric field in the depletion region. When the electric field in the depletion region is sufficiently high, it leads to the high velocity of the minority carriers entering the depletion region. The velocity of the minority carriers is sufficient to break covalent bonds from the crystal atoms or ions in the depletion region. Breaking of the covalent bond results in the release of electron –hole pairs which again break covalent bonds from the other crystal atoms or ions in the depletion region forming a cumulative chain reaction. Thus, results in a very high current in the reverse direction with a nearly constant diode voltage drop. Thus, when the voltage across the diode is higher than the reverse break down voltage, it will lead to a very high negative current flow through the PN junction diode, damaging it completely. Thus, the PN junction diode in faulty condition will act as short-circuit and may be harmful to the circuit where short circuit condition is not allowed.

Example 1: Calculate the thermal voltage drop on the diode which is operating at temperatures 30° C, 50° C and 70° C. Comment on the voltage drop across the diode for the current of I_D =2 A and I_o= 20u A. Assume an ideality factor of 1.1.

Solution: Temperature in Kelvin; For 30°C, the thermal voltage can be computed using:

$$V_T = \frac{KT}{q}$$

$$V_{T@30} = \frac{1.3\times10^{-23}\times323}{1.6\times10^{-19}} = 0.026\ V$$

$$V_{T@50} = \frac{1.3\times10^{-23}\times323}{1.6\times10^{-19}} = 0.028\ V$$

$$V_{T@30} = \frac{1.3\times10^{-23}\times343}{1.6\times10^{-19}} = 0.029\ V$$

Voltage drop for the diode is given by,

$$V_D = \ln\frac{I_d}{I_o} - \propto V_T$$

$$V_{D@30} = \ln\frac{2}{20\times10^{-6}} - 1.1(.026)$$
$$= 13.81 - 1.1(.026)$$
$$=13.789 \text{ V}$$

$$V_{D@50} = \ln\frac{2}{20\times10^{-6}} - 1.1(.028)$$
$$= 13.81 - 1.1(.028)$$
$$=13.779 \text{ V}$$

$$V_{D@70} = \ln\frac{2}{20\times10^{-6}} - 1.1(.029)$$
$$= 13.81 - 1.1(.029)$$
$$=13.778 \text{ V}$$

1.6 OPERATING POINT CALCULATION IN PN JUNCTION DIODE

Till now we have discussed I-V characteristics of the PN junction diode. It was also noted that the I-V characteristics of the PN junction diode is non-linear, due to the presence of the exponential term. So, in a circuit having a PN junction diode, one needs to compute the operating point of the PN junction diode using its I-V characteristics. To determine the operating point of the PN junction diode, one needs to compute the point of intersection of the load and the PN junction diode I-V characteristics. Thus, the PN junction diode when forward-biased conducts with a small drop across it. And in the reverse-biased condition, it blocks the applied voltage with a little current (in μA or still lesser) flowing through it. Below gives a simple example of the PN junction diode circuit with a series resistive load and input source of V_{in}.

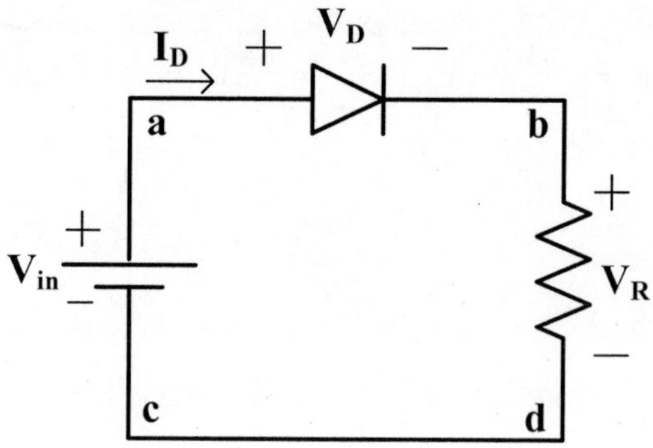

Fig. 1.6.1 PN junction diode circuit with a series resistive load and input source of V_{in}.

As the input voltage V_{in} is greater than PN junction diode threshold voltage, the PN junction diode being forward-biased leading to voltage drop of V_D for current I_D through it. From Shockley equation as given below:

$$I_D = I_o(e^{\frac{qV_d}{\alpha KT}} - 1) \tag{1.6.1}$$

As there are two variables V_D and I_D, so another equation is required. Applying KVL in loop abcd we have,

$$V_{in} - V_D - I_D R = 0$$
$$V_{in} = V_D + I_D R$$
$$V_D = V_{in} - I_D R$$
$$I_D = \frac{V_{in} - V_D}{R} \tag{1.6.2}$$

Substituting I_D we have,

$$\frac{V_{in} - V_D}{R} = I_o\left(e^{\frac{qV_d}{\alpha KT}} - 1\right) \tag{1.6.3}$$

The above equation is non-linear and can be solved using the Newton Raphson method. Another graphical method would be plotting graphs for both the I-V characteristics of the PN junction diode and load line. The load line of the PN junction diode network can be obtained by using

$$V_D = V_{in} + I_D R \quad (1.6.4)$$

The above equation is linear with V_D and I_D as variables. Substituting V_D=0 gives the interaction of the load line in the current (I) axis,

$$V_{in} = V_D + I_D R$$
$$0 = V_{in} + I_D R$$
$$I_D = \frac{V_{in}}{R} \tag{1.6.5}$$

Substituting I_D=0 gives the interaction of the load line in the voltage (V) axis,

$$V_{in} = V_D + I_D R$$
$$V_{in} = V_D + 0R$$
$$V_{in} = V_D \quad (1.6.6)$$

Using graph paper and drawing the load line and I-V characteristics of the PN junction diode to obtain the operating point of the PN junction diode. The point of the intersection of the load line and the I-V characteristics of the PN junction diode represent the operating point of the PN junction diode.

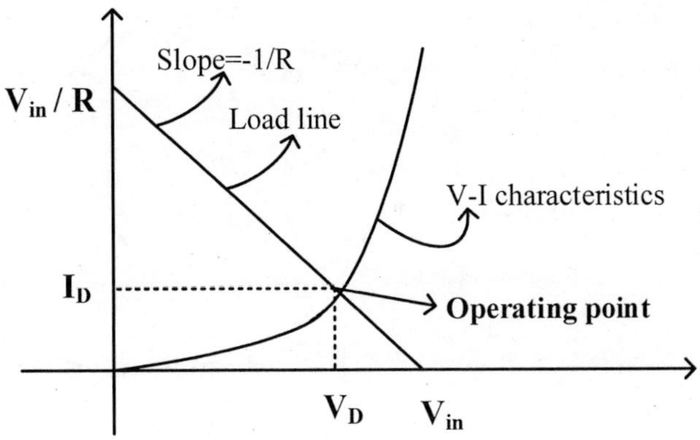

Fig. 1.6.2 Load line and I-V characteristics of the PN junction diode.

1.6.1 Piecewise Linearization of the PN Junction Diode Characteristics

Based on above example it is clear that solving non-linear equation is cumbersome. Further, analysis becomes more complex when the load also has non-linear characteristics. For load like resistive load, the I-V characteristic is the straight line and the operating point can be computed easily by estimating or approximating the diode voltage drop. Also, here one can see that PN junction diode I-V characteristics can be divided into two parts as shown in Fig. 1.6.1.1. One part is fixed voltage drop and the other is resistive drop.

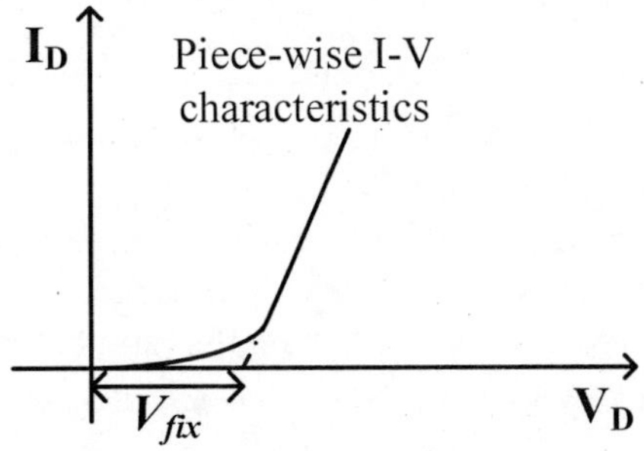

Fig. 1.6.1.1 Piecewise linearization of the PN junction diode characteristics.

The PN junction diode I-V characteristic can be divided into two regions: (i) horizontal line representing threshold voltage or fixed voltage drop in the PN junction diode characteristics, and (ii) the other is inclined straight line representing a resistive drop in the PN junction diode characteristics. The partition of I-V characteristics of the PN junction diode into two parts is also termed as piecewise linearization of the PN junction diode characteristics. With the given simplification the computation of the diode operating point becomes simple linear equation without any non-linear exponential term. Thus, it simplifies the computation involved for solving the circuits having diode. Thus, the electrical equivalent circuit model of the PN junction diode can be represented as a fixed voltage drop 'V_{fix}' with a series resistance 'R_s' during the forward-biased condition as shown in Fig. 1.6.1.2. Thus, the voltage drop across the PN junction diode can be represented as,

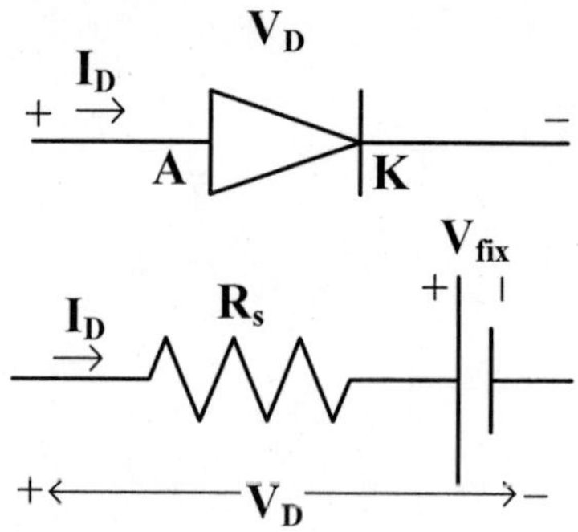

Fig. 1.6.1.2 Electrical equivalent circuit model of the PN junction diode.

Thus, voltage drop across the PN junction diode can be represented as,

$$V_D = V_{fix} + I_D R_s \tag{1.6.1.1}$$

Now substituting above equation in

$$V_{in} - V_D - I_D R = 0$$

$$V_{in} - V_{fix} - I_D R_s - I_D R = 0$$

$$I_D = \frac{V_{in} - V_{fix}}{R_s + R} \tag{1.6.1.2}$$

Once I_D is computed, V_D can be calculated.

The given solution has nearly the same accuracy as that for the exact operating point computed by solving non-linear equation. Thus, the operating point of the PN junction diode in the forward-biased condition can be easily computed using piecewise linearization. The accuracy of the solution depends on the preciseness in the value of fixed voltage drop V_{fix} and series resistance R_s. The value of V_{fix} and series resistance R_s can be derived using the I-V characteristics of the PN junction diode at the given temperature. Thus, series resistance R_s needs to be evaluated based on the operating point of the PN junction diode. The operating point can be fixed or vary with time. For a fixed operating point, the PN junction diode resistance can be simply computed using the ratio of the PN junction diode drop by the current flowing through it. The measured resistance is also termed static resistance R_{static} and is given by,

$$R_{dynamic} = \frac{\Delta V_d}{\Delta I_d} \tag{1.6.1.3}$$

If the operating point varies with time, then the resistance of the PN junction diode can be measured by computing the ratio of the change in the operating voltage to the change in the operating current. Fig. 1.6.1.3 shows the operating characteristics of the PN junction diode with a time-varying voltage source. The measured resistance is also termed as dynamic or ac resistance in such case is given by,

$$R_{dynamic} = \frac{\Delta V_d}{\Delta I_d} \tag{1.6.1.4}$$

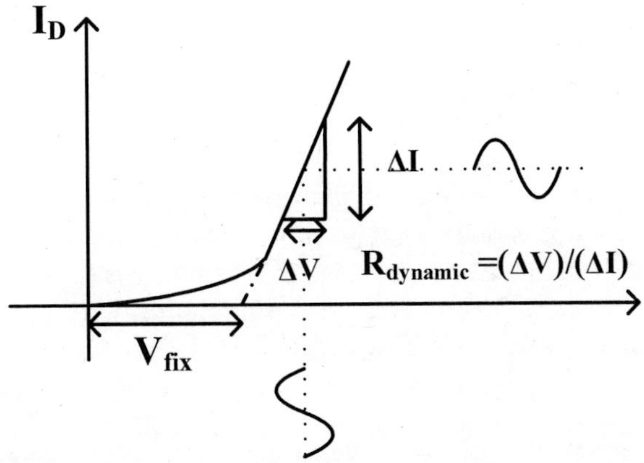

Fig. 1.6.1.3 Operating characteristics of the PN junction diode with a time-varying voltage source

The dynamic resistance can also be measured using a derivate of the I-V characteristics for a given operating current. The expression for the dynamic resistance based on the operating point is derived as follows,

$$R_{dynamic} = \frac{1}{\frac{dI_d}{dV_d}} \tag{1.6.1.5}$$

$$\frac{dI_d}{dV_d} = \frac{d(I_O(e\frac{V_d}{\alpha V_T - 1}))}{dV_d} \tag{1.6.1.6}$$

$$\frac{dI_d}{dV_d} = I_O \frac{V_d}{(e^{\alpha V_T})} \frac{1}{\alpha V_T} \tag{1.6.1.7}$$

$$\frac{dI_d}{dV_d} = \frac{(I_O + I_d)}{\alpha V_T} = \frac{I_d}{\alpha V_T} \tag{1.6.1.8}$$

$$R_{dynamic} = \frac{\alpha V_T}{I_d} \tag{1.6.1.9}$$

At room temperature, the thermal drop in the PN junction diode is given by,

$$R_{dynamic} = \frac{0.26}{I_d} \tag{1.6.1.10}$$

1.7 APPLICATION OF THE PN JUNCTION DIODES

The PN junction diodes mainly find their application in:

1. Rectifiers
2. Voltage multipliers
3. Clipper and clampers

1.8 RECTIFIER

The voltage rectifier is another important application of the diode where the AC input voltage is converted into a DC voltage. The PN junction diode normally acts as closed in forward-biased conditions and acts as an open switch in reverse-biased conditions. When an AC voltage is applied to the PN junction diode, the PN junction diode will be in the forward-biased condition during the positive half cycle of the AC voltage and the PN junction diode will be in the reverse biased condition during the negative half cycle of the AC voltage. If a load is connected at the output of the PN junction diode, then the voltage across the load will be only a positive half cycle of the AC voltage. During the negative half cycle of the AC signal, the PN junction diode will act as an open switch and zero voltage will be applied across the load. Fig. 1.8.1 shows the operation of the PN junction diode for an AC input signal. The output voltage will appear across the load during the positive half cycle only. Therefore, the output voltage across the load is unidirectional. The converter used for converting the alternating input signal into a unidirectional output signal is called a rectifier. There are two types of rectifiers: half wave and full wave rectifiers.

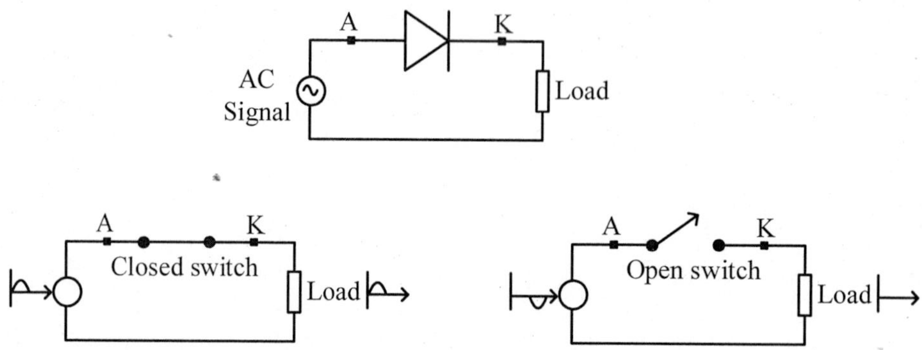

(a) For positive half cycle of AC signal **(b) For negative half cycle of AC signal**

Fig. 1.8.1 Operation of the PN junction diode for an AC input signal.

1.8.1 Half Wave Rectifier

Half wave rectifiers, as the name indicates, allow only one-half cycle (say positive cycle) of the AC input voltage and blocks the other half (say negative cycle) of the AC input voltage. Fig. 1.8.1.1 shows the circuit diagram of a half wave rectifier connected to a resistive load R_L. The AC input signal is applied to the primary winding of the step-down transformer having a turns ratio of N_1:N_2. If V_{PP} is the peak value of the primary voltage of the step-down transformer having turns ratio of N_1:N_2 then the peak value of the secondary voltage V_{PS} of the transformer is given by:

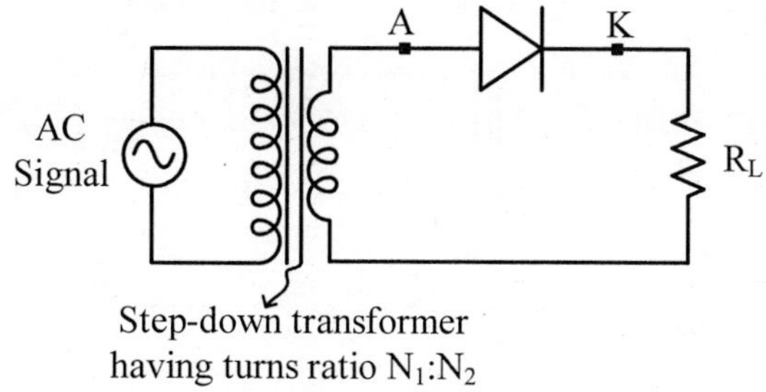

Fig. 1.8.1.1 Circuit diagram of a half wave rectifier connected to a resistive load R_L.

$$V_{PS} = \frac{V_{PP}N_2}{N_1} \tag{1.8.1.1}$$

One end of the second winding of the step-down transformer is connected to the anode terminal of the PN junction diode. The cathode terminal of the PN junction diode is connected to one end of the resistive load. The other end of the resistive load is connected to the other end of the secondary winding of step-down transformer. Fig. 1.8.1.2 shows the operation with the half wave rectifier connected to a resistive load R_L. During the positive half cycle of the AC input voltage, the PN junction diode is in forward-biased condition. In the forward-biased condition, the PN junction diode acts as a closed switch so that the entire applied voltage to the PN junction diode will appear across the load resistance R_L and a load current I_L will flow through the resistive load R_L. V_O is the output voltage across the resistive load. During the negative half cycle of the AC input voltage, the PN junction diode is in reverse-biased condition. In the reverse-biased condition, the PN junction diode acts as an open switch so that no voltage appears across the load resistance R_L. The waveforms of a half wave rectifier are shown in Fig. 1.8.1.3. From the waveforms of the half wave rectifier, it can be observed that for the positive half cycle of the AC input voltage, the PN junction diode is forward biased and the output voltage V_O will appear across the load resistance R_L. During the negative half cycle of the AC input voltage, the output voltage V_O will across the load resistance R_L is zero since the PN junction diode is reverse biased.

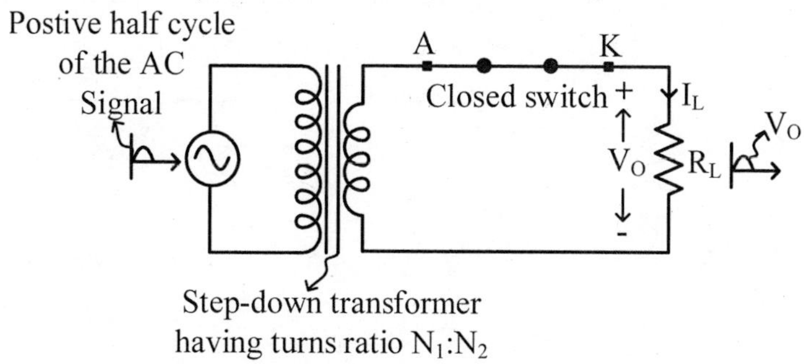

(a) For positive half cycle of AC signal

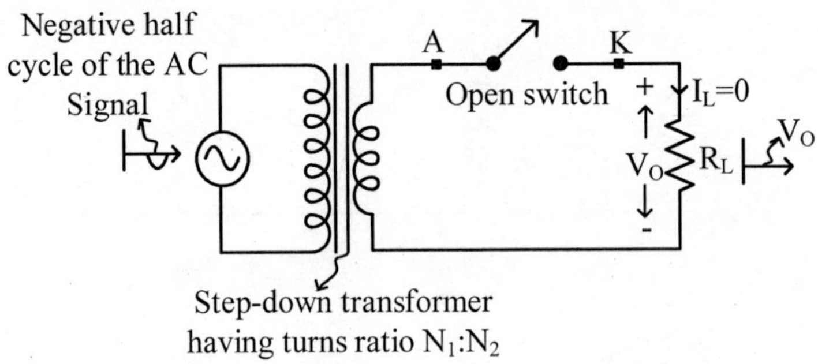

(b) For negative half cycle of AC signal

Fig. 1.8.1.2 Operation of a half wave rectifier connected to a resistive load R_L.

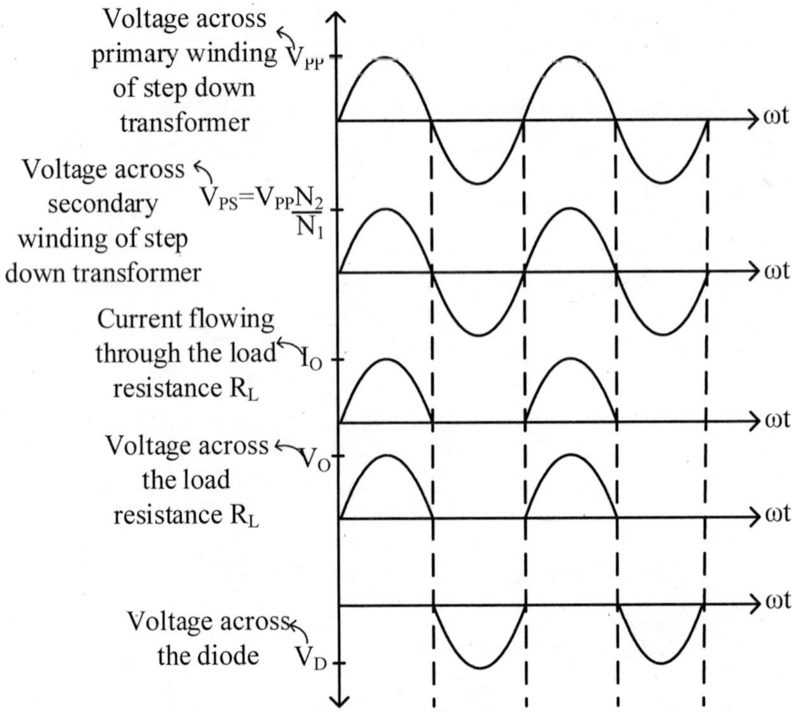

Fig. 1.8.1.3 Waveforms of a half wave rectifier connected to a resistive load R_L.

From the waveforms of half wave rectifier, the average value of output voltage $V_{O,avg}$ for the resistive load can be computed as,

$$V_{o,avg} = \frac{1}{2\pi}\int_0^{\pi} V_{PS}\sin(\omega t)dt \tag{1.8.1.2}$$

$$V_{o,avg} = \frac{V_{PS}}{\pi} \tag{1.8.1.3}$$

From the waveforms of half wave rectifier, the RMS value of output voltage $V_{O,\,RMS}$ for the resistive load can be computed as,

$$V_{o,RMS} = \sqrt{\frac{1}{2\pi}\int_0^{\pi}(V_{PS}\sin\omega t)^2\,d\omega t} \tag{1.8.1.4}$$

$$V_{o,RMS} = \frac{V_{PS}}{2} \tag{1.8.1.5}$$

The DC output power P_{DC} of half wave rectifier can be computed as,

$$P_{DC} = V_{O,avg}I_{L,avg} = \frac{V_{O,avg}^2}{R_L} = \frac{V_{PS}^2}{\pi^2 R_L} \tag{1.8.1.6}$$

The AC input power P_{AC} of half wave rectifier can be computed as

$$P_{AC} = \frac{V_{O,RMS}^2}{R_L} = \frac{V_{PS}^2}{4R_L} \tag{1.8.1.7}$$

The efficiency $\%\eta_{eff}$ of the half wave rectifier can be computed as

$$\%\eta_{eff} = \frac{P_{DC}}{P_{AC}}\times 100 = \frac{4}{\pi^2} = 40.6\% \tag{1.8.1.8}$$

Example 2: A single-phase AC supply having a voltage and frequency of 220 V, 50 Hz is connected to a half wave rectifier using the stepdown transformer having the turns ratio of 5:1. The value of load resistance is 1K Ω. Calculate the average value of the output voltage, RMS value of output voltage, DC output power and AC input power for half wave rectifier.

Solution: The magnitude of voltage across the primary winding of transformer V_{PP}= 220 V.

The magnitude of voltage across the secondary winding of transformer VPS= 220×(1/5) = 44 V.

The average value of output voltage $V_{o,RMS} = \frac{V_{PS}}{\pi} = 44/3.14 = 14\ V$

The RMS value of output voltage $V_{o,RMS} = \frac{V_{PS}}{2} = \frac{44}{2} = 22\ V$

The DC output power PDC of half wave rectifier $= \frac{V_{PS}^2}{\pi^2 R_L} = \frac{44^2}{3.14^2 \times 1000} = 0.196\ W$

The AC input power PAC of half wave rectifier $= \frac{V_{PS}^2}{4R_L} = \frac{44^2}{4\times 1000} = 0.48\ W$

1.8.2 Full Wave Rectifier

Full wave rectifiers as the name indicate allow both cycles (positive and negative cycles) of the AC input voltage to give the output voltage. Fig. 1.8.2.1 shows the circuit diagram of a full wave rectifier connected to a resistive load R_L. The full wave rectifier consists of a step-down centre tapped transformer. One end of the centre tapped step down transformer is connected to the anode of PN junction diode D_1. The other end of a centre tapped step down transformer is connected to the anode of PN junction diode

D_2. The cathode terminal of PN junction diodes D_1 and D_2 are connected to one end of load resistance R_L. The other end of the load resistance R_L is connected centre terminal of the centre tapped step down transformer.

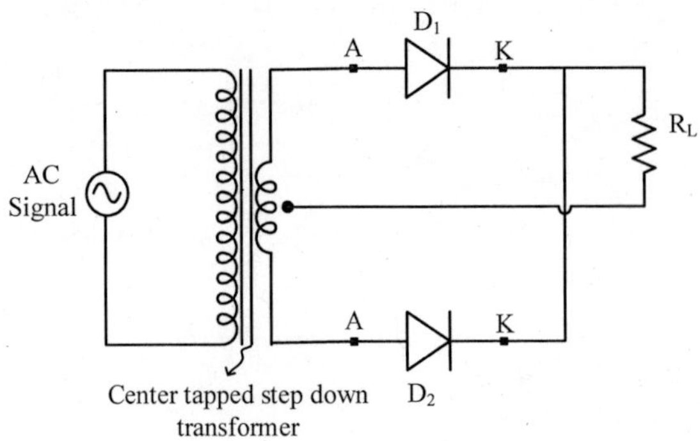

Fig. 1.8.2.1 Circuit diagram of a full wave rectifier connected to a resistive load R_L.

Fig. 1.8.2.2 shows the operation of a full wave rectifier connected to a resistive load R_L during the positive and negative half cycles of the AC input voltage During the positive half cycle of AC input voltage, the PN junction diode D_1 is forward-biased so that it acts as a closed switch. As a result, the positive half cycle of AC input voltage is applied across the load resistance R_L. Due to this, a current I_L flows through the load resistance R_L and a voltage V_L will appear across the load resistance R_L. During the positive half cycle of AC input voltage, the PN junction diode D_2 is reverse-biased so that it acts as an open switch and will not conduct. During the negative half cycle of AC input voltage, the PN junction diode D_2 is forward-biased so that it acts as a closed switch.

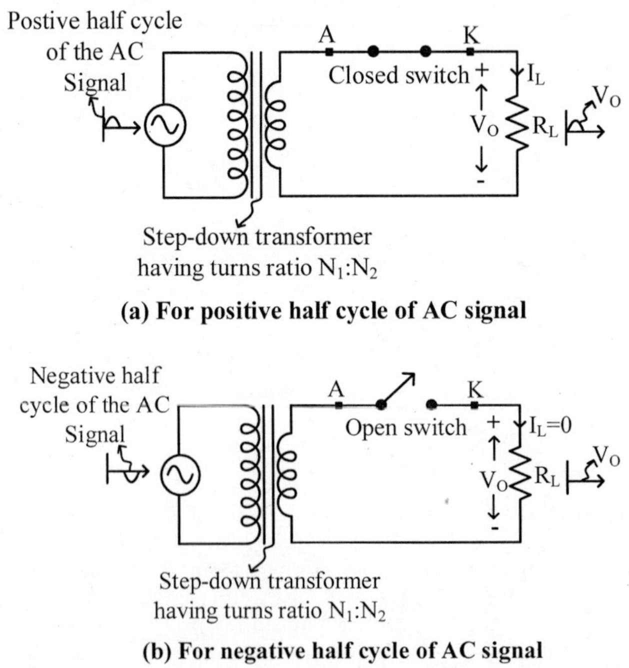

Fig. 1.8.2.2 Operation of a full wave rectifier connected to a resistive load RL.

As a result, the negative half cycle of AC input voltage is applied across the load resistance R_L in the reverse direction so that a positive voltage will appear across the load resistance R_L. Due to this, a current I_L flows through the load resistance R_L and a voltage V_L will appear across the load resistance R_L. During the negative half cycle of AC input voltage, the PN junction diode D_1 is reverse biased so that it acts as an open switch and will not conduct. Therefore, a full wave rectifier allows both half cycles of the AC input voltage and also a unidirectional output voltage will also appear across the load resistance R_L.

The waveforms of a full wave rectifier are shown in Fig. 1.8.2.3. From the waveforms of the full wave rectifier, it can be observed that for the positive half cycle of the AC input voltage, the PN junction diode D_1 is forward biased and the output voltage V_O will across the load resistance R_L. During the negative half cycle of the AC input voltage, the PN junction diode D_2 is forward biased and the positive output voltage V_O will across the load resistance R_L.

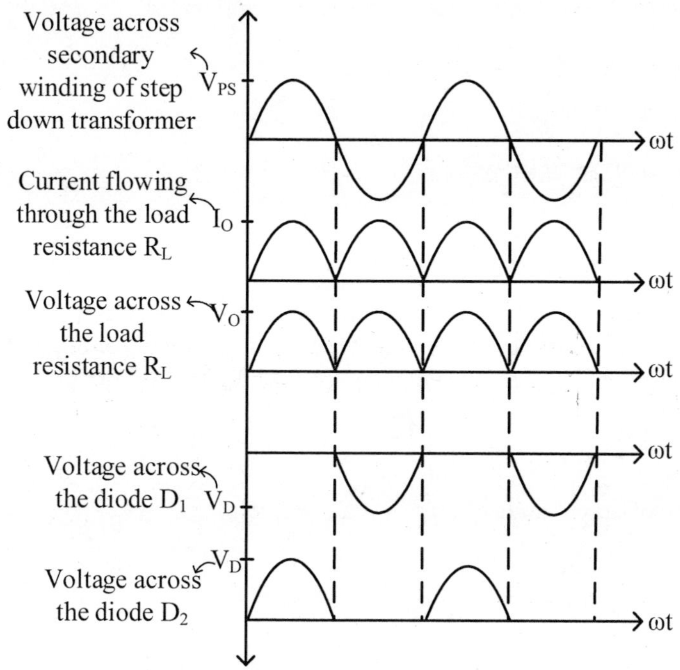

Fig. 1.8.2.3 Waveforms of a full wave rectifier connected to a resistive load RL.

From the waveforms of full wave rectifier, the average value of output voltage $V_{O,avg}$ for the resistive load can be computed as,

$$V_{o,avg} = \frac{1}{2\pi}\int_0^{\pi} V_{PS}\sin(\omega t)dt \tag{1.8.2.1}$$

$$V_{o,avg} = \frac{2V_{PS}}{\pi} \tag{1.8.2.2}$$

From the waveforms of full wave rectifier, the RMS value of output voltage $V_{O,\,RMS}$ for the resistive load can be computed as,

$$V_{o,RMS} = \sqrt{\frac{1}{\pi}\int_0^{\pi}(V_{PS}\sin\omega t)^2\, d\omega t} \qquad (1.8.2.3)$$

$$V_{o,RMS} = \frac{V_{PS}}{\sqrt{2}} \qquad (1.8.2.4)$$

The DC output power P_{DC} of full wave rectifier can be computed as,

$$P_{DC} = V_{O,avg} I_{L,avg} = \frac{V_{O,avg}^2}{R_L} = \frac{4V_{PS}^2}{\pi^2 R_L} \qquad (1.8.2.5)$$

The AC input power P_{AC} of full wave rectifier can be computed as

$$P_{AC} = \frac{V_{O,RMS}^2}{R_L} = \frac{V_{PS}^2}{2R_L} \qquad (1.8.2.6)$$

The efficiency $\%\eta_{eff}$ of the full wave rectifier can be computed as

$$\%\eta_{eff} = \frac{P_{DC}}{P_{AC}}\times 100 = \frac{8}{\pi^2} = 81.2\% \qquad (1.8.2.7)$$

Example 3: A full wave rectifier is fed from a centre-tapped step-down transformer. The peak value of voltage at the secondary winding of the transformer is 30 V. The load resistance is 1K Ω. Calculate the average value of the output voltage, RMS value of output voltage, DC output power and AC input power for full wave rectifier.

Solution: The magnitude of voltage across the secondary winding of transformer VPS= 30 V.

The average value of output voltage $V_{o,avg} = \frac{2V_{PS}}{\pi} = 2\times 30/3.14 = 19.10\ V$

The RMS value of output voltage $V_{o,RMS} = \frac{V_{PS}}{\sqrt{2}} = \frac{30}{\sqrt{2}} = 21.21\ V$

The DC output power PDC of half wave rectifier= $\frac{4V_{PS}^2}{\pi^2 R_L} = \frac{4\times 30^2}{3.14^2\times 1000} = 0.36\ W$

The AC input power PAC of half wave rectifier = $\frac{V_{PS}^2}{2R_L} = \frac{30^2}{2\times 1000} = 0.45\ W$

From the above, it can be observed that the efficiency of full wave rectifier is double the efficiency of half wave rectifier. The above-discussed full wave rectifier circuit is simple with minimum PN junction diodes; however, the above circuit suffers from the following disadvantages:

i. Requirement of centre tapped transformer
ii. Diode of twice blocking voltage
iii. Transformer with higher rating VA due its underutilization of half winding in each cycle of output voltage.

In order to overcome the above-discussed drawbacks, the full wave bridge rectifier is used. Fig. 1.8.2.4 shows the circuit diagram of a full wave bridge rectifier connected to a resistive load R_L.

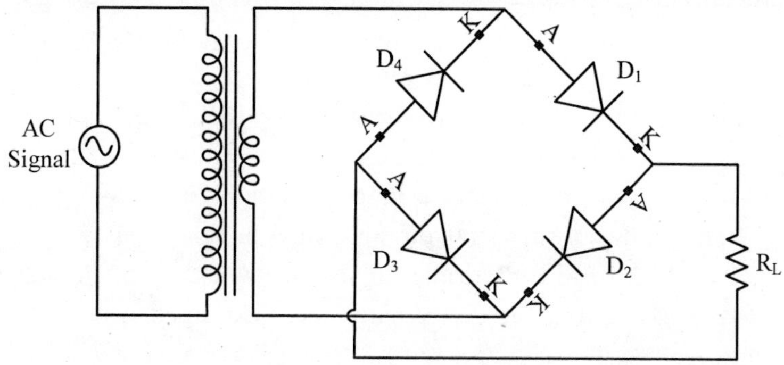

Fig. 1.8.2.4 Circuit diagram of a full wave bridge rectifier connected to a resistive load RL.

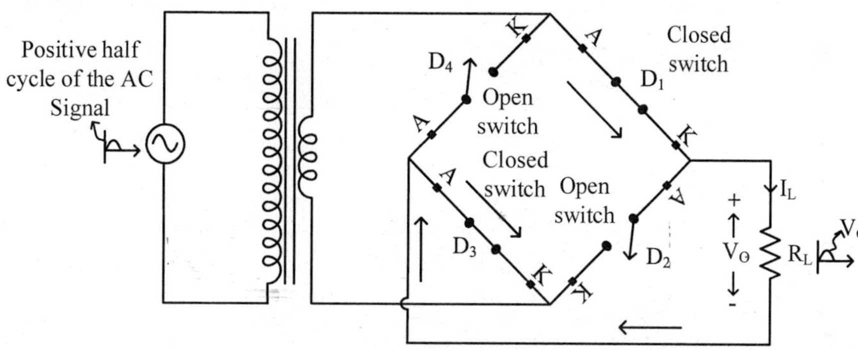

(a) For positive half cycle of AC signal

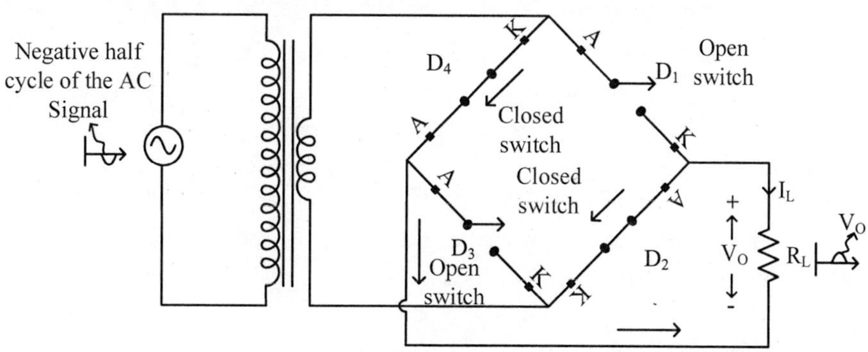

(b) For negative half cycle of AC signal

Fig. 1.8.2.5 Operation of a full wave bridge rectifier connected to a resistive load RL.

The full wave bridge rectifier consists of a stepdown transformer and four PN junction diodes connected in the form of a bridge as shown in Fig. 1.8.2.4. The secondary winding of the stepdown transformer is connected to one diagonal of the bridge and the load resistance R_L is connected in the other diagonal of the bridge. Fig. 1.8.2.5 shows the operation of a full wave bridge rectifier connected to a resistive load R_L. During the positive half cycle of AC input voltage, the PN junction diodes D_1 and D_3 are forward-biased so that they act as a closed switch. As a result, the positive half cycle of AC input voltage is applied across the load resistance R_L. Due to this, a current I_L flows through the load resistance R_L and a voltage V_L will appear across the load resistance R_L. During the positive half cycle of AC input voltage, the PN junction diodes D_2 and D_4 are reverse-biased so that it acts as an open switch and will not conduct. During the negative half cycle of AC input voltage, the PN junction diodes D_2 and D_4 are forward-biased so that it acts as a closed switch. As a result, the negative half cycle of AC input voltage is applied across the load resistance R_L in the reverse direction so that a positive voltage will appear across the load resistance R_L. Due to this, a current I_L flows through the load resistance R_L and a voltage V_L will appear across the load resistance R_L. During the negative half cycle of AC input voltage, the PN junction diodes D_1 and D_3 are reverse-biased so that it acts as an open switch and will not conduct. The expressions for the average value of output voltage, RMS value of output voltage, DC output power, AC input power and efficiency will be the same as that of the full wave rectifier. As the full wave bridge rectifier uses four diodes, the total voltage drop will be double the full wave rectifier.

1.9 ZENER DIODE

A special type of diode that is typically operated in reverse biased zener break down region. Such diodes are typically used for voltage protection and are connected in parallel to the load. The symbol for zener diode is given in Fig. 1.9.1. It is the same as the diode except for letter z instead of the horizontal line in the cathode of the diode. The voltage regulation capability of the zener diode typically

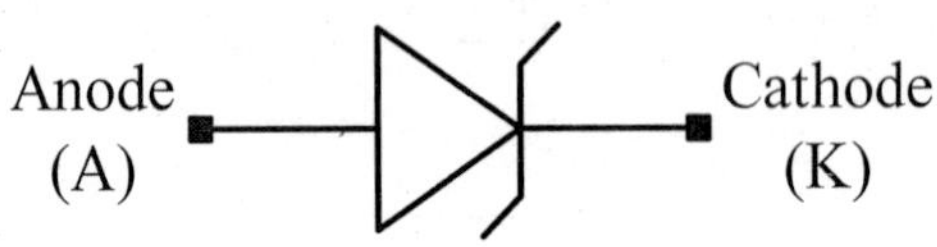

Fig. 1.9.1 Symbol of zener diode.

depends on the doping level of the p-n junction diode. So, for manufacturing zener diode doping of the p-n junction needs to be done in an accurate and controlled manner. These are heavily doped diodes which result two breakdown regions during their operation. The first one is zener breakdown region occurring at the low operating voltage. During operation of the zener breakdown region intense electric field is generated in the small depletion region leading to the tunnelling out of electrons from the valance band. This results in the flow of current in the reverse biased condition while maintaining the low zener breakdown voltage across it. Zener diode basically exploits the I-V characteristics of the diode in the reverse biased condition. It can be noted in the reverse biased condition that diodes maintain the voltage across with the flow of high current. The given property is utilized with the flow of controlled current depending on the doping level in the zener diode. Thus, the voltage protection feature of zener diode can be supported by the fact that during the operation in the zener break down region, it regulates the voltage across it with the flow controlled current through it. Zener diode with 5V zener break down voltage are easily available. However, zener diode with zener break down voltage between 1V to 200V are commercially available. The I-V characteristics of the zener diode is given in Fig. 1.9.2.

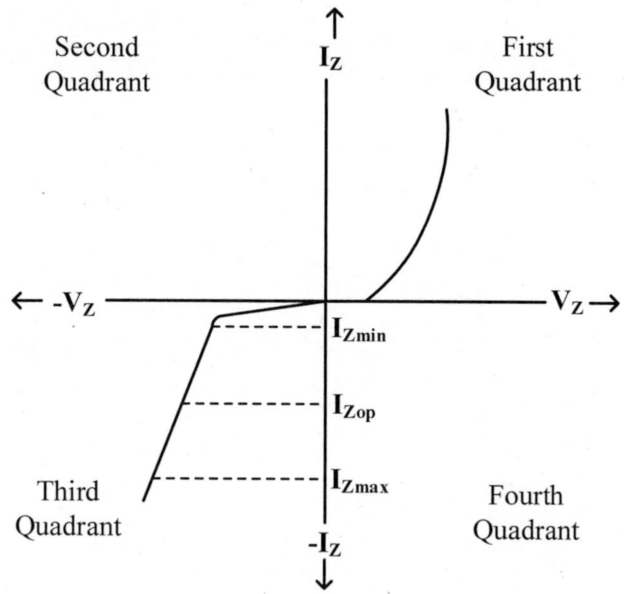

Fig. 1.9.2 I-V characteristics of the zener diode.

It can be noted from the I-V characteristics that in reverse biased condition zener diode enters into break down region when the current through it is higher than I_{zmin}. Further, I_{zop} in I-V characteristics represents the recommended or typical current through zener diode during operation in the zener break down region. And the last I_{zmax} in I-V characteristics represents the maximum value of current through zener diode which it can withstand during operation in the zener break down region. For the design of the zener diode with voltage protection, it should be noted that current through the zener diode should be maintained around I_{z_op}. Further, the zener diode is connected in parallel with the load across which voltage should not exceed zener break down voltage value. Further, circuit zener current limiter resistor R_z which limits the current through the zener diode below I_{zmax}.The circuit using zener diode for voltage protection can have the following cases:

i. Input voltage and load resistance at output is fixed.
ii. Input voltage has variation with load resistance at output is fixed.
iii. Input voltage is fixed while the load resistance varies.

Typically, the zener circuit has the series resistance which needs to be designed to maintain the zener diode to operate near the zener break down region. The circuit schematic for the typical zener diode used for voltage protection is given in Fig. 1.9.3. The given circuit consists of input voltage source $v_{in}(t)$, current limiter resistor R_z, zener diode and output load resistance R_L.

Case I: Take a case when input voltage source V_{IN} and output load resistance R_L are fixed. It is required to operate the zener diode near zener breakdown region. Thus, it is required to obtain the minimum, maximum and typical value of the current limiter resistor R_z. For the minimum value of current limiter resistor R_{z_min}, the current through the zener diode should not exceed I_{zmax}. Assuming the zener diode

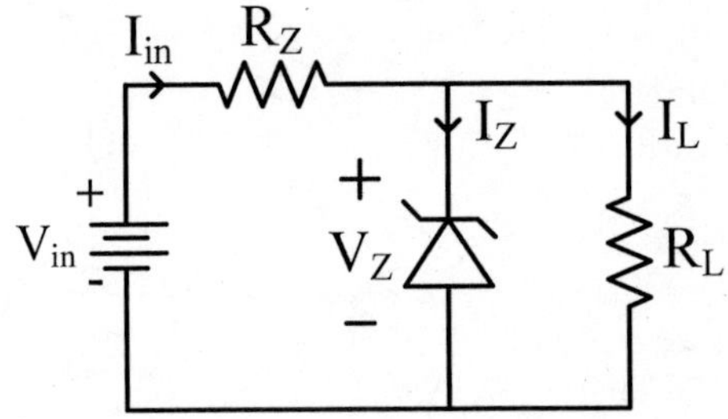

Fig. 1.9.3 Circuit schematic for the typical zener diode used for voltage protection.

maintains the zener break down voltage of V_z across it, the current through load resistance is given by,

$$I_L = \frac{V_z}{R_L} \tag{1.9.1}$$

Therefore, input current I_{in} is given by,

$$I_{in} = I_{z\max} + I_L$$

$$I_{in} = I_{z\max} + \frac{V_z}{R_L}$$

$$R_{z\_\min} = \frac{V_{in} - V_z}{I_{in}}$$

$$R_{z\_\min} = \frac{V_{in} - V_z}{I_{z\max} + \frac{V_z}{R_L}} \tag{1.9.2}$$

Similarly, for the maximum value of current limiter resistor R_{z_max}, the current through the zener diode should not be less than exceed I_{zmin}. Assuming the zener diode maintains the zener break down voltage of V_z across it, the current through load resistance is given by,

$$I_L = \frac{V_z}{R_L} \tag{1.9.3}$$

Therefore, input current I_{in} is given by,

$$I_{in} = I_{z\min} + I_L$$

$$I_{in} = I_{z\min} + \frac{V_z}{R_L}$$

$$R_{z\_\max} = \frac{V_{in} - V_z}{I_{in}}$$

$$R_{z\_\max} = \frac{V_{in} - V_z}{I_{z\min} + \frac{V_z}{R_L}} \tag{1.9.4}$$

Thus, the value of the current limiter resistor R_z should lie between R_{z_min} and R_{z_max} to operate zener diode in zener breakdown region. For zener diode to operate in typical operating value of current limiter resistor R_z is given by,

$$R_{z\_op} = \frac{V_{in} - V_z}{I_{zop} + \frac{V_z}{R_L}} \quad (1.9.5)$$

Now when the current limiter resistor R_z is fixed, then it is required to compute whether zener diode is operating zener breakdown region or not?

Or in other words the voltage across the load resistor R_L should be greater than or equal to zener break down voltage V_Z. Using the voltage divider rule, the output voltage across the load resistor is given by,

$$V_o = \frac{V_{in}}{R_L + R_z} \quad (1.9.6)$$

For the zener diode operating in zener breakdown region the output voltage should be $V_o \geq V_Z$

$$V_o = \frac{V_{in}}{R_L + R_z} \geq Vz \quad (1.9.7)$$

Case II: Take a case when input voltage source V_{in} is variable and output load resistance R_L is fixed as shown in Fig. 1.9.4. It is required to operate zener diode near zener breakdown region.

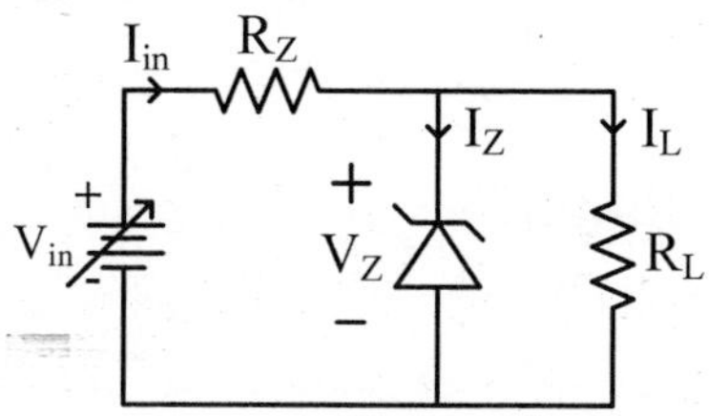

Fig. 1.9.4 Circuit schematic for the typical zener diode used for voltage protection with variable input voltage.

Let input varies between V_{in_min} and V_{in_max}. For fixed R_L, it is required to obtain the minimum, maximum and typical value of current limiter resistor R_Z. For the zener diode operating zener break down region, the voltage across the load resistor R_L should be greater than or equal to zener break down voltage V_Z. Using the voltage divider rule, the output voltage across the load resistor is given by,

$$V_o = \frac{V_{in\_min}}{R_L + R_z} \geq Vz \quad (1.9.8)$$

Assuming the above condition is satisfied then the minimum, maximum and typical value of current limiter resistor R_z taking into account input voltage variation.

For the minimum value of the current limiter resistor R_{z_min}, the current through the zener diode should not exceed I_{zmax}. Assuming the zener diode maintains the zener break down voltage of V_z across it, the current through load resistance is given by,

$$I_L = \frac{V_z}{R_L} \quad (1.9.9)$$

Therefore, input current I_{in} with V_{in_max} is given by,

$$I_{in} = I_{z\max} + I_L$$

$$I_{in} = I_{z\max} + \frac{V_z}{R_L}$$

$$R_{z\_\min} = \frac{V_{in\_\max} - V_z}{I_{in}}$$

$$R_{z\_\min} = \frac{V_{in\_\max} - V_z}{I_{z\max} + \frac{V_z}{R_L}} \tag{1.9.10}$$

The designed value of the current limiter resistor R_z should not be less than R_{z_min} computed above. Similarly, for a maximum value of current limiter resistor R_{z_max}, the current through the zener diode should not be less than exceed I_{zmin}. Assuming the zener diode maintains the zener break down voltage of V_z across it, the current through load resistance is given by,

$$I_L = \frac{V_z}{R_L} \tag{1.9.11}$$

Therefore input current I_{in} with a minimum value of the input voltage V_{in_min} is given by,

$$I_{in} = I_{z\min} + I_L$$

$$I_{in} = I_{z\min} + \frac{V_z}{R_L}$$

$$R_{z\_\max} = \frac{V_{in\_\min} - V_z}{I_{in}}$$

$$R_{z\_\max} = \frac{V_{in\_\min} - V_z}{I_{z\min} + \frac{V_z}{R_L}} \tag{1.9.12}$$

Thus, the value of the current limiter resistor R_z should lie between R_{z_min} and R_{z_max} to operate the zener diode in zener breakdown region. For zener diode to operate in typical operating value of current limiter resistor R_z is given by,

$$R_{z\_op} = \frac{V_{in} - V_z}{I_{zop} + \frac{V_z}{R_L}} \tag{1.9.13}$$

Case III: Take a case when input voltage source V_{in} is fixed and output load resistance R_L is variable as shown in Fig. 1.9.5. It is required to operate the zener diode near the zener breakdown region. Let output load resistance R_L varies between R_{L_min} and R_{L_max}. For input voltage source V_{in}, it is required to obtain the minimum, maximum and typical value of current limiter resistor R_Z.

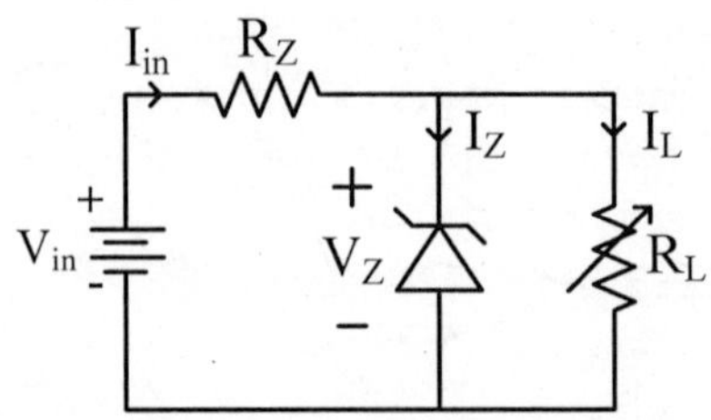

Fig. 1.9.5 Circuit schematic for the typical zener diode used for voltage protection with variable load resistance.

For the zener diode operating zener break down region, the voltage across the load resistor R_L should be greater than or equal to the zener break down voltage V_z. Using the voltage divider rule, the output voltage across the load resistor is given by,

$$V_o = \frac{V_{in}}{R_{L\_min} + R_z} \geq Vz \tag{1.9.14}$$

Assuming the above condition is satisfied then the minimum, maximum and typical value of current limiter resistor R_z taking into account input voltage variation. For the minimum value of current limiter resistor R_{z_min}, the current through the zener diode should not exceed I_{zmax}. Assuming the zener diode maintains the zener break down voltage of V_z across it, the current through load resistance is given by,

$$I_L = \frac{V_z}{R_{L\_min}} \tag{1.9.15}$$

Therefore, input current I_{in} with R_{L_min} is given by,

$$I_{in} = I_{z\max} + I_L$$

$$I_{in} = I_{z\max} + \frac{V_z}{R_{L\_min}}$$

$$R_{z\_min} = \frac{V_{in} - V_z}{I_{in}}$$

$$R_{z\_min} = \frac{V_{in} - V_z}{I_{z\max} + \dfrac{V_z}{R_{L\_min}}} \tag{1.9.16}$$

The designed value of the current limiter resistor R_z should not be less than R_{z_min} computed above. Similarly, for the maximum value of current limiter resistor R_{z_max}, the current through the zener diode should not be less than exceed I_{zmin}. Assuming the zener diode maintains the zener break down voltage of V_z across it, the current through load resistance is given by,

$$I_L = \frac{V_z}{R_{L\_max}} \tag{1.9.17}$$

Therefore input current I_{in} with a minimum value of the input voltage V_{in_min} is given by,

$$I_{in} = I_{z\min} + I_L$$

$$I_{in} = I_{z\min} + \frac{V_z}{R_{L\_max}}$$

$$R_{z\_max} = \frac{V_{in\_min} - V_z}{I_{in}}$$

$$R_{z\_max} = \frac{V_{in\_min} - V_z}{I_{z\min} + \dfrac{V_z}{R_{L\_max}}} \tag{1.9.18}$$

Thus, the value of the current limiter resistor R_z should lie between R_{z_min} and R_{z_max} to operate the zener diode in zener breakdown region. For zener diode to operate in typical operating value of current limiter resistor R_z is given by,

$$R_{z\_op} = \frac{V_{in} - V_z}{I_{zop} + \frac{V_z}{R_L}} \quad (1.9.19)$$

1.10 CLIPPING CIRCUIT

One of the important applications of the PN junction diode is the clipping or clipper circuit. Clipper circuit as its name implies clips or chops the portion of the signal waveform without changing the other portion. Basically, there are two types of clipper circuit-series and parallel type clipper circuits. The term series and parallel comes from the position of the PN junction diode with respect to the load. When the PN junction diode is in series with the load, then the given clipper circuit represents the series type clipper circuit. Similarly, when the PN junction diode is in parallel with the load, then the given clipper circuit represents the parallel type clipper circuit. DC supply or battery in the clipper circuit is typically connected in series with the PN junction diode.

For analyzing the clipper circuits below are the steps to be followed while solving the circuits having a PN junction diode and battery which clips the portion of the input signal waveform as follows:

i. First obtain the Thevenin equivalent circuit across two terminals of the PN junction diode. In other words, remove the PN junction diode or open the terminals in the circuit across the PN junction diode.
ii. Then find the Thevenin voltage across the PN junction diode assuming the anode and cathode terminal of the diode as positive and negative respectively.
iii. As the input signal waveform varies with time, derive the condition for the Thevenin voltage greater than and less than zero. Here, assume zero current through the resistor resulting in zero voltage drops across it.
iv. For the positive or greater than zero Thevenin voltage across the PN junction diode is considered forward-biased. Thus, the PN junction diode can be replaced by a short circuit or a drop of 0.7V in the circuit. The circuit then can be solved for determining the output voltage using Kirchoff's voltage or current law.
v. For the negative or less than zero Thevenin voltage, the PN junction diode is considered reversed-biased. Thus, the PN junction diode can be replaced by an open circuit. The circuit then can be solved for determining the output voltage using Kirchoff's voltage or current law.

1.10.1 Series Type Clipper Circuit

Fig. 1.10.1.1 shows the circuit of a series type clipper. The series type clipper circuit consists of an AC supply $V_{in}(t)$, PN junction diode D_1, voltage source V_{dc} and a load resistance R_L. The voltage across the load resistance is V_O and the current flowing through the diode D_1 is I_D.

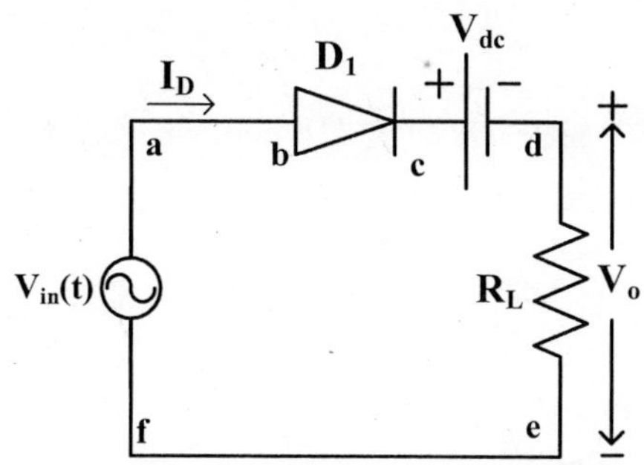

Fig. 1.10.1.1 Circuit of a series type clipper.

i. For the analyzing the above circuit, the steps discussed previously must be used. Define the Thevenin circuit by removing diode in the given clipper circuit as shown in Fig. 1.10.1.2.

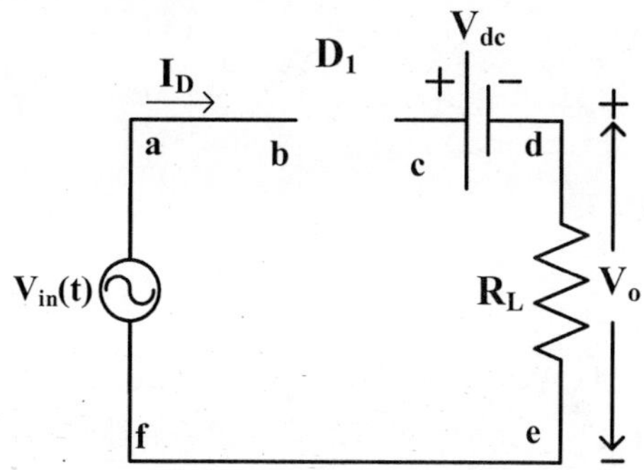

Fig. 1.10.1.2 Circuit of a series type clipper by replacing the PN junction diode by open circuit (Step-I).

ii. Carry out second step by applying Thevenin theorem across terminal 'b' and 'c' as shown in Fig. 1.10.1.3. Applying KVL in the loop abcdef we have,

$$V_{in} - V_{dc} - V_{th} - V_O = 0 \tag{1.10.1.1}$$

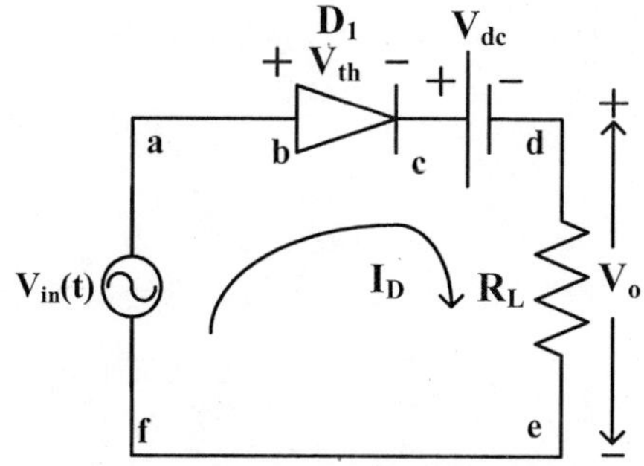

Fig. 1.10.1.3 Circuit of a series type clipper by applying Thevenin voltage across the PN junction diode (Step-II).

iii. Assuming voltage across resistor R_L to be zero (step-III) as the current through it is zero so that V_O is zero. Thus, terminal 'd' and 'e' are having same potential one can compute the biasing condition for the PN junction diode.

$$V_{in} - V_{dc} - V_{th} = 0 \tag{1.10.1.2}$$

iv. For PN junction diode to operate in forward-biased condition V_{th} should be greater than zero.

$$V_{th} \geq 0$$

$$V_{in} - V_{dc} - V_{th} = 0$$

$$V_{in} - V_{dc} = V_{th}$$

$$V_{in} - V_{dc} = V_{th} \geq 0$$

$$V_{in} - V_{dc} \geq 0$$

$$V_{in} \geq V_{dc} \tag{1.10.1.3}$$

Thus, the PN junction diode will be in a forward-biased condition when above condition is satisfied. Thus, assuming the given condition is satisfied and PN junction diode is in forward-biased condition the clipper circuit becomes,

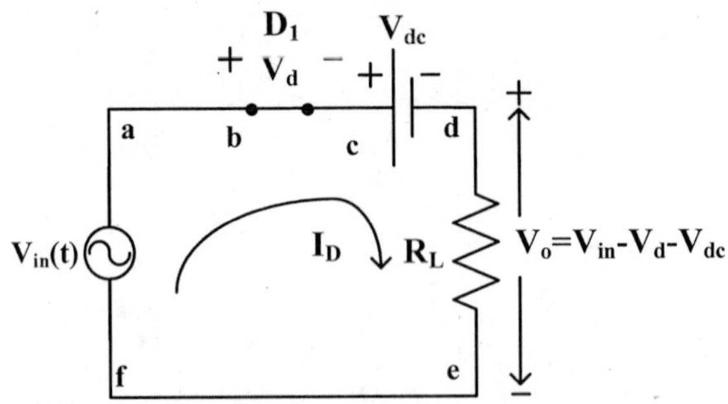

Fig. 1.10.1.4 Series type clipper in which the PN junction diode is replaced the closed switch (Step-IV).

Applying KVL in the loop abcdef, to solve for the output voltage we have,

$$V_{in} - V_{dc} - V_{o} - V_{d} = 0$$

$$V_{o} = V_{in} - V_{dc} - V_{d} \tag{1.10.1.3}$$

v. During the reverse biased condition of PN junction diode, the PN junction diode is replaced by an open switch. This results in zero voltage across the load resistance R_L.

$$V_{o} = 0 \tag{1.10.1.4}$$

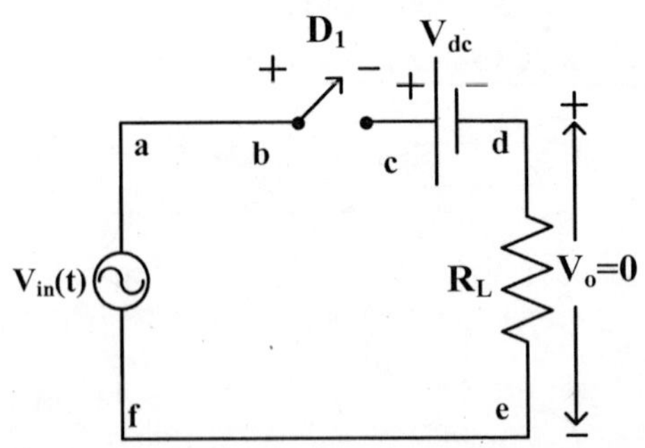

Fig. 1.10.1.5 Series type clipper in which the PN junction diode is replaced open switch (Step-V).

Fig. 1.10.1.6 shows the waveforms of the series type clipper. Whenever the applied input signal magnitude is greater than the voltage V_{dc}, the diode is forward biased and the output voltage will appear across the load resistance R_L which is a part of the input voltage. Whenever, the applied input voltage is less than the voltage V_{dc}, the diode is reverse biased and the output voltage will have zero value. In this way, the series clipping circuit clips the part of input signal.

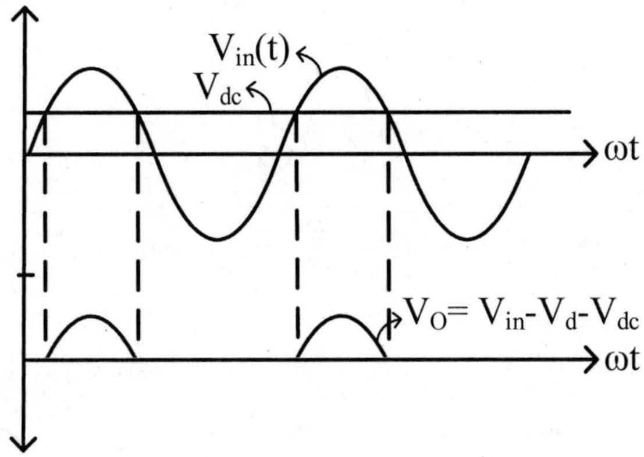

Fig. 1.10.1.6 Waveforms of the series type clipper.

1.10.2 Parallel Type Clipper Circuit

Fig. 1.10.2.1 shows the circuit of a parallel type clipper. The parallel type clipper circuit consists of an AC supply $V_{in}(t)$, PN junction diode D_1, resistance R_s, voltage source V_{dc}. The voltage across the branch c and e is V_O and the current flowing through the diode D_1 is I_D.

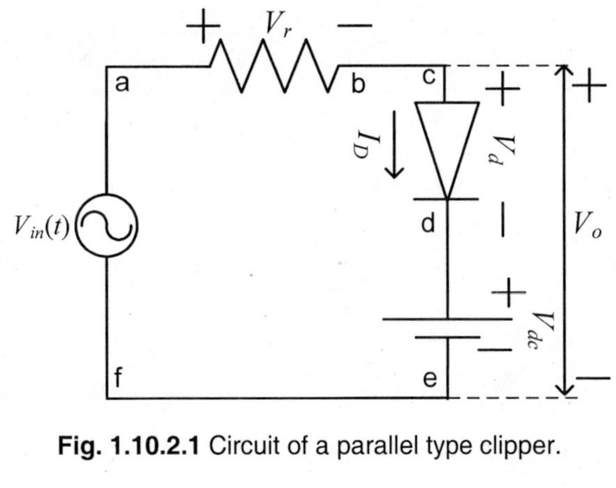

Fig. 1.10.2.1 Circuit of a parallel type clipper.

i. Carry out first step by applying Thevenin theorem across terminal 'c' and 'd'. Applying KVL in the loop abcdef we have,

$$V_{in} - V_{dc} - V_{th} - V_r = 0 \tag{1.10.2.1}$$

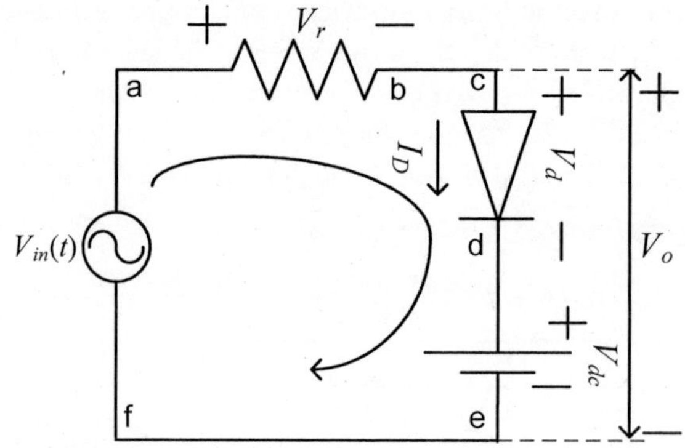

Fig. 1.10.2.2 Circuit of a parallel type clipper used in Step-I.

ii. Assuming voltage across resistor 'R' to be zero as the current through it is zero. Thus, terminal 'a' and 'b' are having same potential one can compute the biasing condition for the diode. For the diode to operate in forward biased condition V_{th} should be greater than zero.

$$V_{th} \geq 0$$
$$V_{in} - V_{dc} - V_{th} = 0$$
$$V_{in} - V_{dc} = V_{th}$$
$$V_{in} - V_{dc} = V_{th} \geq 0$$
$$V_{in} - V_{dc} \geq 0$$
$$V_{in} \geq V_{dc} \tag{1.10.2.2}$$

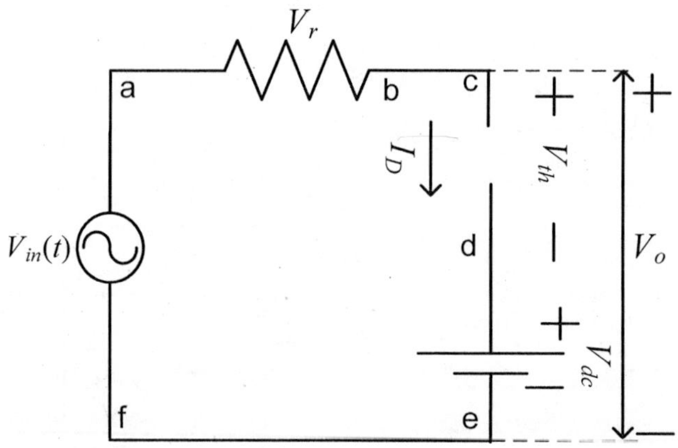

Fig. 1.10.2.3 Circuit of a parallel type clipper used in Step-II.

iii. Thus, diode will be in forward biased condition when above condition is satisfied. Thus, assuming the given condition is satisfied and diode is in forward biased condition the clipper circuit becomes, Applying KVL in the loop abcdef, to solve for output voltage we have,

$$V_o - V_{dc} + V_d = 0$$
$$V_o = V_{dc} - V_d \tag{1.10.2.3}$$

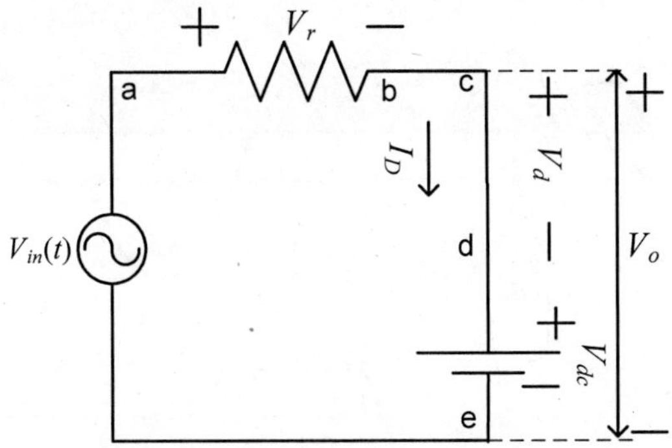

Fig. 1.10.2.4 Circuit of a parallel type clipper used in Step-III.

iv. Thus, diode will be in reverse biased condition when condition is satisfied. Thus, assuming the given condition is satisfied and diode is in forward biased condition the clipper circuit becomes,

Applying KVL in the loop abcdef, to solve for output voltage we have,

$$V_{in} - V_r - V_o = 0$$

$$V_{in} - IR - V_o = 0 \qquad (I = 0)$$

$$V_o = V_{in} \tag{1.10.2.4}$$

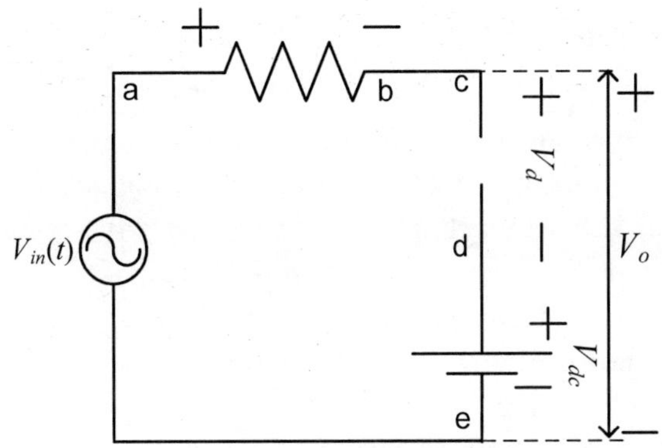

Fig. 1.10.2.5 Circuit of a parallel type clipper used in Step-IV.

Fig. 1.10.2.6 shows the waveforms of the parallel type clipper. Whenever the applied input signal magnitude is greater than the voltage V_{dc}, the diode is forward biased and the output voltage will be clamped to DC bus voltage. Whenever, the applied input voltage is less than the voltage V_{dc}, the diode is reverse biased and the output voltage will have the same magnitude as the input voltage. In this way, the parallel clipping circuit clips the part of the input signal.

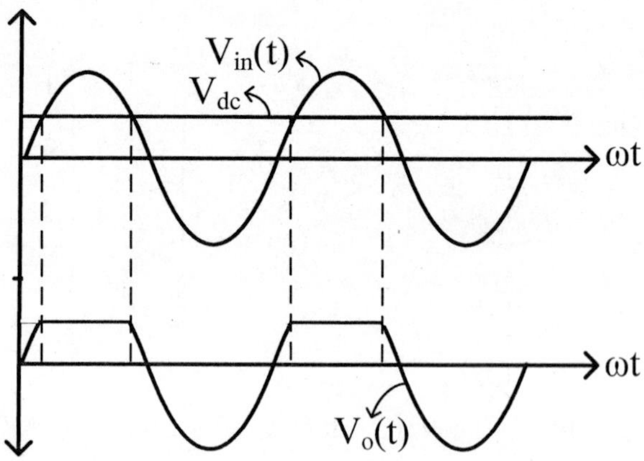

Fig. 1.10.2.6 Waveforms of the parallel type clipper.

1.11 CLAMPING CIRCUIT

Clamper circuit as its name implies clamps the AC signal with DC offset. During the clamping process, the nature of the AC signal is not changed but only there is change in the offset of the signal. The clamper circuit normally uses a capacitor to clamp the output voltage. In general there are two types of clamper circuits: positive clamper and negative clamper

1.11.1 Positive Clamper Circuit

In the positive clamper circuit, the DC offset is provided such that the input is shifted above 0 V. The schematic of the positive clamper circuit is shown in Fig. 1.11.1.1. The circuit consists of an AC input signal $V_{in}(t)$, capacitor C, PN junction diode D_1 and the load resistance R_L. During the first positive half cycle of the input signal, the PN junction diode D_1 is reverse-biased such that the input voltage will appear across the output load resistance. In the first negative half cycle of the input signal, the PN junction diode D_1 will get forward-biased and the capacitor will get charged to the peak reverse voltage. In the second positive half cycle of the input signal, the capacitor voltage will get added up with the input voltage and the signal will get level shifted to double the peak voltage of the input signal. Fig. 1.11.1.2 shows the waveforms of the positive clamper circuit. From the waveforms, it can be observed that the dc offset is added to the input signal such that the output voltage Vo(t) is shifted above 0 V.

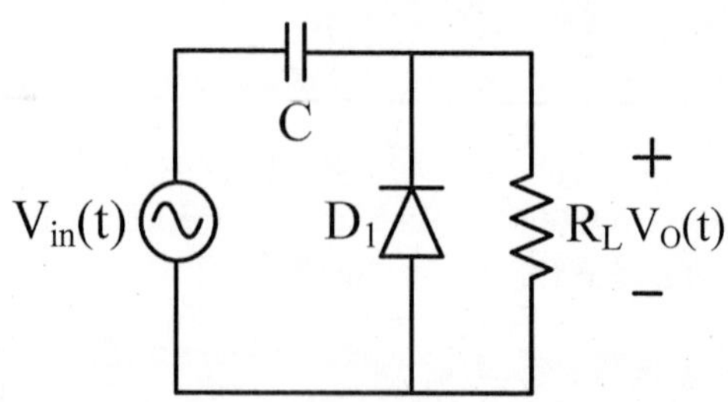

Fig. 1.11.1.1 Circuit of a positive clamper circuit.

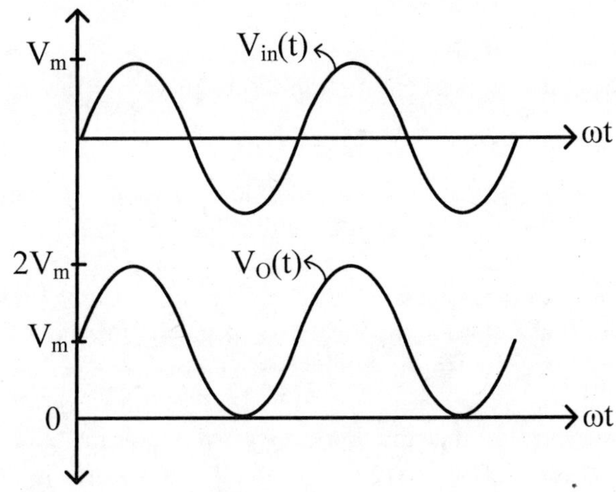

Fig. 1.11.1.2 Waveforms of the positive clamper circuit.

1.11.2 Negative Clamper Circuit

In the negative clamper circuit, the DC offset is provided such that the input is shifted below 0 V. The schematic of the negative clamper circuit is shown in Fig. 1.11.2.1. The circuit consists of an AC input signal $V_{in}(t)$, capacitor C, PN junction diode D_1 and the load resistance R_L. During the first positive half cycle of the input signal, the PN junction diode D1 is forward-biased such that the capacitor will get charged to the peak positive voltage. In the first negative half cycle of the input signal, the PN junction diode D1 will get reverse-biased and the capacitor will get charged to the peak reverse voltage. In the second positive half cycle of the input signal, the capacitor voltage will get added up with the input voltage and the signal will get level shifted to double the peak voltage of the input signal. Fig. 1.11.2.2 shows the waveforms of the negative clamper circuit. From the waveforms, it can be observed that the dc offset is added to the input signal such that the output voltage Vo(t) is shifted below 0 V.

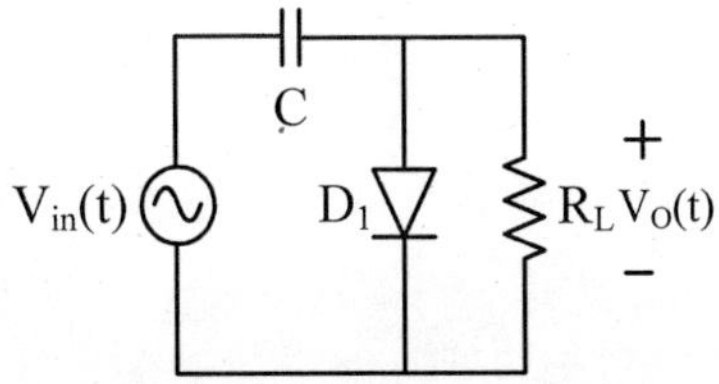

Fig. 1.11.2.1 Circuit of a negative clamper circuit.

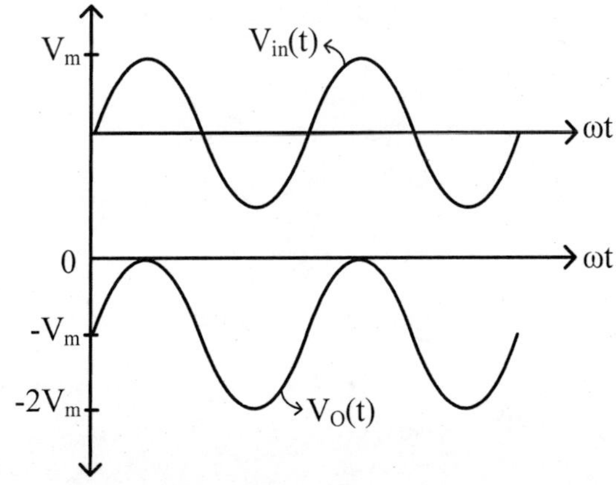

Fig. 1.11.2.2 Waveforms of the negative clamper circuit.

UNIT SUMMARY

The chapter presents the detail of the PN junction diode and zener diode. The complete analysis of the PN junction diode which includes its I-V characteristics and biasing modes are discussed in the chapter. Different applications like clipper, clamper and rectifier are also discussed in detail. iV is given., tunnel diode helps students to get a primary idea about the operation of the PN junction diode under both unbiased and biased operating conditions. Once the operation of the PN junction diode is clear, then the of the PN junction diode are discussed to understand the electrical behaviour of the PN junction diode. Then the I-V characteristic equation of the PN junction diode is explained to determine the relationship between the current and voltage in the PN junction diode. This characteristics equation is necessary to identify the behaviour of the various operating conditions. All these concepts are required for obtaining the operating point of the PN junction diode. The operating point will always ensure that the diode always operates in stable and desired operating conditions. Next, the practical applications of the PN junction diode are discussed to identify its real-time applications like rectifiers, clipping circuits and clamping circuits. The unit also explains the construction, working and characteristics of the zener diode All these concepts are essential to in understanding the operation of various electronic circuits.

The laboratory experiments related to the I-V characteristics of the PN junction diode, zener diode, clipping and clamping circuits, and design of the rectifiers were presented in the appendix of the book.

EXERCISES

Multiple Choice Questions

1.1 When PN junction is in forward bias, by increasing the battery voltage

a. Circuit resistance increases
b. Current through P-N junction increases
c. Current through P-N junction decreases
d. None of the above happen

1.2 When a PN junction is reverse-biased

a. Holes and electrons tend to concentrate towards the junction
b. The barrier tends to break down
c. Holes and electrons tend to move away from the junction
d. None of the above

1.3 Zener diodes are also known as

a. Voltage regulators
b. Forward bias diode
c. Breakdown diode
d. None of the above

1.4 For the output in below figure Value of the V_1 and V_2 is

a. 6, 8
b. 8, 6
c. 9, 9
d. 8, -6

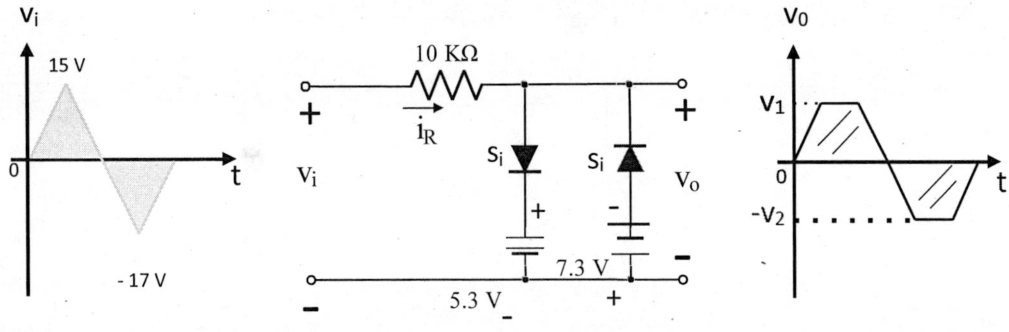

Answers for Multiple Choice Questions			
1.1 **a**	1.2 **c**	1.3 **a**	1.4 **a**

Short Answer Type Questions

1.1 Give details of the formation of p-n junction with p and n type materials?

1.2 Explain the P-N junction diode operation in forward biased condition.

1.3 Give the circuit diagram using diode for, Gate using negative logic, and explain its working?

1.4 Explain two types of breakdown in Zener diode.

Long Answer Type Questions

1.1 For the network, determine the range of RL and IL that will result in VRL being maintained at 10 V. Also, determine the maximum wattage rating of the zener diode in below figure:

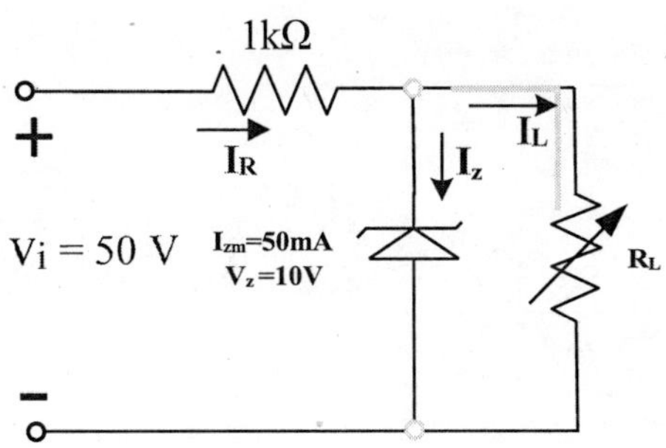

1.2 For the network determine the range of Vi that will maintain VL at 8V and not exceed the maximum power rating of the Zener diode.

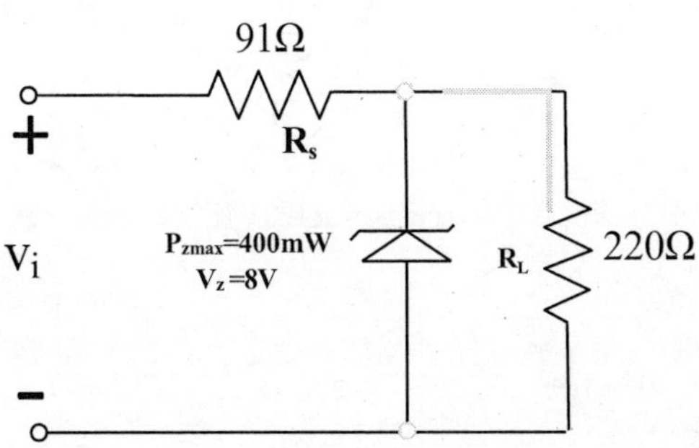

1.3 Design a clamper circuit for the given input and output waveform.

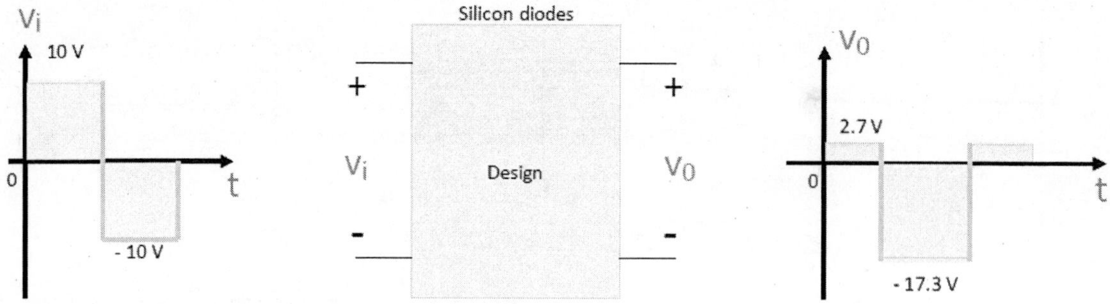

1.4 Design a clamper to perform the function indicated in

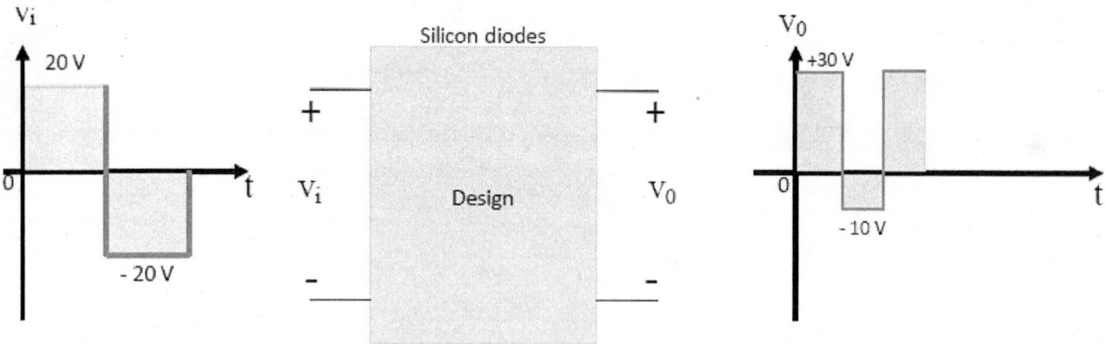

KNOW MORE

For Additional Info Scan Me

REFERENCES

1. Robert Boylestad, Louis Nashelsky- Electronic Devices and Circuit Theory, Prentice Hall Upper Saddle River, New Jersey Columbus, Ohio
2. Donald L. Schilling, Charles Belove -Electronic Circuits: Discrete and Integrated, McGraw-Hill Book Company New York St. Louis San Francisco Toronto London Sydney.
3. Integrated Electronics Analog Digital Circuits, Jacob Millman and D. Halkias, McGraw Hill.

02 BIPOLAR JUNCTION TRANSISTOR

UNIT SPECIFICS

Through this unit we have discussed the following aspects:

- *Introduction to the BJT;*
- *Operation of the transistor;*
- *Various configurations of the transistor;*
- *Biasing of the transistor;*
- *Quiescent point (Q-Point) of the transistor;*
- *Selection of quiescent point (Q-Point) of the transistor;*
- *Biasing stabilization;*
- *Biasing circuits for the transistors;*
- *BJT as a switch;*
- *BJT as an amplifier;*
- *Hybrid parameter model of the transistor;*
- *Small signal analysis of the amplifier using the h-parameters;*
- *Small signal analysis of the amplifier using the simplified h-parameters;*
- *Current mirror;*
- *High-frequency response of the amplifiers.*

RATIONALE

This fundamental unit on the bipolar junction transistor helps students to get a primary idea about the operation of the bipolar junction transistor under both unbiased and biased operating conditions. Next various configurations of the transistor like common base, common emitter and common collector were discussed. After obtaining a good knowledge on these topics, next the crucial aspect that decides the operation of bipolar junction transistor in the electronic circuit needs to be studied. Biasing of the bipolar junction transistor is a crucial aspect that decides the operation of the bipolar junction transistor in any electronic circuit. Next, the operating point of the Q-point of the bipolar junction transistor is discussed. The operating point will ensure that the bipolar junction transistor always operates in stable and desired operating conditions. The various biasing circuits of the bipolar junction transistor are discussed next. After the biasing, the operation of BJT as a switch and amplifier is discussed. This includes the operation of BJT as closed switch and open switch. Along with this, the amplifier configurations like common base configuration, common emitter configuration and common collector configuration with their input and output characteristics are discussed. The parameters of the amplifiers like input impedance, voltage gain, current gain and output admittance need to be obtained. For which a small signal analysis is required. This small signal analysis uses the h-parameter model of the bipolar junction transistor. The detailed steps for obtaining the amplifier parameters like input impedance, voltage gain, current gain and output admittance using the small signal analysis is presented. To minimize the complexity in obtaining the amplifier parameters the small signal analysis using a simplified h-parameter model is also presented. Next the high-frequency response of the amplifier is presented to analyze the variation of the amplifier gain with frequency. All these concepts are essential to in understanding the operation of various electronic circuits. The laboratory experiment related to input and output characteristics of the bipolar junction transistor in common emitter configuration is presented in the appendix of the book.

PRE-REQUISITES

Physics: Semiconductor (Class XII)

UNIT OUTCOMES

List of outcomes of this unit is as follows:

U2-O1: Understand the basics of bipolar junction transistor along with its operation

U2-O2: Realize various biasing circuits required the bipolar junction transistor along with the selection of the operating point

U2-O3: Realize various operations of the bipolar junction transistor in the electronic circuits

U2-O4: Determine the amplifier parameters by using the small signal analysis

U2-O5: Understand the high-frequency response of the amplifier

Unit-2 Outcomes	EXPECTED MAPPING WITH COURSE OUTCOMES (1- Weak Correlation; 2- Medium correlation; 3- Strong Correlation)				
	CO-1	CO-2	CO-3	CO-4	CO-5
U2-01	3	1	2	1	-
U2-02	3	1	2	1	-
U2-03	3	1	2	1	-
U2-04	3	1	3	1	-
U2-05	3	1	3	1	-

2.1 INTRODUCTION TO THE BJT

The transistor is a basically three terminal devices. In general, there are many types of transistors. The bipolar junction transistor (BJT) is one of the transistors. In the BJT both the majority and minority charge carriers (both electrons and holes) will contribute to the current conduction. Fig. 2.1.1 shows the construction of BJT. The BJT is a three-terminal two junction device that has emitter (E), base (B) and collector (C) terminals. The BJT is generally used for amplification and switching applications in electronic circuit.

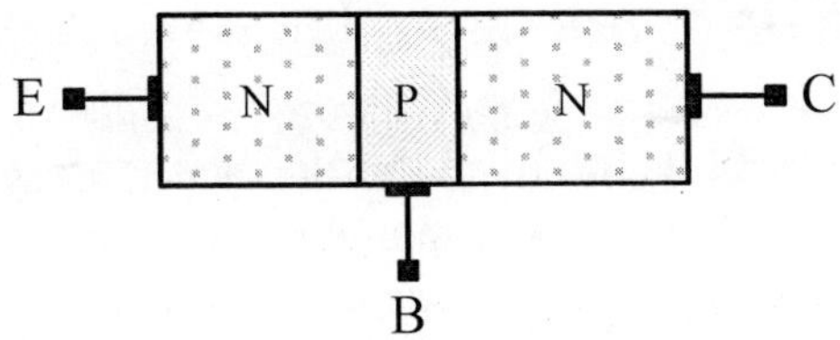

Fig. 2.1.1 Construction of BJT.

The BJT is commonly referred as bi-polar transistor. Based on the construction, there are two types of BJTs: (a) N-P-N type and (b) P-N-P type. The N-P-N type BJT is constructed by sandwiching a P-type semiconductor between two N-type semiconductors. Similarly, the P-N-P type BJT is constructed by sandwiching an N-type semiconductor between two P-type semiconductors. Fig. 2.1.2 shows the construction of N-P-N and P-N-P transistors.

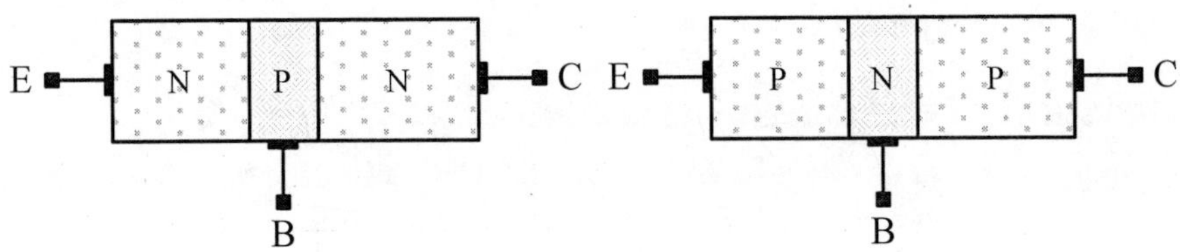

(a) N-P-N type **(b) P-N-P type**

Fig. 2.1.2 Construction of N-P-N type and P-N-P type transistors.

The middle region of the transistor is normally called the base (B). The base region will be generally very narrow in area and lightly doped. While the other two regions are normally called emitter (E) and collector (C). The emitter region is heavily doped and the collector region is moderately doped. The collector region is normally made wide to dissipate a larger amount of power while the emitter region is moderately wide. With the sand-witch arrangement and presence of three semi-conductor material, it results in the formation of two junctions in the device. Due to these three regions with semiconductor material, there are two junctions in the transistor. The first junction is between the emitter and the base which is normally called as emitter junction (J_E). The second region is between the collector and the base which is normally called as collector junction (J_C). Fig. 2.1.3 shows the symbol of the N-P-N and P-N-P transistors. The direction of the arrow mark is always in the direction of conventional current which is opposite to the flow of electrons between the emitter to the base. In the case of N-P-N type transistor, the electrons will flow from the emitter to the base so that the direction of the arrow mark is towards the emitter terminal. In case of P-N-P type transistor, the electrons will flow from the base to the emitter so that the director of arrow mark is towards the base terminal.

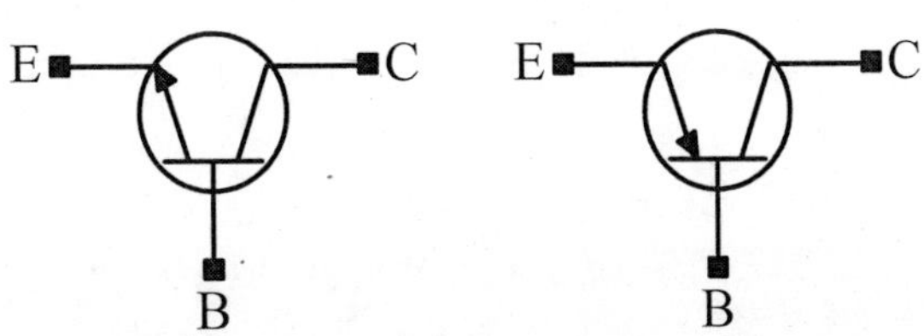

(a) N-P-N type **(b) P-N-P type**

Fig. 2.1.3 Symbol of N-P-N type and P-N-P type transistors.

2.2 OPERATION OF THE TRANSISTOR

As discussed above the transistor has two junctions: the emitter junction (J_E) and the collector junction (J_C). Presence of junctions results in the formation of the depletion layers, between the emitter-base terminals and the collector-base terminals as shown in Fig. 2.2.1. The formation of the depletion layers is due to the diffusion process, where the charge carriers in the depletion region will penetrate more into the lightly doped region compared to the heavily doped regions. The depletion layer formed between the emitter-base junction will penetrate more into the base region compared to the emitter region. Similarly, the depletion layer formed between the collector-base junction will penetrate more into the base region compared to the collector region. As a result, the effective area of the base region will reduce as shown in Fig. 2.2.1. Due to these depletion regions, there will be a barrier potential across these junctions which is called as barrier potential. It should be noted that for the current conduction through the transistor, a voltage greater than the barrier voltage must be applied. For typical practical doping conditions, the barrier voltage will be approximately 0.7V for the silicon transistor whereas for the germanium transistor, the barrier voltage will be approximately 0.3V.

2.2.1 Operation of the Biased Transistor

During the biasing, the external voltages are applied to the transistor. Due to these external voltages, the emitter junction (J_E) and the collector junction (J_C) can be forward or reverse-biased. Depending on the nature of these junctions the transistor can be operated in different regions. The four possible bias combinations of the transistor along with the region and application is shown in Table 2.2.1.1

For amplifier applications, the transistor must be operated in the active region. In order to operate the transistor in the active region, the emitter junction must be forward biased and the collector junction must be reverse biased. The biasing for the transistor to operate in the active region is shown in Fig. 2.2.1.1. The emitter junction is forward-biased by using an external power supply V_{EE}. Similarly, the collector junction is reverse-biased by using the external power supply V_{CC}.

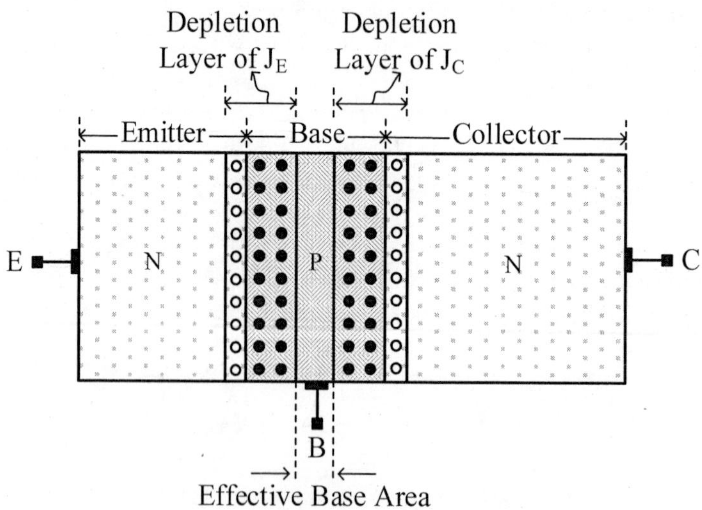

(a) N-P-N type

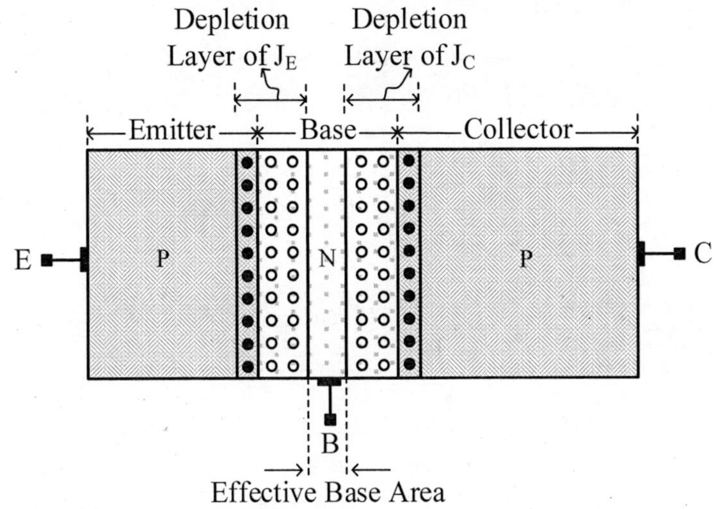

(b) P-N-P type

Fig. 2.2.1 N-P-N type and P-N-P type transistors at equilibrium condition.

Table 2.2.1.1

Emitter Junction (J_E)	Collector Junction (J_C)	Region	Application
Forward-Biased	Reverse-Biased	Active	Amplifier
Forward-Biased	Forward-Biased	Saturation	Closed Switch
Reverse-Biased	Reverse-Biased	Cut-off	Open Switch

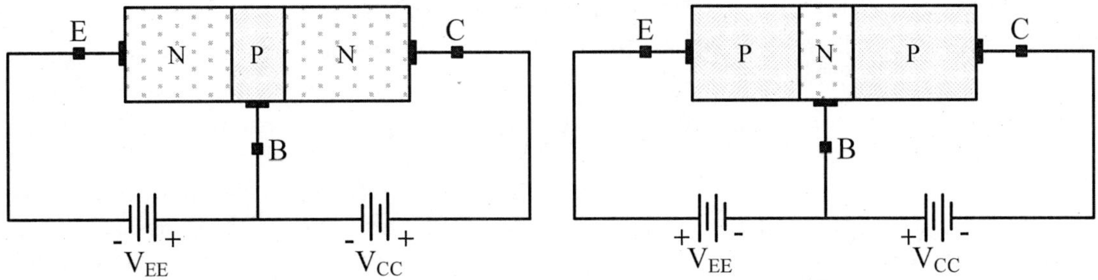

Fig. 2.2.1.1 Biasing for the transistor to operate in the active region.

In order to operate the transistor as a closed switch, the transistor must operate in the saturation region. For operation in the saturation region, the emitter junction and the collector junction must be forward-biased. The biasing for the transistor to operate in the saturation region is shown in Fig. 2.2.1.2. The emitter junction and the collector junction are forward-biased by using the external power supplies V_{EE} and V_{CC}.

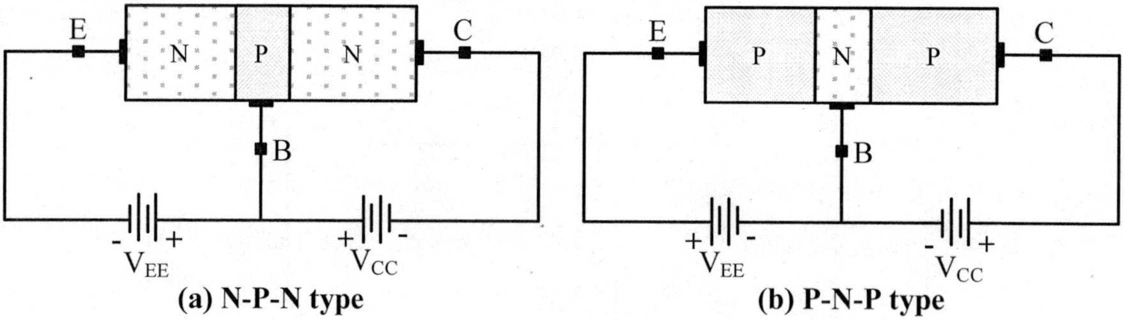

Fig. 2.2.1.2 Biasing for the transistor to operate in the saturation region.

In order to operate the transistor as an open switch, the transistor must be in the cutoff region. For the cut-off region, the emitter junction and the collector junction must be reverse-biased. The biasing for the transistor to operate in the cut-off region is shown in Fig. 2.2.1.3. The emitter junction and the collector junction are reverse-biased by using the external power supplies V_{EE} and V_{CC}.

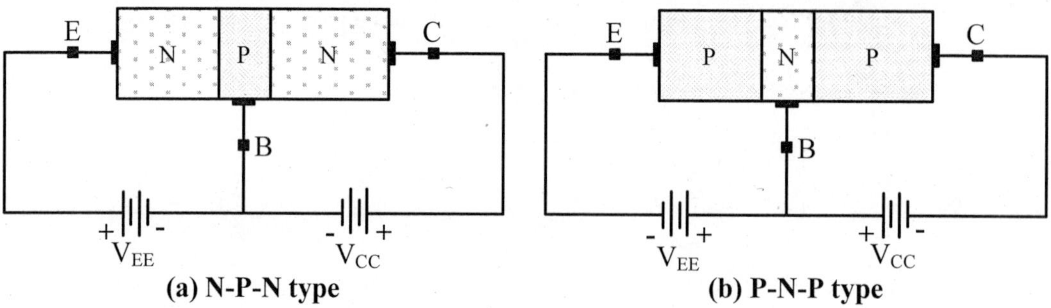

Fig. 2.2.1.3 Biasing for the transistor to operate in the cut-off region.

2.2.1.1 Operation of the Transistor in the Active Region

For the N-P-N transistor to operate in the active region, the emitter junction must be forward-biased and the collector junction must be reverse-biased. The emitter junction is forward-biased by using an external power supply V_{EE} as shown in Fig. 2.2.2.1.1. Similarly, the collector junction is reverse-biased by using the external power supply V_{CC} as shown in Fig. 2.2.2.1.1. The resistors R_E and R_C at the emitter and collector terminals respectively are normally used to limit the current and also biasing the transistors. As the emitter junction is forward-biased, the width of the depletion at the emitter junction reduces. Similarly, the collector junction is reverse-biased so that the width of the depletion at the collector junction increases. Due to the external power supply V_{EE}, the electrons in the emitter region crosses the junction J_E and enter the base region. Since the base region is lightly doped, only some of the electrons coming from the emitter region combine with the holes in the base region. Due to this

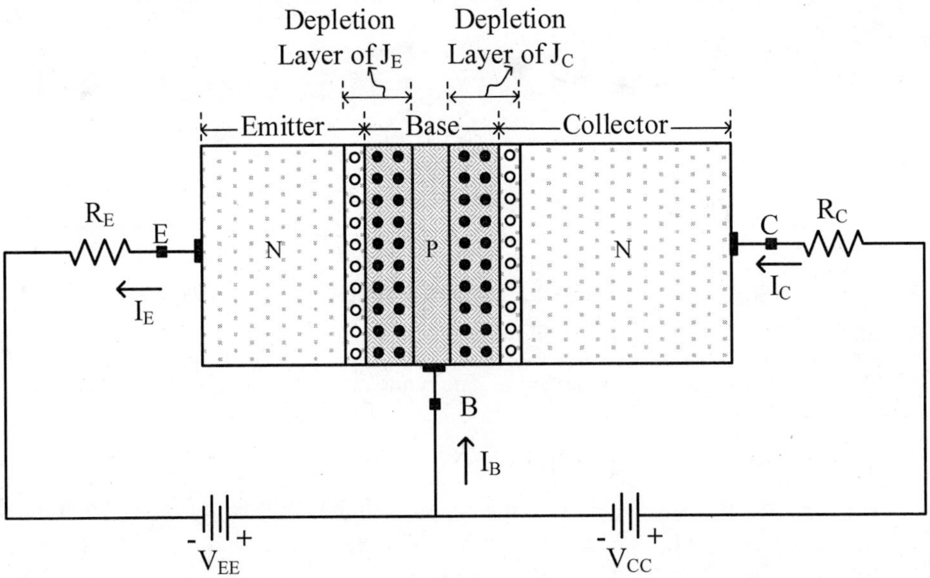

Fig. 2.2.2.1.1 Operation of N-P-N transistor in the active region.

recombination, there will be a current I_B in the base region. Base current also has a part when carriers from the base region are also injected into the emitter region due to forward bias. The remaining electrons in the base region cross the junction J_C and enter into the collector region to the positive of the power supply V_{CC}. This results in the collector current I_C and this current is due to the majority charge carriers (electrons). Therefore, the emitter current I_E is given by

$$I_E = I_B + I_C \tag{2.2.2.1.1}$$

Similarly, because of the reverse-biased collector junction, a reverse leakage current I_{CO} due to minority charge carriers (holes) will flow in the collector region. Therefore, the collector current I_C is given by

$$I_C = I_{C(majority)} + I_{CO} \tag{2.2.2.1.2}$$

Same strategy holds true for P-N-P transistor for its operation in the active region.

2.3 VARIOUS CONFIGURATIONS OF THE TRANSISTOR

In general, most of electronic circuits are two-port networks which means there are two input terminals and two output terminals. The transistor is basically a three-terminal device. In order to construct an electronic circuit using the transistor, one terminal of the transistor needs to be common for the input and output terminals. This leads to a common base, common emitter and common collector configuration. Fig. 2.3.1.1 shows the common base, common emitter and common collector configurations. In the common base configuration, the base terminal is common to both the input and output terminals of the circuit. In the common emitter configuration, the emitter terminal is common to both the input and output terminals of the circuit. Similarly, in the common collector configuration, the collector terminal is common to both the input and output terminals of the circuit.

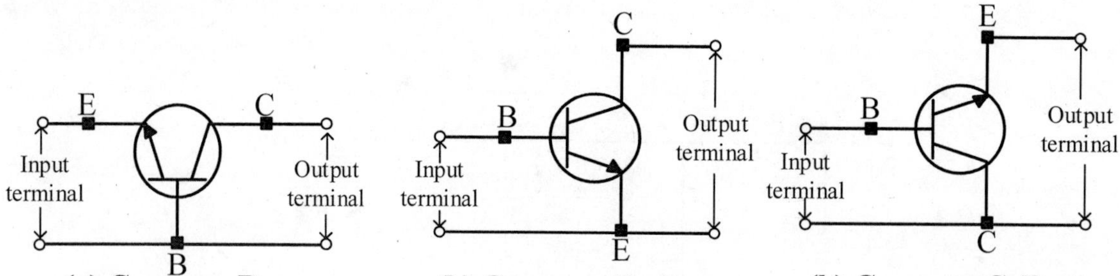

(a) Common Base Configuration **(b) Common Emitter Configuration** **(b) Common Collector Configuration**

Fig. 2.3.1.1 Common base, common emitter and common collector configurations of the transistor.

2.3.1 Common Base Configuration

In the case of the common base configuration, the base terminal is kept common for both the input and output terminals of the circuit. Fig. 2.3.1.1.1 shows the common base configuration of the transistor. The resistors R_C and R_E are placed to limit the current. The voltage sources V_{XX} and V_{CC} are given for the biasing of the transistor. The emitter and base terminals act as input terminals and the collector and base terminals act as output terminals. The voltage V_{EB} is the input voltage to the transistor and the

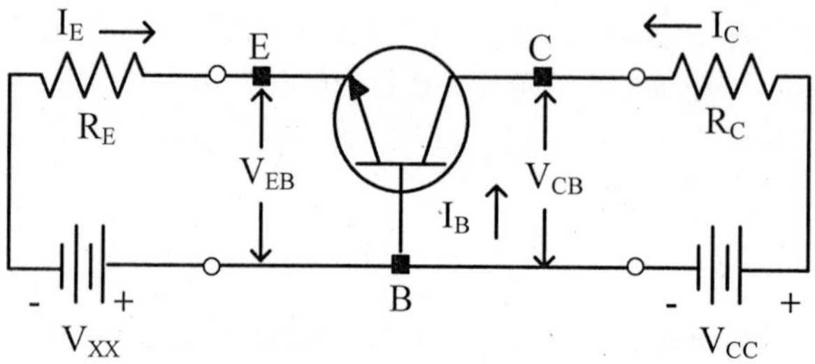

Fig. 2.3.1.1.1 Common base configuration of the transistor.

current I_E is the input current of the transistor. The voltage V_{CB} is the output voltage to the transistor and the current I_C is the output current of the transistor.

In the common base configuration, I_E is the input current of the transistor and I_C is the output current of the transistor. The current gain α in the common base configuration is given by

$$\alpha = \frac{I_C}{I_E} \tag{2.3.1.1.1}$$

The collector current I_C is again equal to

$$I_C = I_{C(majority)} + I_{CBO} \tag{2.3.1.1.2}$$

where $I_{C(majority)}$ is the current injected due to the majority charge carriers and I_{CBO} is the reverse current flowing due to minority charge carriers between collector and base with emitter open. This current is very negligible when compared to $I_{C(majority)}$. Therefore, the collector current I_C is almost equal to $I_{C(majority)}$.

$$I_C = I_{C(majority)} = \alpha I_E \tag{2.3.1.1.3}$$

The emitter current I_E is almost equal

$$I_E = \frac{I_C}{\alpha} \tag{2.3.1.1.4}$$

Again, the emitter current IE is given by

$$I_E = I_C + I_B \tag{2.3.1.1.5}$$

$$I_B = I_E - I_C = I_E - \alpha I_E = (1-\alpha) I_E \tag{2.3.1.1.6}$$

The collector current I_C is always less than the emitter current I_E. Therefore, the current gain α in the common base configuration is always less than unity. The typical values of current gain α will be from 0.95 to 0.995.

• Next, the input characteristics need to be plotted. The input characteristics are plotted between the input voltage V_{EB} and the input current I_E at a constant output voltage V_{CB}. Fig. 2.3.1.1.2 shows the input characteristics of the common base configuration. After the barrier potential voltage (i.e., 0.7V for the silicon and 0.3V for germanium), the emitter current increases rapidly with the increase in the input voltage V_{EB}. This results in very low dynamic input resistance. With the increase in the voltage V_{CB}, the width of the depletion layer of collector junction J_C increases, as a result, the effective base area reduces. This slightly increases the emitter current I_E. This effect is called early effect. The output characteristics are plotted between the output voltage V_{CB} and the output current I_C at a constant input current I_E. Fig. 2.3.1.1.3 shows the output characteristics of the common base configuration. There are three regions in the output characteristics: active, saturation and cut-off region. During the active region, the emitter junction is forward-biased and the collector junction is reverse- biased. In this region, the collector current I_C is almost constant for various values of V_{CB} and approximately equal to the emitter current I_E. The collector current I_C increases with the increase in emitter current I_E. The dynamic output

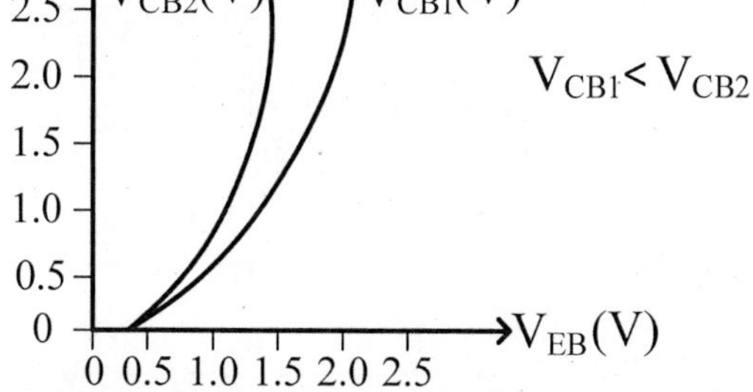

Fig. 2.3.1.1.2 Input characteristics of common base configuration.

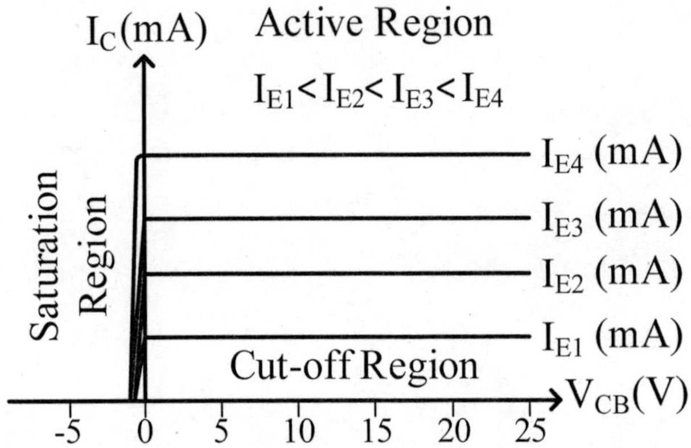

Fig. 2.3.1.1.3 Output characteristics of common base configuration.

resistance is very high due to the constant collector current I_C. In the saturation region, both the emitter junction and the collector junctions are forward-biased. In the saturation region, with the decrease in the negative value of V_{CB}, the width of the depletion layer increases, the results in an increase in the collector current I_C. In the cut-off region, both the emitter junction and the collector junctions are reverse-biased. This results in the zero-collector current I_C for various values of V_{CB} for a finite value of I_E.

2.3.2 Common Emitter Configuration

In the case of the common emitter configuration, the emitter terminal is kept common for both the input and output terminals of the circuit. Fig. 2.3.1.2.1 shows the common emitter configuration of the transistor. The resistors R_C and R_B are placed to limit the current. The voltage sources V_{XX} and V_{CC} are given for the biasing of the transistor. The base and emitter terminals act as input terminals and the collector and emitter terminals act as output terminals. The voltage V_{BE} is the input voltage to the transistor and the current I_B is the input current of the transistor. The voltage V_{CE} is the output voltage to the transistor and the current I_C is the output current of the transistor.

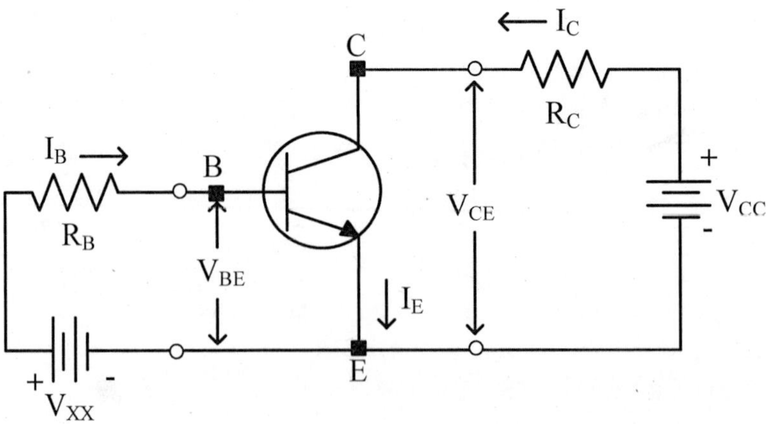

Fig. 2.3.1.2.1 Common emitter configuration of the transistor.

In the common emitter configuration, I_B is the input current of the transistor and I_C is the output current of the transistor. The current gain β in the common emitter configuration is given by

$$\beta = \frac{I_C}{I_B} \quad (2.3.1.2.1)$$

The emitter current I_E is given by

$$I_E = I_C + I_B \quad (2.3.1.2.2)$$

$$I_B = I_E - I_C \quad (2.3.1.2.3)$$

Substituting equation (2.3.1.2.3) in equation (2.3.1.2.1) gives

$$\beta = \frac{I_C}{I_E - I_C} \quad (2.3.1.2.4)$$

Now dividing RHS terms with I_E on numerator and denominator gives

$$\beta = \frac{\frac{I_C}{I_E}}{1 - \frac{I_C}{I_E}} \quad (2.3.1.2.5)$$

The term I_C/I_B is equal to the current gain α in the common base configuration

$$\beta = \frac{\alpha}{1-\alpha} \quad (2.3.1.2.6)$$

From the equation (2.3.1.1.2) and (2.3.1.1.3),

$$I_C = \alpha I_E + I_{CBO} \quad (2.3.1.2.7)$$

Substituting (2.3.1.2.2) in (2.3.1.2.7) gives

$$I_C = \alpha\left(I_C + I_B\right) + I_{CBO} \quad (2.3.1.2.8)$$

The above equation can be rewritten as

$$I_C\left(\frac{1-\alpha}{\alpha}\right) = I_B + \frac{I_{CBO}}{\alpha} \quad (2.3.1.2.9)$$

$$I_C = I_B\left(\frac{\alpha}{1-\alpha}\right) + \frac{I_{CBO}}{1-\alpha} \quad (2.3.1.2.10)$$

Substituting (2.3.1.2.5) in (2.3.1.2.10) gives

$$I_C = I_B\beta + \left(1+\beta\right)I_{CBO} \quad (2.3.1.2.11)$$

and

$$(1+\beta)I_{CBO} = I_{CEO} \tag{2.3.1.2.12}$$

where I_{CEO} is the reverse leakage current between collector and emitter while base is open. The current I_{CBO} cannot be ignored because it is amplified by a factor 1+β. The current gain β in the common emitter configuration is very high. Next, the input characteristics need to be plotted. The input characteristics are plotted between the input voltage V_{BE} and the input current I_B at a constant output voltage V_{CE}. Fig. 2.3.1.2.2 shows the input characteristics of the common emitter configuration. After the barrier potential voltage (i.e., 0.7V for the silicon and 0.3V for germanium), the base current I_B increases rapidly with the increase in the input voltage V_{BE}. This results in very low dynamic input resistance. The output characteristics are plotted between the output voltage V_{CE} and the output current I_C at a constant input current I_B. Fig. 2.3.1.2.3 shows the output characteristics of the common emitter configuration. There are three regions in the output characteristics: active region, saturation region and cut-off region. During the active region, the emitter junction is forward-biased and the collector junction is reverse-biased. In this region, the collector current I_C slightly increases with the increase in V_{CE}. The dynamic output resistance is high. In the saturation region, both the emitter junction and the collector junctions are forward-biased. In the saturation region, the collector current I_C is independent of input current I_B. In the cut-off region, both the emitter junction and the collector junctions are reverse-biased. The region below I_B equal to zero is the cut-off region.

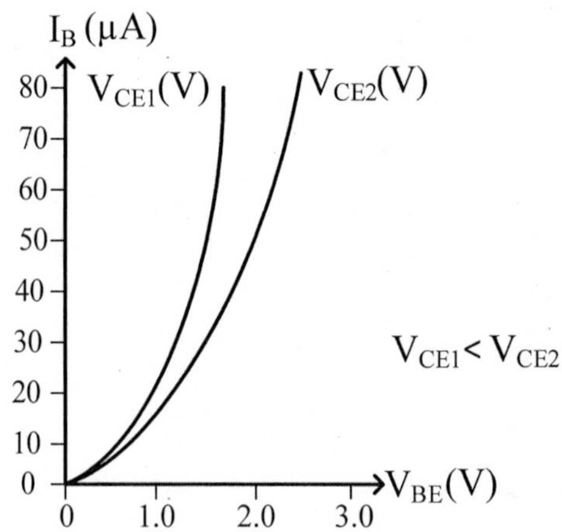

Fig. 2.3.1.2.2 Input characteristics of common emitter configuration.

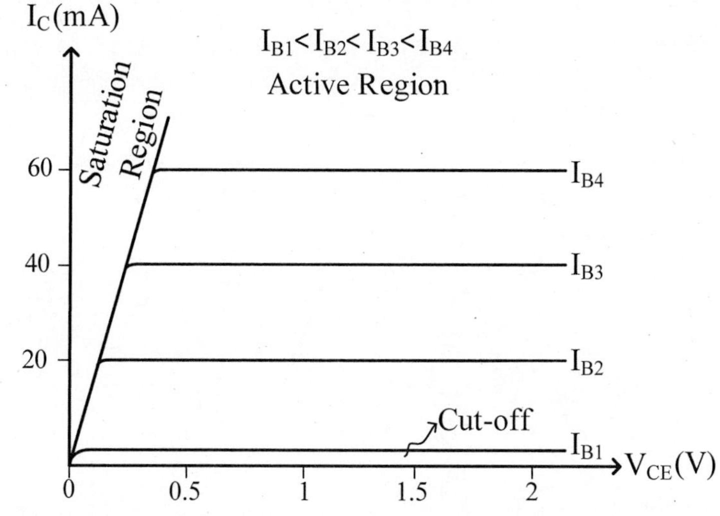

Fig. 2.3.1.2.3 Output characteristics of common emitter configuration.

2.3.3 Common Collector Configuration

In the case of the common collector configuration, the collector terminal is kept common for both the input and output terminals of the circuit. Fig. 2.3.1.3.1 shows the common collector configuration of the transistor. The resistors R_E and R_B are placed to limit the current. The voltage sources V_{XX} and V_{CC} are given for the biasing of the transistor. The base and collector terminals act as input terminals and the emitter and collector terminals act as output terminals. The voltage V_{BC} is the input voltage to the transistor and the current I_B is the input current of the transistor. The voltage V_{EC} is the output voltage to the transistor and the current I_E is the output current of the transistor. In the common collector configuration, I_B is the input current of the transistor and I_E is the output current of the transistor. The current gain γ in the common collector configuration is given by

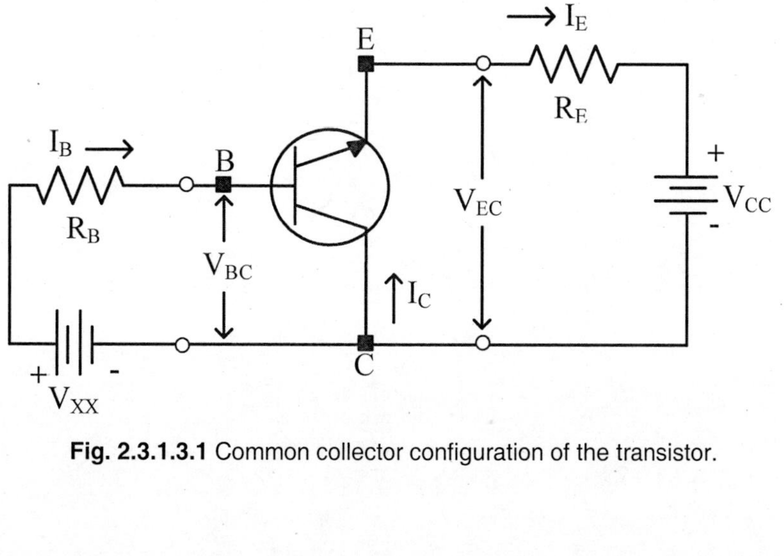

Fig. 2.3.1.3.1 Common collector configuration of the transistor.

$$\gamma = \frac{I_E}{I_B} \tag{2.3.1.3.1}$$

The emitter current I_E is given by

$$I_E = I_C + I_B \tag{2.3.1.3.2}$$

$$I_E = I_B + \alpha I_E + I_{CBO} \tag{2.3.1.3.3}$$

$$I_E = \frac{I_B}{1-\alpha} + \frac{I_{CBO}}{1-\alpha} \tag{2.3.1.3.4}$$

$$I_E = I_B(1+\beta) + I_{CBO}(1+\beta) \tag{2.3.1.3.5}$$

Neglecting I_{CBO} we get

$$I_E = I_B(1+\beta) \tag{2.3.1.3.6}$$

$$(1+\beta) = \frac{I_E}{I_B} = \gamma \tag{2.3.1.3.7}$$

Next, the input characteristics need to be plotted. The input characteristics are plotted between the input voltage V_{CB} and the input current I_B at a constant output voltage V_{CE}. Fig. 2.3.1.3.2 shows the input characteristics of the common collector configuration. The input resistance of the common collector configuration is very high. The output characteristics are plotted between the output voltage V_{CE} and the output current I_E at a constant input current I_B. Fig. 2.3.1.3.3 shows the output characteristics of the common collector configuration. The output characteristics of the common collector configuration is almost the same as that of the common emitter configuration since the value of the collector current I_C is almost equal to the emitter current I_E. That is why the common collector configuration is also called as emitter follower.

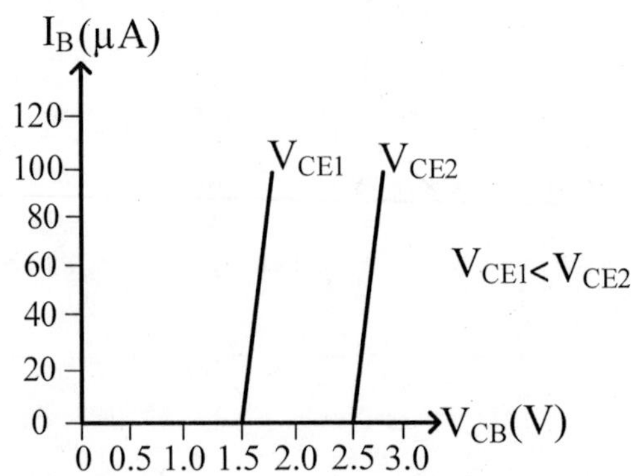

Fig. 2.3.1.3.2 Input characteristics of common collector configuration.

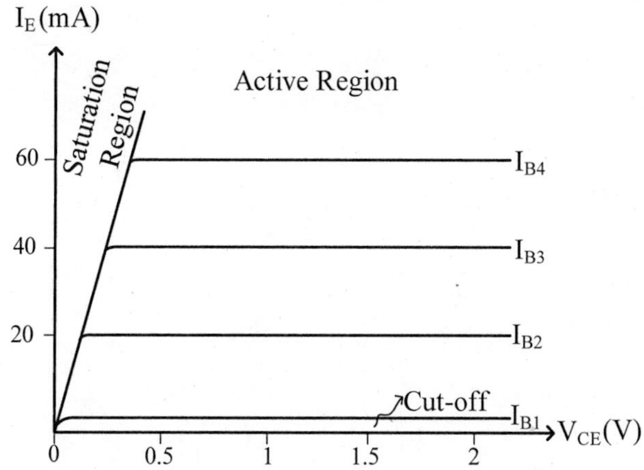

Fig. 2.3.1.3.3 Output characteristics of common collector configuration.

Among the three configurations of the transistor, the common base configuration has the current gain less than 1 and a high voltage gain. The common emitter configuration has high current gain and voltage gain. The common collector configuration has high current gain and a voltage gain of less than 1. For the amplifier application common emitter configuration is preferred over common base and common collector configurations. The common collector configuration is normally used for impedance matching applications. The common base configuration is used in the input stage of multistage amplifier because of low input impedance.

2.4 BIASING OF THE TRANSISTOR

The biasing of the transistor is generally required for two purposes:

1. As discussed earlier, the transistor will normally operate in the active region, saturation region, cut-off region and inverse active region. The region of operation of the transistor is selected by biasing the transistor. By biasing the transistor, the emitter junction and the collector junction can be forward-biased or reverse-biased so that desired operating region will be selected.
2. One of the major applications of the transistor is amplification. During amplification, the weak input signal is transformed into the strong output signal. The biasing of the transistor will normally provide the additional power to convert the weak input signal is transformed to the strong output signal.

The biasing of the transistor can be of four types: fixed bias circuit, collector-to-base bias, voltage divider bias and emitter stabilized bias.

2.5 QUIESCENT POINT (Q-POINT) OF THE TRANSISTOR

In order to explain the Q-point or operating point of the transistor, the circuit shown in Fig. 2.5.1 is considered. The circuit shown in Fig. 2.5.1 consists of one external power supply V_{CC}, one transistor and two biasing resistors R_B and R_C. The circuit shown in Fig. 2.5.1 is a common emitter configuration. The biasing resistor R_B is connected between the positive terminal of V_{CC} to the base terminal of the transistor. Similarly, the biasing resistor R_C is connected between the positive terminal of V_{CC} to the collector terminal of the transistor. I_B is the current flowing into the base terminal and I_C is the current flowing into the collector terminal.

Applying the KVL on the collector network

$$V_{CC} - I_C R_C - V_{CE} = 0 \tag{2.5.1}$$

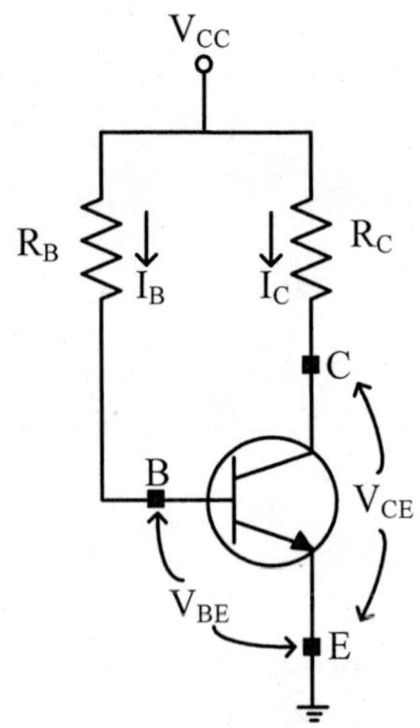

Fig. 2.5.1 Circuit considered for explaining the Q-point of transistor.

$$I_C = \frac{-V_{CE} + V_{CC}}{R_C} \tag{2.5.2}$$

The expression (2.5.2) is a straight line where $1/R_C$ is the slope of the line and V_{CC}/R_C is the y-intercept. The above straight line is called the DC load line. Plotting the DC load line on the output characteristics of the common emitter configuration as shown in Fig. 2.5.2 results in the intersection points between the output characteristics and the DC load line. These intersection points are called as the Q-point of the transistors. One X-coordinate of the DC load line is (V_{CC}, 0) and the Y-coordinate of the DC load line is (0, V_{CC}/R_C).

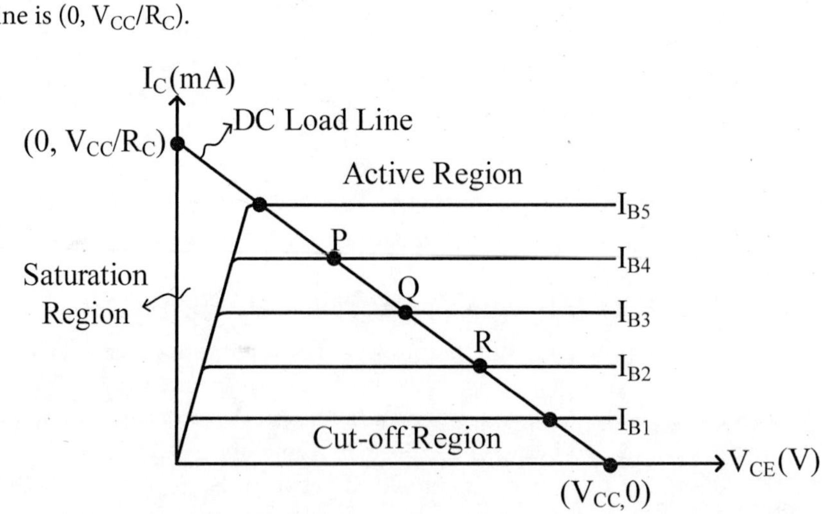

Fig. 2.5.2 DC load line and output characteristics of the common emitter configuration.

2.6 SELECTION OF QUIESCENT POINT (Q-POINT) OF THE TRANSISTOR

The selection of the Q-point depends on the application of the transistor whether to be used as an amplifier or switch. Let us consider the Q-points P, Q and R as shown in Fig. 2.5.2. The Q-point P is near to the saturation region, Q-point Q is at the center of the DC load line and Q-point R is near to the cut-off region.

Case-1: If the Q-point is at P

Let us consider that the Q-point is at location P. Now at this location, if an input signal is applied to the transistor for the amplification, The part of the input signal is in the saturation region. This results in the clipping in the collector current I_C and the output voltage V_{CE} of the corresponding input part that is in the saturation as shown in Fig. 2.6.1. This clipping results in the distortion in the output voltage and current.

Case-2: If the Q-point is at Q

Let us consider that the Q-point is at location Q. Now at this location, if an input signal is applied to the transistor for the amplification, Then the complete input signal is in the active region. This results in the complete output signal without clipping in the collector current I_C and the output voltage V_{CE} as shown in Fig. 2.6.2. Thus, there no distortion of the input signals by clipping for transistor operating in active region for Q-point at location Q.

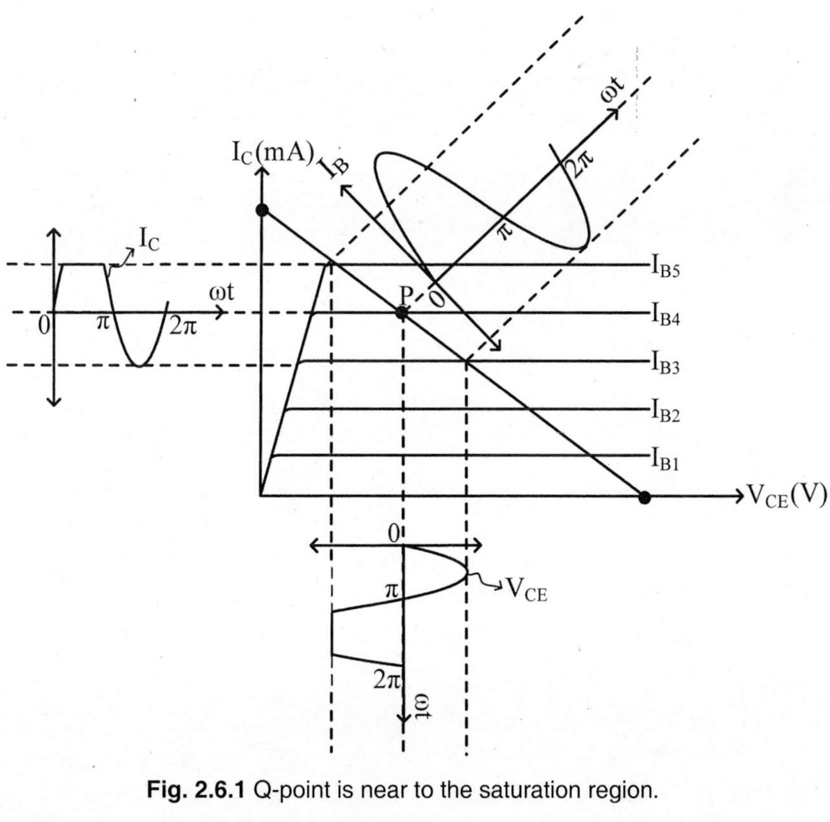

Fig. 2.6.1 Q-point is near to the saturation region.

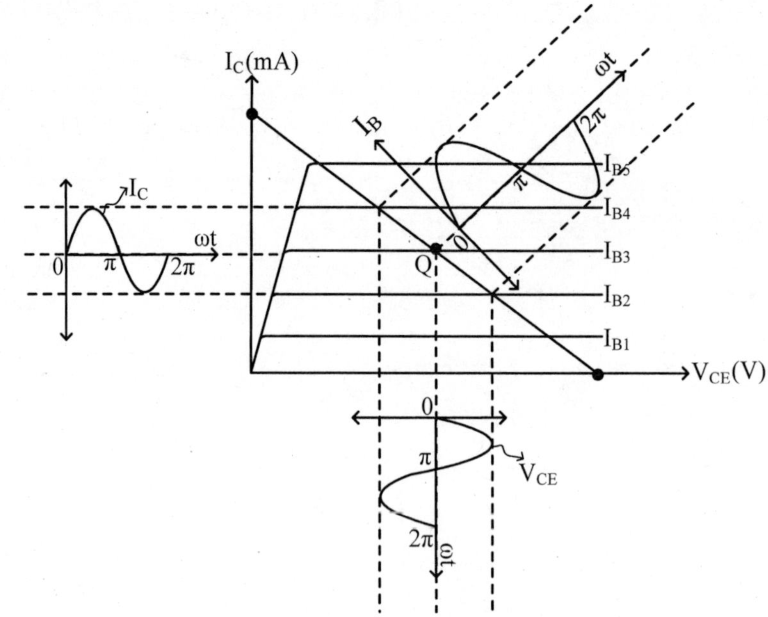

Fig. 2.6.2 Q-point is at the center of the DC load line.

Case-3: If the Q-point is at R

Let us consider that the Q-point is at location R. Now at this location, if an input signal is applied to the transistor for the amplification, The part of the input signal is in the cut-off region. This results in the clipping in the collector current I_C and the output voltage V_{CE} of the corresponding input part that is in the cut-off as shown in Fig. 2.6.3. This clipping results in the distortion in the output voltage and current.

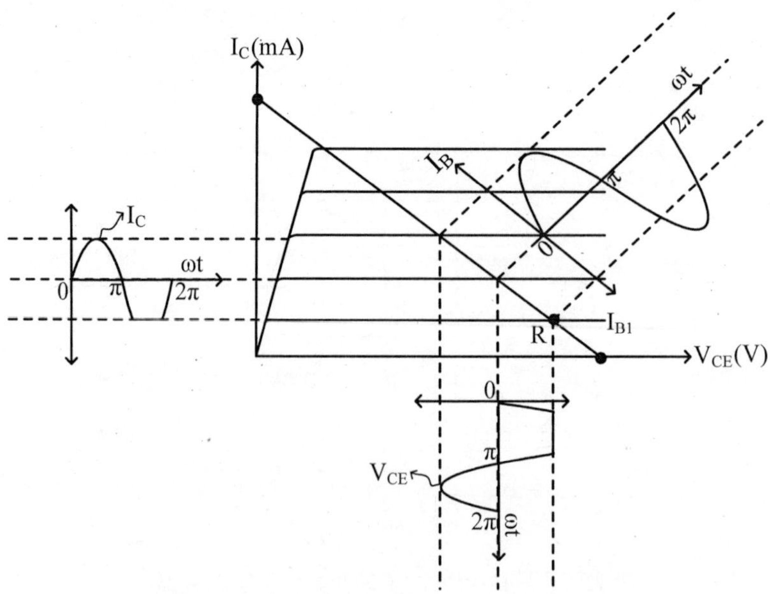

Fig. 2.6.3 Q-point is near to the cut-off region.

From the three cases, it can be observed that whenever the Q-point is at the center of the DC load line, then the output current and output voltage are not distorted and the amplification will happen to the complete input signal. So, the Q-point must be always at the center of the DC load line. The biasing circuit must be designed in such a way that the Q-point should be always at the center of the DC load line.

2.7 BIASING STABILIZATION

The Q-point should be always at the center of the DC load line. But in the real time scenario, the reverse leakage current I_{CO} due to minority charge carriers, base to emitter voltage V_{BE} and the current gain β will vary with the temperature. The collector current that is a function of I_{CO}, V_{BE} and β will also change due to the variation in the temperature. Due to a change in the collector current I_C, the Q-point will shift either toward the saturation region or toward to cut-off region. The biasing circuit should be designed in such a way that the variation of I_{CO}, V_{BE} and β with the variation in the temperature should not affect the Q-point location. This process of stabilizing the Q-point with the biasing circuit is called biasing stabilization.

The biasing circuits should be designed in such a way that the change in the collector current I_C due to changes in the I_{CO}, V_{BE} and β with the variation in temperature should be nullified. The biasing stabilization can be identified from the stability factors S, S', S". These stability factors are nothing but the partial derivatives of the collector current I_C with the I_{CO}, V_{BE} and β. The values of these stability factors should be as minimum as possible. The stability factors normally indicate the deviation in the Q-point because of the variation in the I_{CO}, V_{BE} and β. The minimum value or low value of the stability factors indicates that the deviation in the Q-point is very less.

$$\text{Stability factor } S = \frac{\partial I_C}{\partial I_{CO}} \tag{2.7.1}$$

$$\text{Stability factor } S' = \frac{\partial I_C}{\partial V_{BE}} \tag{2.7.2}$$

$$\text{Stability factor } S'' = \frac{\partial I_C}{\partial \beta} \tag{2.7.3}$$

2.8 BIASING CIRCUITS FOR THE TRANSISTORS

The biasing of the transistor can be of four types: fixed bias circuit, collector-to-base bias and voltage divider bias.

2.8.1 Fixed Bias Circuit

The fixed bias circuit is very simple compared to the other biasing circuits. The fixed biasing circuit generally requires one external power supply V_{CC} and two biasing resistors R_B and R_C as shown in Fig. 2.8.1.1. The biasing resistor R_B is connected between the positive terminal of V_{CC} to the base terminal of the transistor. Similarly, the biasing resistor R_C is connected between the positive terminal of V_{CC} to the collector terminal of the transistor. I_B is the current flowing into the base terminal and I_C is the current flowing into the collector terminal.

Applying KVL on the base network,

$$V_{CC} - I_B R_B - V_{BE} = 0 \tag{2.8.1.1}$$

$$I_B = \frac{V_{CC} - V_{BE}}{R_B} \tag{2.8.1.2}$$

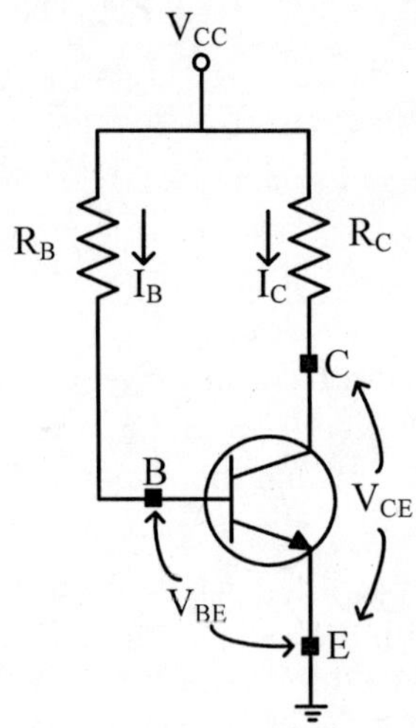

Fig. 2.8.1.1 Fixed bias circuit.

Similarly, applying the KVL on the collector network

$$V_{CC} - I_C R_C - V_{CE} = 0 \tag{2.8.1.3}$$

$$I_C = \frac{V_{CC} - V_{CE}}{R_C} \tag{2.8.1.4}$$

The collector current I_C and the base current I_B are related as

$$I_C = \beta I_B + (1+\beta) I_{CO} \tag{2.8.1.5}$$

The stability factor S is equal to

$$S = \frac{\partial I_C}{\partial I_{CO}} = 1 + \beta \tag{2.8.1.6}$$

The stability factor S' is equal to

$$S' = \frac{\partial I_C}{\partial V_{BE}} = \frac{-\beta}{R_B} \tag{2.8.1.7}$$

The stability factor S" is equal to

$$S'' = \frac{\partial I_C}{\partial \beta} = \frac{V_{CC} - V_{CE}}{R_C \beta} \tag{2.8.1.8}$$

The value of β is very large, so the values of the stability factors are also high. The Q-point will change due to the changes in the I_{CO}, V_{BE} and β with the variation in temperature. Therefore, the fixed bias circuit doesn't have the bias stabilization.

2.8.2 Collector-to-Base Bias Circuit

The collector-to-base bias circuit is shown in Fig. 2.8.2.1. In the case of collector-to-base bias, there is feedback between the input base and output collector terminals. The collector-to-base bias circuit generally requires one external power supply V_{CC} and two biasing resistors R_B and R_C as shown in Fig. 2.8.2.1. The biasing resistor R_B is connected between the collector to the base terminals of the transistor. The biasing resistor R_C is connected between the positive terminal of V_{CC} to the collector terminal of the transistor. I_B is the current flowing into the base terminal and the sum of I_C and I_B is the flowing through the resistor R_C.

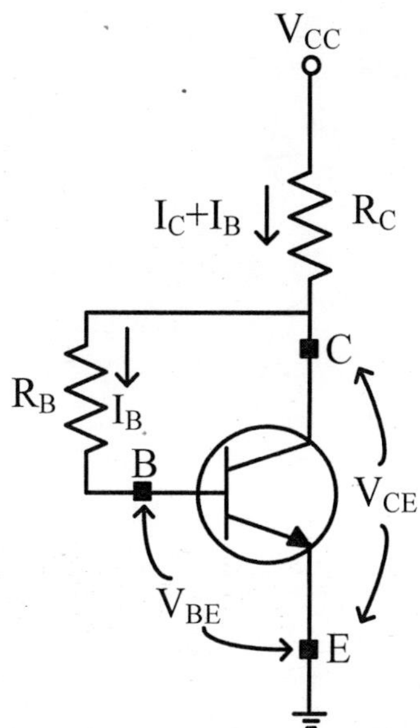

Fig. 2.8.2.1 Collector-to-base bias circuit.

Applying KVL on the base network,

$$V_{CC} - (I_C + I_B)R_C - I_B R_B - V_{BE} = 0 \tag{2.8.1.1}$$

$$V_{CC} = (I_C + I_B)R_C + I_B R_B + V_{BE} \tag{2.8.1.2}$$

$$V_{CC} = (R_C + R_B)I_B + I_C R_C + V_{BE} \Rightarrow V_{CC} = (R_C + R_B)I_B + \beta I_B R_C + V_{BE} \tag{2.8.1.3}$$

$$V_{CC} = (R_C(1+\beta) + R_B)I_B + V_{BE} \tag{2.8.1.4}$$

$$I_B = \frac{V_{CC} - V_{BE}}{R_B + R_C(1+\beta)} \tag{2.8.1.5}$$

Applying KVL on the collector network,

$$V_{CC} - (I_C + I_B)R_C - V_{CE} = 0 \tag{2.8.1.6}$$

$$V_{CC} - \left(I_C + \frac{I_C}{\beta}\right)R_C - V_{CE} = 0 \tag{2.8.1.7}$$

$$I_C = \frac{V_{CC} - V_{CE}}{R_C\left(1 + \frac{1}{\beta}\right)} \tag{2.8.1.8}$$

The collector current I_C and the base current I_B are related as

$$I_C = \beta I_B + (1+\beta)I_{CO} \tag{2.8.1.9}$$

The stability factor S is equal to

$$S = \frac{\partial I_C}{\partial I_{CO}} = \frac{1+\beta}{1+\beta\left(\frac{R_C}{R_C + R_B}\right)} \tag{2.8.1.10}$$

The stability factor S' is equal to

$$S' = \frac{\partial I_C}{\partial V_{BE}} = \frac{-\beta}{R_B + (1+\beta)R_C} \tag{2.8.1.11}$$

The stability factor S" is equal to

$$S'' = \frac{\partial I_C}{\partial \beta} = \frac{I_C}{\beta}\left(\frac{1}{1+\beta\left(\frac{R_C}{R_C + R_B}\right)}\right) \tag{2.8.1.12}$$

Compared to the fixed bias circuit, in the collector-to-base bias circuit, the values of the stability factors S, S', S" are less. So, the collector-to-base bias will provide more bias stabilization compared to the fixed bias circuit.

2.8.3 Voltage Divider Bias Circuit

The voltage divider bias circuit is shown in Fig. 2.8.3.1 The voltage divider bias circuit generally requires one external power supply V_{CC} and four biasing resistors R_1, R_2, R_E and R_C as shown in Fig. 2.8.3.1. The biasing resistor R_1 is connected between the positive of the power supply V_{CC} to the base terminals of the transistor. The biasing resistor R_2 is connected between the base terminals of the transistor to the negative terminal of the power supply V_{CC}. The biasing resistor R_C is connected between the positive terminal of V_{CC} to the collector terminal of the transistor. The biasing resistor R_E is connected between the emitter terminal of the transistor to the negative terminal of the power supply V_{CC}. In order to analyze this circuit, the circuit needs to be simplified using Thevenin's theorem. The Thevenin's equivalent of the voltage divider bias circuit is shown in Fig. 2.8.3.2. The term V_{th} denotes the Thevenin's voltage and the term R_{th} denotes the Thevenin's resistance.

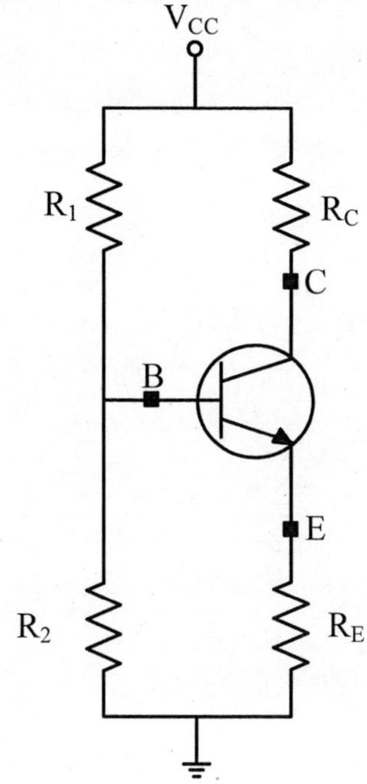

Fig. 2.8.3.1 Voltage divider bias circuit.

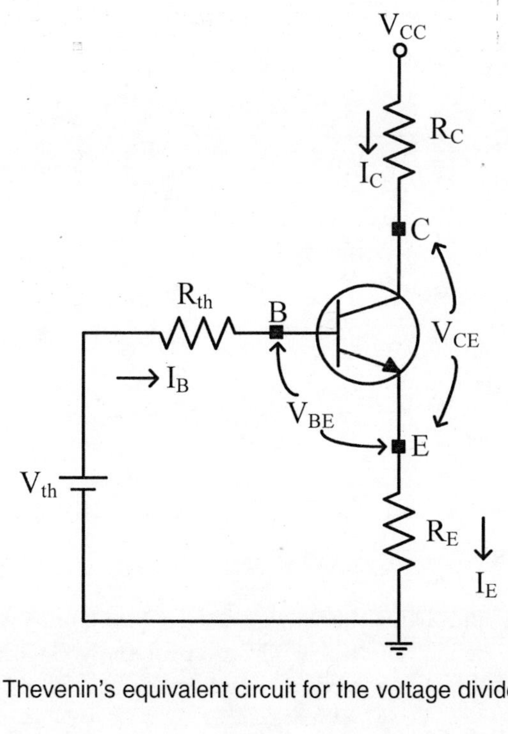

Fig. 2.8.3.2 Thevenin's equivalent circuit for the voltage divider bias circuit.

$$V_{th} = \frac{V_{CC}R_2}{R_1 + R_2} \tag{2.8.3.1}$$

$$R_{th} = \frac{R_1R_2}{R_1 + R_2} \tag{2.8.3.2}$$

Applying KVL on the base network,

$$V_{th} - I_BR_{th} - V_{BE} - I_ER_E = 0 \tag{2.8.3.3}$$

$$V_{th} = I_BR_{th} + V_{BE} + (1+\beta)I_BR_E \tag{2.8.3.4}$$

$$V_{th} = \left(R_{th} + (1+\beta)R_E\right)I_B + V_{BE} \tag{2.8.3.5}$$

$$I_D = \frac{V_{th} - V_{BE}}{R_{th} + R_E(1+\beta)} \tag{2.8.3.6}$$

Applying KVL on the collector network,

$$V_{CC} - I_CR_C - V_{CE} - I_ER_E = 0 \tag{2.8.3.7}$$

$$V_{CC} = I_CR_C + V_{CE} + (1+\frac{1}{\beta})R_EI_C \tag{2.8.3.8}$$

$$I_C = \frac{V_{CC} - V_{CE}}{R_E\left(1+\frac{1}{\beta}\right) + R_C} \tag{2.8.3.9}$$

The collector current I_C and the base current I_B are related as

$$I_C = \beta I_B + (1+\beta)I_{CO} \tag{2.8.3.10}$$

The stability factor S is equal to

$$S = \frac{\partial I_C}{\partial I_{CO}} = \frac{1+\beta}{1+\beta\left(\frac{R_E}{R_E + R_{th}}\right)} \tag{2.8.3.11}$$

The stability factor S' is equal to

$$S' = \frac{\partial I_C}{\partial V_{BE}} = \frac{-\beta}{R_{th} + (1+\beta)R_E} \tag{2.8.3.12}$$

The stability factors S, S' and S" for the voltage divider bias circuit is very less compared to the collector-to-base bias circuit. Hence the bias stabilization for the voltage divider bias circuit is good compared to collector-to-base bias and fixed bias circuit. Hence the voltage divider bias circuit is preferred.

2.9 BJT AS A SWITCH

The BJT can be operated as an open switch or closed switch depending upon the region of operation. When the BJT is operating the cut-off region, it acts as an open switch. When the BJT is operating in the saturation region, it acts as closed switch.

2.9.1 BJT as an Open Switch

For the operation of BJT in the cut-off region, both the emitter junction and collector junction need to be reverse biased. For reverse biasing the emitter junction, a voltage of 0V needs to be applied at the base terminals. Because of this 0V at the base terminal, the value of IB and IC will be equal to 0A. Thus, BJT is acting like a open switch. Fig. 2.9.1.1 shows the operation of BJT as an open switch.

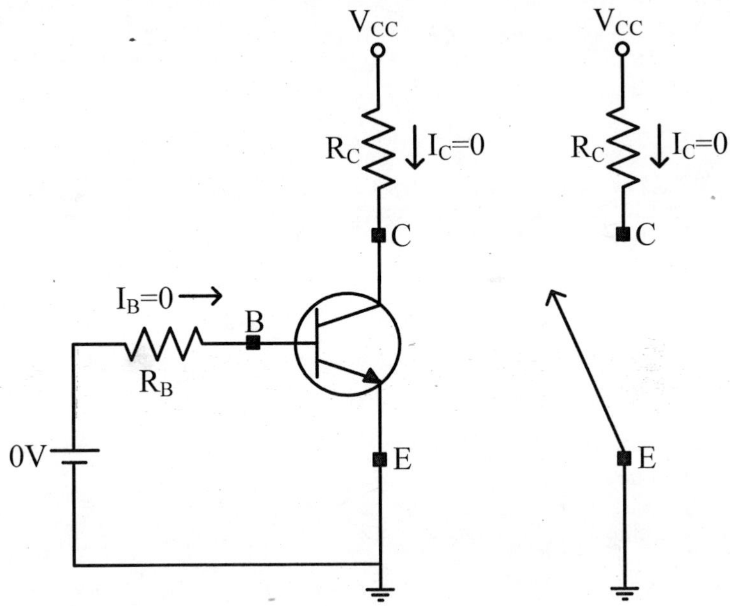

Fig. 2.9.1.1 BJT as an open switch.

2.9.2 BJT as a Closed Switch

For the operation of BJT in the saturation region, both the emitter junction and collector junction need to be forward biased. Normally, there will be a saturation voltage between the collector terminal and emitter terminal which is denoted by the term the $V_{CE(sat)}$. Fig. 2.9.2.1 shows the operation of BJT as a closed switch.

The collector current $I_{C(sat)}$ is equal to

$$I_{C(sat)} = \frac{V_{CC} - V_{CE(sat)}}{R_C} \tag{2.9.2.1}$$

$$I_{B(sat)} = \beta I_{C(sat)} \tag{2.9.2.2}$$

For operating in the saturation region, the current flowing through the base terminal must be greater than the $I_{B(sat)}$.

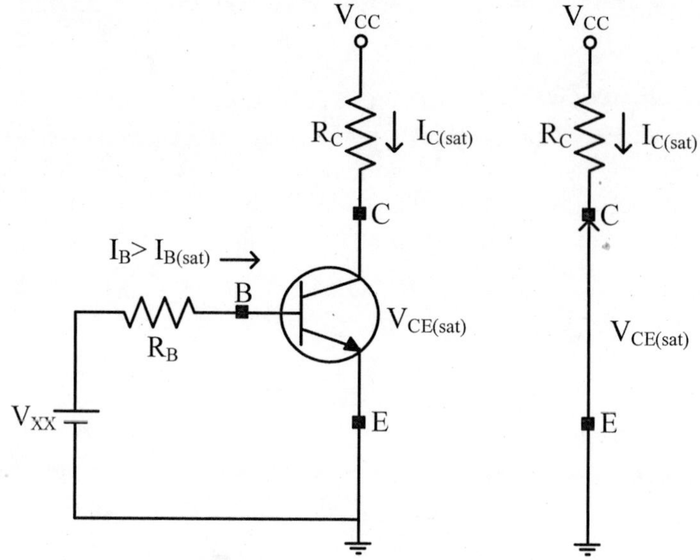

Fig. 2.9.2.1 BJT as a closed switch.

2.10 BJT AS AN AMPLIFIER

Three amplifier configurations are possible using the BJT: common base amplifier, common emitter amplifier and common collector amplifier.

2.10.1 Common Base Amplifier

The circuit schematic of the common base amplifier is shown in Fig. 2.10.1.1. The common base amplifier consists of an input signal V_S which is to be amplified. The term R_S denoted resistance of the input signal. The amplifier consists of two biasing resistors, R_E and R_C. The load resistance is denoted by the term R_L. The voltage across and the current flowing through the load resistance is denoted by the terms V_L and I_L. The capacitors C_i and C_o are used as coupling capacitors. The capacitor C_i is used to block the dc value entering from Vs. Similarly, the C_o is used to block DC reaching to the load.

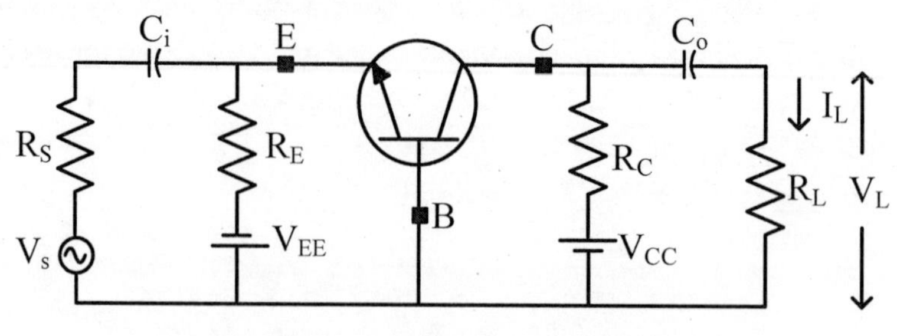

Fig. 2.10.1.1 Common base amplifier

For amplification application, the emitter junction needs to be forward-biased. The external dc source V_{EE} is connected to forward bias the emitter junction. Similarly, the collector junction needs to reverse biased. The external dc source V_{CC} is connected to reverse bias the collector junction. During the positive half cycle of the input voltage, the voltage at the emitter terminal increases. This increase in the voltage reduces the forward biased emitter terminal. This reduction in the forward biased emitter terminal reduces the emitter current I_E. The reduction in the emitter current reduces the collector current I_C. The output voltage V_L is equal to

$$V_L = V_{CC} - I_C R_C \tag{2.10.1.1}$$

The reduction I_C increases the output voltage in positive direction. In this way for the positive half cycle of the input voltage, the output voltage follows the positive half of the input voltage. Similarly, for the negative half cycle of the input voltage, the output voltage follows the negative half cycle of the input voltage. Thus, the input voltage and output voltage are in phase for the common base amplifier.

2.10.2 Common Emitter Amplifier

The circuit schematic of the common emitter amplifier is shown in Fig. 2.10.2.1. The common emitter amplifier consists of an input signal V_S which is to be amplified. The term R_S denotes resistance in series with input signal source. The amplifier consists of four biasing resistors R_1, R_2, R_E and R_C. The load resistance is denoted by the term R_L. The voltage across and the current flowing through the load resistance is denoted by the terms V_L and I_L. The capacitors C_i and C_o are used as coupling capacitors to block the dc value entering the transistor. C_e is the bypass capacitance across the emitter resistance R_E.

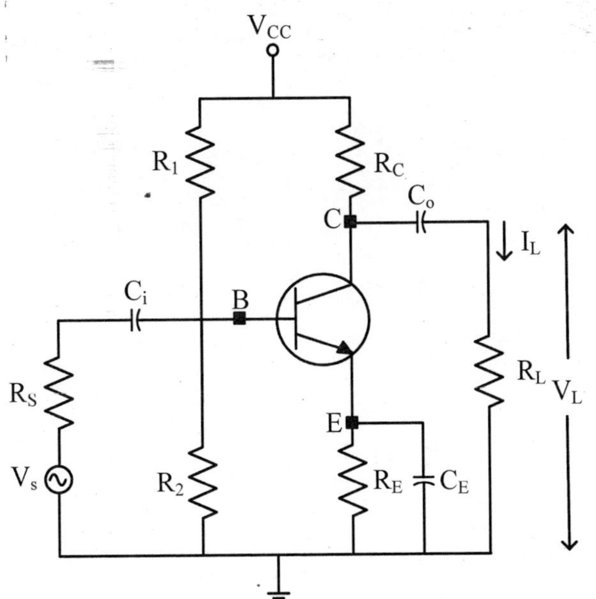

Fig. 2.10.2.1 Common emitter amplifier

The bypass capacitance C_e is connected to avoid the drop in emitter resistance R_E.

During the positive half cycle of the input voltage, the base current I_B increases. This increase in the base current increases the collector current I_C. The output voltage V_L is equal to

$$V_L = V_{CC} - I_C R_C \tag{2.10.2.1}$$

The increase in collector current I_C decreases the output voltage in positive direction. In this way for the positive half cycle of the input voltage, the output voltage will be in negative half cycle. Similarly, for the negative half cycle of the input voltage, the output voltage will be positive half cycle. Thus, the input voltage and output voltage are in out of phase for the common emitter amplifier.

2.10.3 Common Collector Amplifier

The circuit schematic of the common collector amplifier is shown in Fig. 2.10.3.1. The common collector amplifier consists of an input signal V_S which is to be amplified. The term R_S denotes resistance in series with input signal source. The amplifier consists of three biasing resistors R_1, R_2 and R_E. The load resistance is denoted by the term R_L. The voltage across and the current flowing through the load resistance is denoted by the terms V_L and I_L. The capacitors C_i and C_o are used as coupling capacitors to block the dc value entering the transistor.

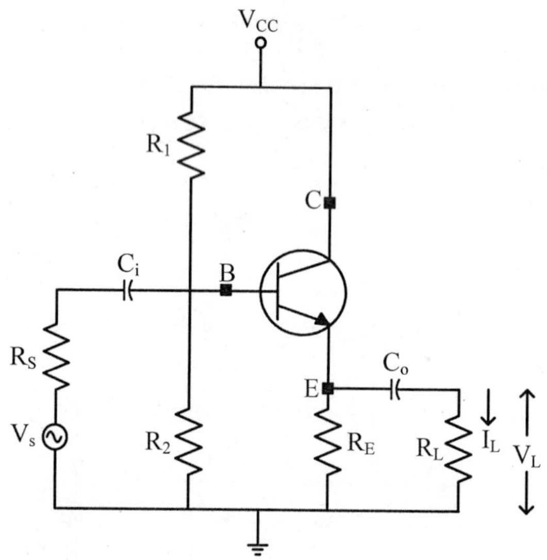

Fig. 2.10.3.1 Common collector amplifier.

During the positive half cycle of the input voltage, the voltage at base terminal increases. For a fixed V_{BE}, the increase in base voltage increases the voltage at the emitter terminal. Since the load is connected at emitter terminal, the output voltage will follow the input voltage. Thus, the input voltage and output voltage are in phase for the common collector amplifier.

2.11 HYBRID PARAMETER MODEL OF THE TRANSISTOR

The hybrid parameter (h-parameter) model of the transistor is normally used to calculate the input impedance, voltage gain, current gain and output admittance of the amplifier. In the case of two port networks, if V_1 is the input voltage of the network, I_1 is the input current, V_2 is the output voltage of the network and I_2 is the output current. Using the h-parameters, the variables V_1, V_2, I_1 and I_2 are related by

$$V_1 = h_{11}I_1 + h_{12}V_2$$
$$I_2 = h_{21}I_1 + h_{22}V_2 \qquad (2.11.1)$$

where

$h_{11} = \left.\frac{V_1}{I_1}\right|_{V_2=0}$ gives the information of input impedance of the network in ohms

$h_{12} = \left.\frac{V_1}{V_2}\right|_{I_1=0}$ gives the information of reverse voltage gain of the network

$h_{21} = \left.\frac{I_2}{I_1}\right|_{V_2=0}$ gives the information of forward current gain of the network

$h_{22} = \left.\frac{I_2}{V_2}\right|_{I_1=0}$ gives the information of output admittance of the network in mhos

In the case of transistor, if V_i is the input voltage of the transistor, V_o is the output voltage of transistor, I_i is the input current of the transistor and I_o us the output current of transistor then

$$V_i = h_iI_i + h_rV_o$$
$$I_o = h_fI_i + h_oV_o \qquad (2.11.2)$$

where

$h_i = \left.\frac{V_i}{I_i}\right|_{V_o=0}$ gives the information of input impedance of the transistor in ohms

$h_r = \left.\frac{V_i}{V_o}\right|_{I_i=0}$ gives the information of reverse voltage gain of the transistor

$h_f = \left.\frac{I_o}{I_i}\right|_{V_o=0}$ gives the information of forward current gain of the transistor

$h_o = \left.\frac{I_o}{V_o}\right|_{I_i=0}$ gives the information of output admittance of the transistor in mhos

Using the equation 2.11.2, the equivalent circuit of the transistor using the h-parameters is given in Fig. 2.11.1.

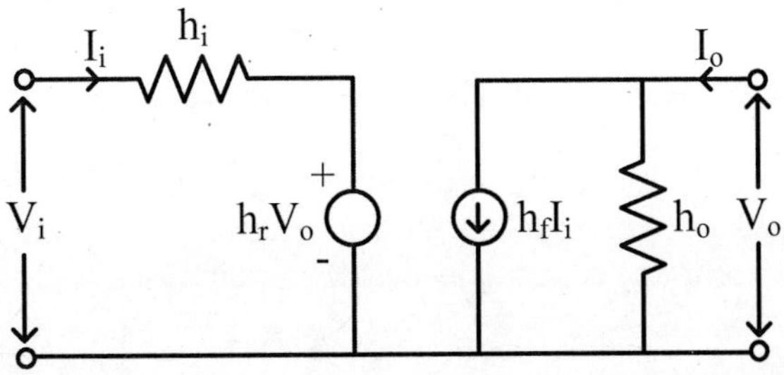

Fig. 2.11.1 h-parameter equivalent circuit of the transistor

In the case of common base configuration, the input voltage to the transistor is V_{EB}, the input current to the transistor is I_E, the output voltage of the transistor is V_{CB} and the output current is I_C. The h-parameters for the transistor in the common base configuration is given by

$$V_{EB} = h_{iB}I_E + h_{rB}V_{CB}$$
$$I_C = h_{fB}I_E + h_{oB}V_{CB} \tag{2.11.3}$$

where h_{IB}, h_{rB}, h_{fB} and h_{oB} are the h-parameters of the transistor in the common base configuration. Using the equation 2.11.3, the equivalent circuit of the transistor in the common base configuration using the h-parameters is given in Fig. 2.11.2.

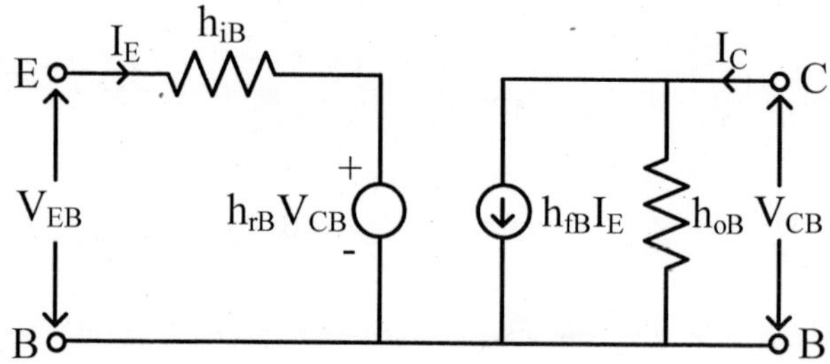

Fig. 2.11.2 h-parameter equivalent circuit of the transistor in the common base configuration.

In the case of common emitter configuration, the input voltage to the transistor is V_{BE}, the input current to the transistor is I_B, the output voltage of the transistor is V_{CE} and the output current is I_C. The h-parameters for the transistor in the common emitter configuration is given by

$$V_{BE} = h_{iE}I_B + h_{rE}V_{CE}$$
$$I_C = h_{fE}I_B + h_{oE}V_{CE} \tag{2.11.4}$$

where h_{iE}, h_{rE}, h_{fE} and h_{oE} are the h-parameters of the transistor in the common emitter configuration. Using the equation 2.11.4, the equivalent circuit of the transistor in the common emitter configuration using the h-parameters is given in Fig. 2.11.3.

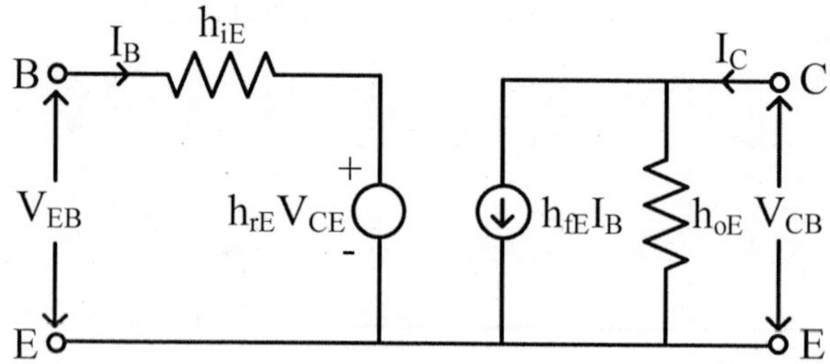

Fig. 2.11.3 h-parameter equivalent circuit of the transistor in the common emitter configuration.

In the case of common collector configuration, the input voltage to the transistor is V_{BC}, the input current to the transistor is I_B, the output voltage of the transistor is V_{EC} and the output current is I_E. The h-parameters for the transistor in the common collector configuration is given by

$$V_{BC} = h_{iC}I_B + h_{rC}V_{EC}$$
$$I_E = h_{fC}I_B + h_{oC}V_{EC} \tag{2.11.5}$$

where h_{iC}, h_{rC}, h_{fC} and h_{oC} are the h-parameters of the transistor in the common collector configuration. Using the equation 2.11.5, the equivalent circuit of the transistor in the common collector configuration using the h-parameters is given in Fig. 2.11.4.

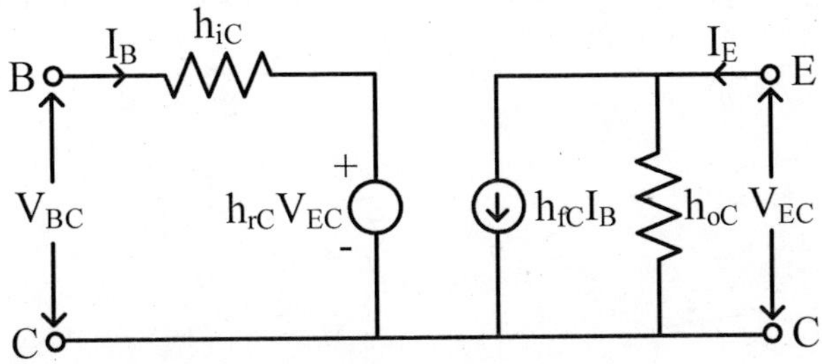

Fig. 2.11.4 h-parameter equivalent circuit of the transistor in the common collector configuration.

Most of the h-parameters for the transistor were given only for common emitter configuration. The h-parameters for the common base configuration and common collector configuration can be calculated from the h-parameters for the common emitter configuration by using the below relations:

$$h_{iB} = \frac{h_{iE}}{1+h_{fE}} \qquad h_{iC} = h_{iE}$$

$$h_{rB} = \frac{h_{iE}h_{oE}}{1+h_{fE}} - h_{rE} \qquad h_{rC} = 1 - h_{rE}$$

$$h_{fB} = -\frac{h_{fE}}{1+h_{fE}} \qquad h_{fC} = -(1+h_{fE})$$

$$h_{oB} = \frac{h_{oE}}{1+h_{fE}} \qquad h_{oC} = h_{oE} \qquad (2.11.6)$$

2.12 SMALL SIGNAL ANALYSIS OF THE AMPLIFIER USING THE H-PARAMETERS

The small signal analysis is required for the amplifier to obtain the parameters like the input impedance of the amplifier, voltage gain of the amplifier, current gain of the amplifier and the output admittance of the amplifier. The small signal analysis gives the linear model of the amplifier in the active region. This linear model helps in minimizing the complex computations. The following steps are required for performing the small signal analysis of the amplifier.

Step-1: In the first step replace all the coupling capacitors and the bypass capacitors with a short circuit.

Step-2: In the second step, replace the external dc power supplies used for the biasing with a short circuit.

Step-3: Replace the transistor with its h-parameter model.

Step-4: Calculate the current gain (A_I) of the amplifier.

Step-5: Calculate the input resistance (R_I) of the amplifier.

Step-6: Calculate the voltage gain (A_V) of the amplifier.

Step 7: Calculate the output admittance (Y_O) of the amplifier.

For performing the small signal analysis let us consider a CE amplifier shown in Fig. 2.12.1. The amplifier consists of coupling capacitors C_i and C_o, bypass capacitors C_E, biasing resistors R_1, R_2, R_E and R_C, the dc power supply V_{CC} for biasing the transistor.

Step-1: In the step-1, replace all the coupling capacitors and the bypass capacitors with a short circuit. The Fig. 2.12.2 shows the circuit of CE amplifier replacing all the coupling capacitors and the bypass capacitors with a short circuit.

Step-2: In the step-2, replace the external dc power supplies used for the biasing with a ground potential. The Fig. 2.12.3 shows the circuit of after replacing the external dc power supplies used for the biasing with a short circuit.

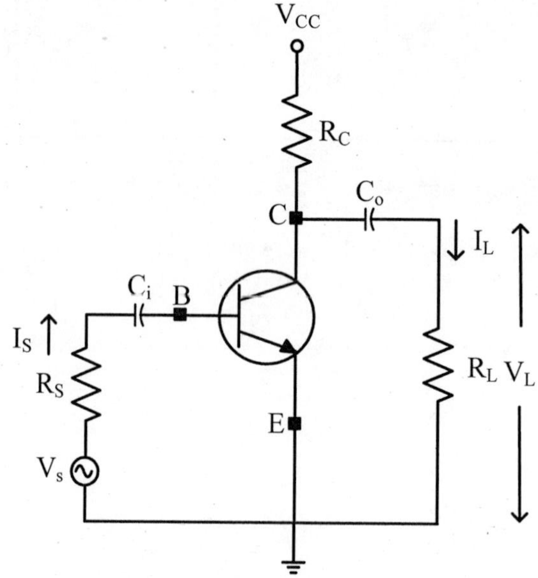

Fig. 2.12.1 CE amplifier.

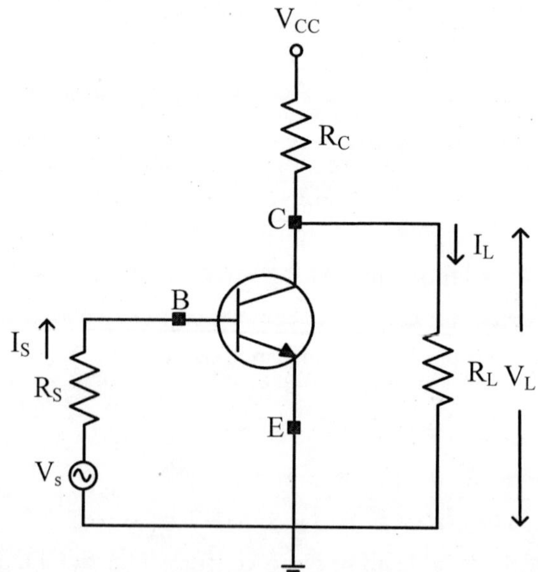

Fig. 2.12.2 CE amplifier replacing all the coupling capacitors and the bypass capacitors with a short circuit.

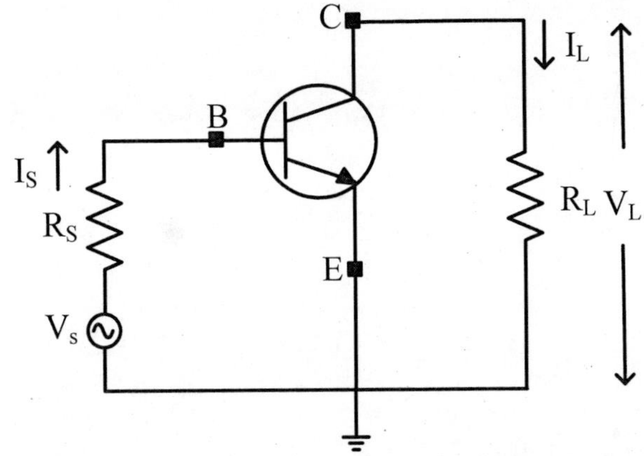

Fig. 2.12.3 Replacing the external dc power supplies used for the biasing with a short circuit.

Step-3: In the step-3, replace the transistor with its h-parameter model. The Fig. 2.12.4 shows the circuit of after replacing the transistor with its h-parameter model.

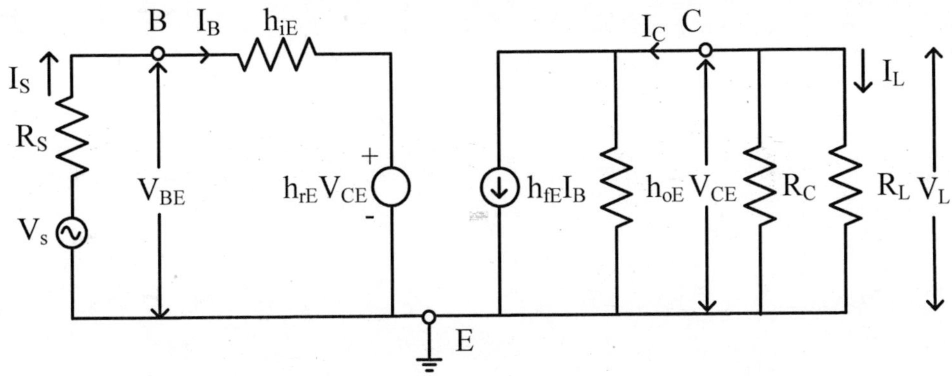

Fig. 2.12.4 Replacing the transistor with its h-parameter model.

Step-4: Calculate the current gain (A_I) of the amplifier.

The current gain of the amplifier A_I is equal to

$$A_I = \frac{I_L}{I_S} = \frac{I_L}{I_B} = -\frac{I_C R_C}{(R_L + R_C) I_B} \tag{2.12.1}$$

From the h-parameters equations for the CE amplifier

$$I_C = h_{fE} I_B + h_{oE} V_{CE} \tag{2.12.2}$$

$$\text{But } V_{CE} = I_L R_L = -\frac{I_C R_C R_L}{R_C + R_L} \; 4 \tag{2.12.3}$$

Substituting (2.12.3) in (2.12.2), we get

$$I_C = h_{fE} I_B - \frac{h_{oE} I_C R_C R_L}{R_C + R_L} \tag{2.12.4}$$

$$I_C + \frac{h_{oE} I_C R_C R_L}{R_C + R_L} = h_{fE} I_B \tag{2.12.5}$$

$$\frac{I_C}{I_B} = h_{fE}\left(\frac{R_C + R_L}{R_C + R_L + h_{oE} R_C R_L}\right) \tag{2.12.6}$$

Multiplying $\frac{R_C}{(R_C + R_L)}$ on both sides, we get

$$\frac{I_C R_C}{I_B (R_C + R_L)} = h_{fE}\left(\frac{R_C}{R_C + R_L + h_{oE} R_C R_L}\right) \tag{2.12.7}$$

Therefore, the current gain *AI* is equal to

$$A_I = \frac{I_L}{I_B} = -\frac{I_C R_C}{(R_L + R_C) I_B} = -h_{fE}\left(\frac{R_C}{R_C + R_L + h_{oE} R_C R_L}\right) \tag{2.12.8}$$

Step-5: Calculate the input resistance (RI) of the amplifier.
The input impedance RI is given by

$$R_I = \frac{V_{BE}}{I_B} \tag{2.12.9}$$

From the h-parameters equations for the CE amplifier

$$V_{BE} = h_{iE} I_B + h_{rE} V_{CE} \tag{2.12.10}$$

and $V_{CE} = I_L R_L = -\frac{I_C R_C R_L}{R_C + R_L} = A_I I_B R_L$ (2.12.11)

substituting (2.12.11) in (2.12.10) gives

$$V_{BE} = h_{iE} I_B + h_{rE} A_I I_B R_L \tag{2.12.12}$$

Therefore, the input impedance

$$R_I = \frac{h_{iE} I_B + h_{rE} A_I I_B R_L}{I_B} = h_{iE} + h_{rE} A_I R_L \tag{2.12.13}$$

Step-6: Calculate the voltage gain (AV) of the amplifier.
The voltage gain AV is given by

$$A_V = \frac{V_{CE}}{V_{BE}} \tag{2.12.14}$$

Substituting (2.12.9) and (2.12.11) in (2.12.14) gives

$$A_V = \frac{A_I I_B R_L}{V_{BE}} = \frac{A_I R_L}{R_I} \tag{2.12.15}$$

By including the source resistance R_S, the voltage gain A_{VS} will be

$$A_{VS} = \frac{A_V R_I}{R_I + R_S} \tag{2.12.16}$$

Step-7: Calculate the output admittance (Y_O) of the amplifier.

The output admittance Y_O is given by

$$Y_O = \left.\frac{I_C}{V_{CE}}\right|_{V_S=0} \tag{2.12.17}$$

From the h-parameters equations for the CE amplifier

$$I_C = h_{fE} I_B + h_{oE} V_{CE} \tag{2.12.18}$$

Substituting (2.12.18) in (2.12.17) gives

$$Y_O = \frac{h_{fE} I_B}{V_{CE}} + h_{oE} \tag{2.12.19}$$

Applying KVL in the input loop we get

$$(R_S + h_{iE})I_B + h_{rE}V_{CE} = V_S \tag{2.12.20}$$

But for calculating Y_O, V_S needs to be taken as 0. Therefore,

$$(R_S + h_{iE})I_B + h_{rE}V_{CE} = 0 \tag{2.12.21}$$

Simplifying (2.12.20), we get

$$\frac{I_B}{V_{CE}} = -\frac{h_{rE}}{(R_S + h_{iE})} \tag{2.12.22}$$

Substituting (2.12.21) in (2.12.19), we get

$$Y_O = -\frac{h_{rE} h_{fE}}{(R_S + h_{iE})} + h_{oE} \tag{2.12.23}$$

2.13 SMALL SIGNAL ANALYSIS OF THE AMPLIFIER USING THE SIMPLIFIED H-PARAMETERS

The computation of the amplifier parameters like the input impedance, voltage gain, current gain and the output admittance using h-parameters involves complex computations. In order to minimize these computations, simplified h-parameter model can be used. In the simplified h-parameter model, the parameters like h_r and h_o can be neglected. Neglecting these parameters, the transistor equations will be

$$V_i = h_i I_i$$
$$I_o = h_f I_i \tag{2.13.1}$$

Using the equation (2.13.1), the simplified h-parameter equivalent circuit will be

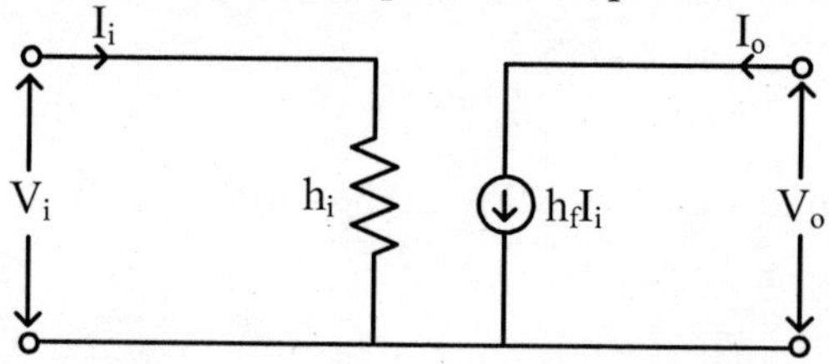

Fig. 2.13.1 Simplified h-parameter equivalent circuit of the transistor.

Using the above equivalent circuit, the analysis of amplifier parameters can be simplified. The condition for choosing between the exact analysis and the simplified analysis is $h_oR_L \geq 0.1$. If this condition is satisfied then exact analysis needs to be used. If the above condition failed, then the simplified analysis needs to be used.

The following steps are required for performing the simplified small signal analysis of the amplifier.

Step-1: Compute whether $h_oR_L < 0.1$ or not. If this condition satisfied then proceed for simplified analysis

Step-2: In the first step replace all the coupling capacitors and the bypass capacitors with a short circuit.

Step-3: In the second step, relace the external dc power supplies used for the biasing with a short circuit.

Step-4: Replace the transistor with its h-parameter model.

Step-5: Calculate the current gain (A_I) of the amplifier.

Step-6: Calculate the input resistance (R_I) of the amplifier.

Step-7: Calculate the voltage gain (R_I) of the amplifier.

Step-8: Calculate the output admittance (R_O) of the amplifier.

2.14 CURRENT MIRROR

Current mirror is the realization current dependent current source circuit. Thus, the output current at the load point is function of particular current at a given point in the network circuit. The output current at the load point is the function of the current at the particular point in the network circuit. The output current increases or decreases as the current at the particular point in the network circuit. In order to fabricate current mirror circuit one need two identical transistors with same β value. The two transistors are connected back-to-back as shown in Fig. 2.14.1

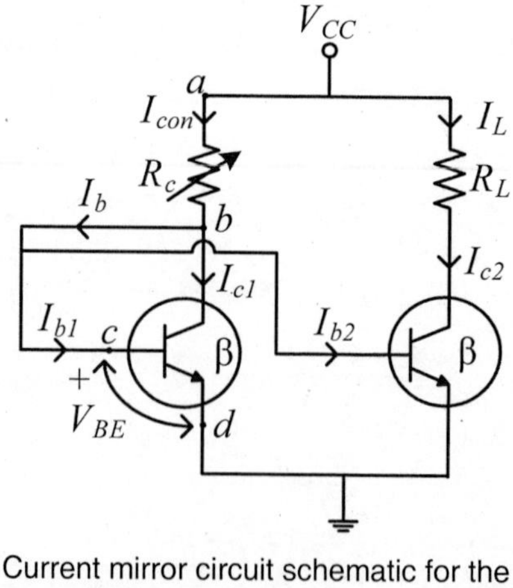

Fig. 2.14.1 Current mirror circuit schematic for the using transistor.

As two transistors are identical, it results in same base to emitter drop V_{BE} for the equal value of base currents (I_{B1} and I_{B2} for transistor 1 and 2 respectively) two flowing through them. For the circuit shown in Fig. 2.14.1, collector terminal of the transistor 1 is connected to variable resistor, while the collector terminal of the transistor 2 is connected to load. The current in the load resistor is controlled by the variable resistor. The control current I_{con} in the variable resistor is given by,

Applying KVL in loop abcd we have,

$$V_{CC} - I_{CON}R_C - V_{BE} = 0$$

$$I_{CON} = \frac{V_{CC} - V_{BE}}{R_C} \tag{2.14.1}$$

As drop V_{BE} is typically constant (diode drop of 0.7V), so increase and decrease of control resistor R_C will define the control current I_{con} assuming V_{CC} is constant. As identical transistors are used so the control current gets equally shared between two transistors and is given by,

$$I_{b1} = I_{b2} = \frac{I_b}{2} \tag{2.14.2}$$

Transistor operating in active region and considering ($\beta >> 2$) we have,

$$I_{con} = I_{c1} + I_b$$

$$I_b = I_{b1} + I_{b2}$$

$$I_{b2} = I_{b1}$$

$$I_{con} = I_{c1} + 2I_{b1}$$

$$I_{con} = \beta I_{b1} + 2I_{b1} = (\beta + 2)I_{b1} \approx \beta I_{b1} \tag{2.14.3}$$

Now the output load current I_{c2} is given by,

$$I_{c2} = \beta I_{b2}$$

$$I_{c2} = \beta I_{b1}$$

$$I_{c2} = I_{con}$$

$$I_L = I_{c2} = \frac{V_{cc} - V_{BE}}{R_c} \tag{2.14.4}$$

Thus, current in the output load is controlled by control resistor R_c or current in the output load is the mirror image of the current through control resistor R_c.

2.15 HIGH-FREQUENCY RESPONSE OF THE AMPLIFIERS

Before starting the study of the high-frequency response of the BJT circuit, let us understand some basic terms employed in the computation of the audio levels in the amplifier circuits. The unit "bel" is typically employed to measure audio levels. The unit "bel" is derived from the surname of Alexander Graham Bell. Alexander Graham Bell discovered that the audio level of any circuit is related to power on a logarithmic basis. Thus, audio levels of any circuit is given by,

$$G = \log_{10}\frac{P_2}{P_1} \text{ bel} \tag{2.15.1}$$

Where P_2 is the output power (power absorbed by the amplifier) and P_1 is the reference power associated with the standard audio levels. The reference power P_1 has the typical value of 1mW with a resistor of 600Ω. The resistance 600Ω is the standard characteristics impedance of the audio transmission line. The logarithmic scale of the amplifier simplifies the computation. For e.g. the increase in power from 4W to 16W will result in a change in audio level by two times not four times. Thus, if one needs to increase the audio level by three times with reference power, P is given by $(P)^3$, not 3P. Hence, using a logarithmic scale simplifies the computation. Thus, the logarithmic scale in decibels (dB) is used as given below,

$$G_{dB} = 10\log_{10}\frac{P_2}{P_1} \text{ dB} \tag{2.15.2}$$

Apart from the given computation is also given on the basis of voltage and current in the audio circuit. Assuming the impedance employed is the same as that for the reference power in the audio circuit, the gain equation becomes,

$$G_{dB} = 10\log_{10}\frac{P_2}{1mW} \text{ dB} \tag{2.15.3}$$

Where P_2 is the power in mW. Further, the given eq (2.15.3) is also applied to the circuit with an impedance not equal to or other than 600Ω. In a circuit with an impedance not equal to or other than 600Ω, the effect of the resistance on power is neglected. The voltage or current values typically define the power. So, the gain in terms of voltage or current can be computed as,

$$G_{dB} = 10\log_{10}\frac{P_2}{P_1} \text{ dB}$$

$$G_{dB} = 10\log_{10}\frac{\frac{(V_2)^2}{R_2}}{\frac{(V_1)^2}{R_1}} \text{ dB}$$

$$G_{dB} = 10\log_{10}\frac{(V_2)^2}{(V_1)^2} \text{ dB } (R_1 \neq R_2, \text{their effect is ignored})$$

$$G_{dB} = 20\log_{10}\frac{(V_2)}{(V_1)} \text{ dB}$$

$$G_{dB} = 20\log_{10}\frac{(i_2)}{(i_1)} \text{ dB} \tag{2.15.4}$$

Thus, the gain for the given audio circuit can be computed in terms of voltage or current ratio while neglecting the effect of the resistance. Further, the given value of gain varies with frequency of the source signal. Thus, it is necessary to compute the gain and its variation with change in frequency. Thus, it is necessary to study the frequency response of the transistor circuit. The frequency response of the transistor circuit has three bands based on the frequency value. The three frequency bands are defined as follows:

i. Low-frequency band
ii. Medium frequency band
iii. High-frequency band

The frequency range between 10Hz to 100Hz comes under the low-frequency band. The next frequency range between 100Hz to 100 kHz comes under the medium frequency band. And the frequency band greater than 100 kHz comes under high frequency. Further, a typical transistor circuit can be divided into three categories based on input source coupling with the transistor circuit. They are:

i. RC coupled
ii. Transformer coupled
iii. Direct coupled

The response of the circuit in the different frequency band depends on the circuit category or the type of elements in the circuit. Let us discuss one by one the effect of the each frequency band on each categories of the circuit. Typical response of the given three categories of the transistor circuit is given in Fig. 2.15.1.

In the case of the low-frequency band, the input and output coupling parameters like coupling capacitors and transformers have significant effect in the response of the system. At low frequency coupling capacitor offers infinite impedance and it results in complete attenuation of the signal across the coupling capacitor. As the frequency increases, the coupling capacitor impedance decreases, thus the attenuation of input source signal decreases. In other words, amplitude of the signal increases. Thus, the coupling capacitors which typically come in series with input signal acts as a high pass filter as can be observed in Fig. 2.15.1 (a). Direct coupled does not have any effect on the signal as can be observed in Fig. 2.15.1(c). While, transformer coupled provides zero impedance for zero frequency and its value increases proportionally with the frequency. The transformer provides the galvanic isolation

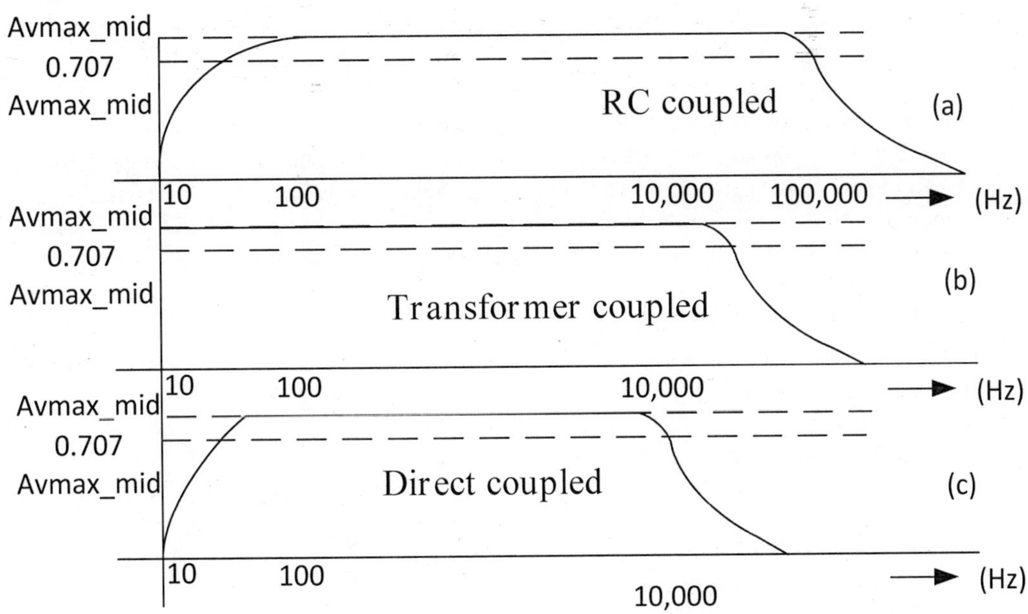

Fig. 2.15.1 Frequency response for (a) RC coupled; (b) Transformer coupled; (c) Direct coupled amplifier

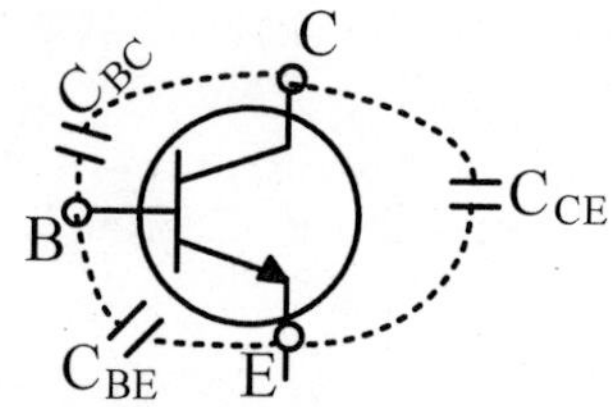

Fig. 2.15.2 Parasitic capacitance within the transistor device.

between input signal and transistor circuit apart from transferring the signal information. At low frequency the transformer magnetizing inductance between input source and ground offers zero impedance and it results in complete attenuation of the signal across the magnetizing inductance as can be observed in Fig. 2.15.1(b). As the frequency increases, the magnetizing inductance impedance increases, thus the attenuation of the input source signal decreases. In other words, the amplitude of the signal increases. Thus, the magnetizing inductance which typically comes in parallel with output acts as a high pass filter. Therefore, the signal's amplitude depends upon the transformer magnetizing inductance between input source and ground. Thus, the given coupling (both transformer and coupling capacitor coupled) acts as a high pass filter where signal amplitude increases as the frequency of the input source increases. The impedance of the coupling capacitors is nearly comparable for the frequency greater than low frequency band. Thus, for the frequency above the low frequency band, the impedance offered by coupling capacitor and transformer does not attenuate the input signal and the frequency band comes under mid frequency band region. In this zone, there is no attenuation of the input source signal and gain is almost same for all the three categories. In this zone of mid frequency region input signal has maximum amplitude. However, when source frequency reaches near high frequency region the amplitude of the input signal again starts attenuating in all the three types of transistor coupling circuit. This high frequency zone starts at the frequency of 10 kHz, slightly less than 10kHz and 100kHZ for the direct, transformer and RC coupled transistor circuit respectively. High frequency region comes into existence due to the presence of parasitic capacitance within the transistor device as shown in Fig. 2.15.2, apart from the parasitic resistance in the circuit due wiring in the input and output circuit. The given parasitic capacitance exist typically between ground and terminal forms a low pass filter. Thus, signal starts attenuating when frequency enters high frequency zone due to dominant effect of the parasitic capacitance. Fig. 2.15.2 shows different parasitic capacitances which exist in the transistor circuitry by dotted lines. The various parasitic capacitance are C_{BC} (transistor parasitic capacitance between base and collector), C_{BE} (transistor parasitic capacitance between base and emitter), C_{CE} (transistor parasitic capacitance between collector and emitter), C_{Wi} (parasitic capacitance between ground and input wiring circuit), and C_{Wo} (parasitic capacitance between ground and output wiring circuit). It can be noted that the parasitic capacitance C_{BC} exists between input base and output collector terminal. Thus, it has the effect on both input and output capacitance. The Miller theorem is employed to get effective capacitance at input and output. Before discussing the response of the transistor circuit in the high frequency region let us understand the effect in low frequency region. Take a simple case of voltage divider transistor circuit having input source v_i and is couple to the transistor circuit by a coupling capacitor C_i as shown in Fig. 2.15.3. The ac equivalent of the given transistor circuit is given in Fig. 2.15.3. The input circuit with coupling capacitor C_i forms high pass filter as shown in Fig. 2.15.3 (b). The actual input voltage V_i' given to the transistor circuit is given by,

$$V_i' = \frac{V_i z_i}{xc_i + z_i}$$

$$xc_i = \frac{1}{\omega_i C_i}; z_i = R_2 \parallel R_1 \parallel \beta R_i \tag{2.15.5}$$

where, V_i is input signal with angular frequency ω_i, R_1 and R_2 are voltage divider resistor as shown in Fig. 2.15.3. Thus, near zero frequency magnitude of the input signal to transistor circuit V_i' is near to zero, as impedance of the coupling capacitor is near to infinite.

Now as the frequency input signal increases, the magnitude of the input signal to transistor circuit V_i' also increases as shown in Fig. 2.15.1 (a) and (c). Thus, in low frequency range the transistor circuit behaves as high pass filter and the amplitude of the signal increases from zero to maximum value. The magnitude of the input signal can be normalized based on frequency of the input source as,

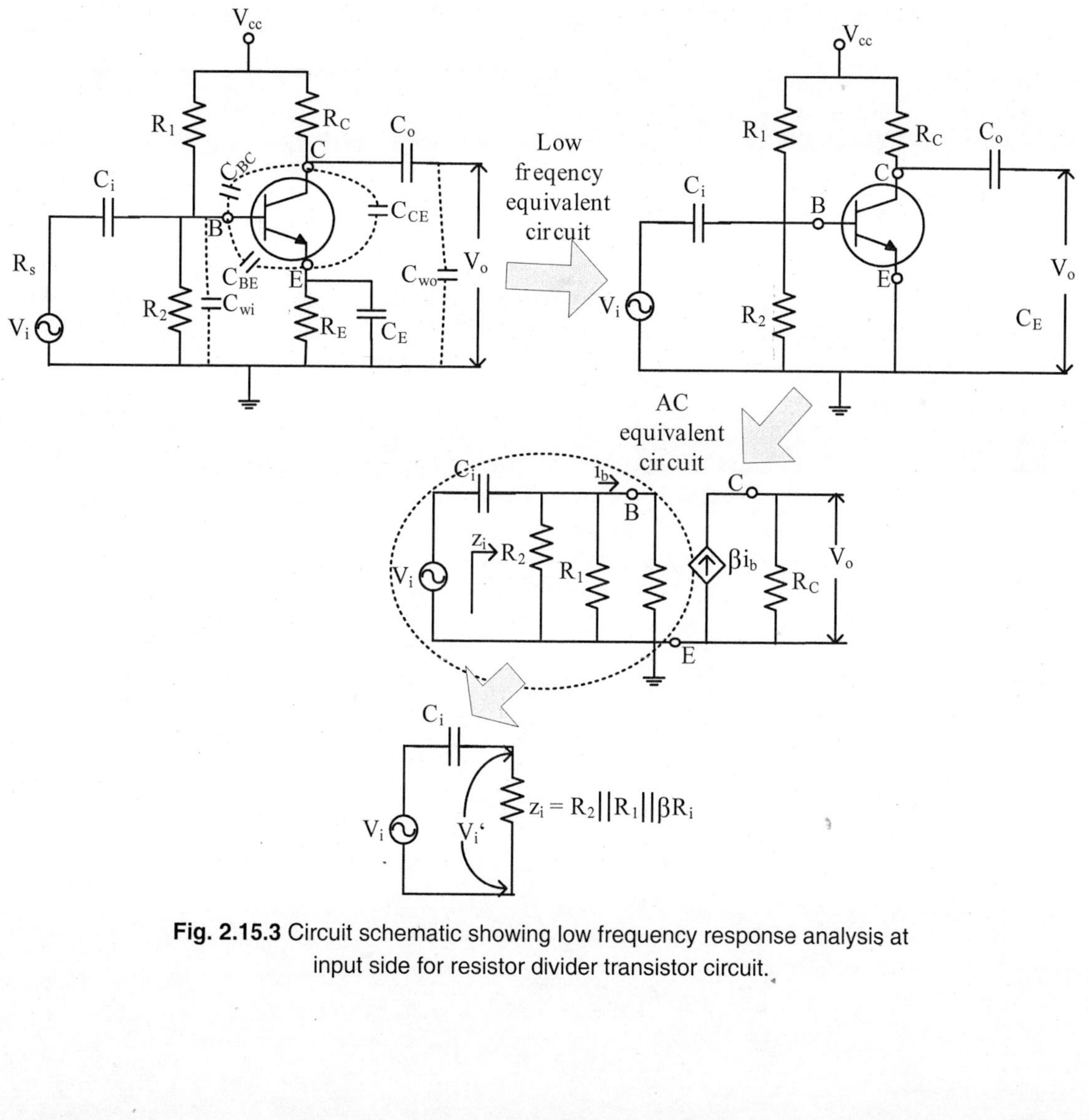

Fig. 2.15.3 Circuit schematic showing low frequency response analysis at input side for resistor divider transistor circuit.

$$V_i' = \frac{V_i z_i}{\sqrt{(xc_i)^2 + (z_i)^2}}$$

$$V_i' = \frac{V_i \frac{z_i}{xc_i}}{\sqrt{(1)^2 + (\frac{z_i}{xc_i})^2}}$$

$$V_i' = \frac{V_i \frac{f}{f_r}}{\sqrt{(1)^2 + (\frac{f}{f_r})^2}} \text{ where } f_r = \frac{1}{2\pi z_i C_i} \tag{2.15.6}$$

The above expression clearly shows that for input source frequency $f >> f_r$, there is no attenuation of the input signal and region of frequency operation is defined as medium frequency region. The frequency value at which signal amplitude is 0.707 times of its maximum value is the starting frequency value for the range of medium frequency response. In medium frequency range the signal maintains its maximum strength for all transistor circuit. The medium frequency response range ends at frequency the where gain is 0.707 times of its maximum value as shown in Fig. 2.15.1. High frequency response region starts from the end frequency value medium frequency range. In high frequency region signal strength starts attenuating and transistor circuit acts as a low pass filter as shown in Fig. 2.15.1. In high frequency region, the transistor device parasitic capacitance has the comparative impedance. Or in other words, high frequency region corresponds to the frequency zone where transistor device parasitic capacitance as shown in Fig. 2.15.2 comes into picture. Before starting analysis for the response of the transistor circuit in high frequency region let us understand Miller's effect which applies to the parasitic capacitor connecting input and output terminal i.e. capacitance are C_{BC} (transistor parasitic capacitance between base and collector). This capacitor has its effect on the performance of the transistor both at input and output. Thus, it becomes necessary to determine the input and the output capacitance i.e. Millers capacitance at input and output respectively. Let us understand the Millers effects at input and output capacitance C_M ($X_m=1/(\omega C_M)$, ω angular frequency of the input source)as shown in Fig. 2.15.4. Assume the gain between the input voltage (V_i) and output voltage (V_o) is given by A_v as shown in Fig. 2.15.4. The effective Millers capacitance at the input of the transistor circuit can be computed as,

$$i_i = i_{i2} + I_{Mi}$$

$$\frac{V_i}{z_{ieq}} = \frac{V_i}{z_i} + \frac{V_i - V_o}{X_m}$$

$$\frac{1}{z_{ieq}} = \frac{1}{z_i} + \frac{1 - \frac{V_o}{V_i}}{X_m}$$

$$\frac{1}{z_{ieq}} = \frac{1}{z_i} + \frac{1 - A_V}{X_m}$$

$$\frac{1}{z_{ieq}} = \frac{1}{z_i} + \frac{1}{X_{mi}} \text{ where } X_{mi} = \frac{X_m}{1 - A_V} \tag{2.15.7}$$

Where, z_i and z_{ieq} are the equivalent input resistance at transistor terminals and over-all equivalent input resistance at input source terminals as shown in Fig. 2.15.3. Thus, the over-all equivalent input resistance at input source terminals z_{ieq} is parallel combination of z_i and effective input Miller capacitance X_{mi} as shown in Fig. 2.15.4 .

Similarly, for the output circuit the over-all equivalent output impedance at output source terminals z_{oeq} is given by,

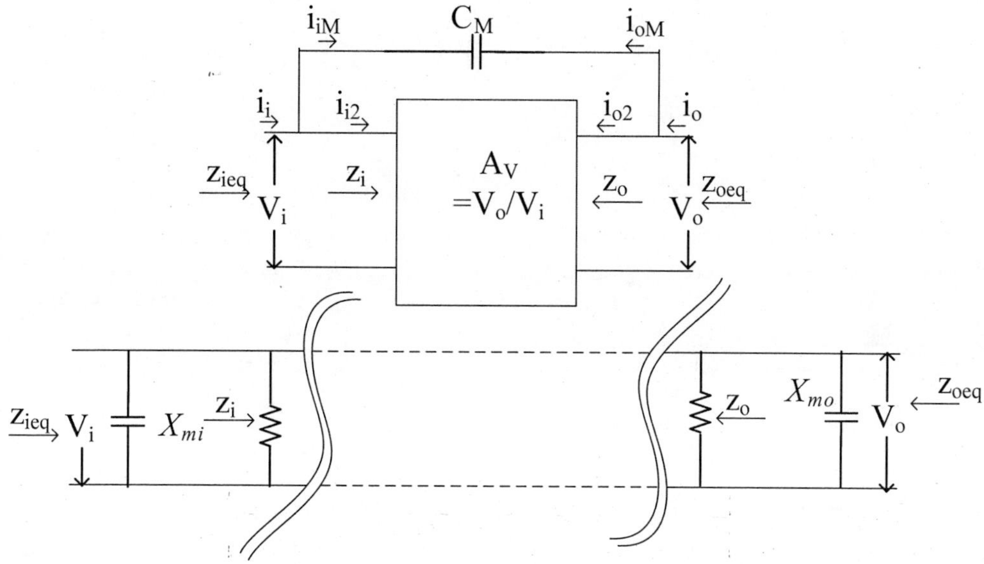

Fig. 2.15.4 Circuit schematic for Miller capacitance derivation.

$$i_o = i_{o2} + I_{oM}$$

$$\frac{V_o}{z_{oeq}} = \frac{V_o}{z_o} + \frac{-V_i + V_o}{X_m}$$

$$\frac{1}{z_{oeq}} = \frac{1}{z_o} + \frac{-\frac{V_i}{V_o} + 1}{X_m}$$

$$\frac{1}{z_{oeq}} = \frac{1}{z_i} + \frac{-\frac{1}{A_V} + 1}{X_m}$$

$$\frac{1}{z_{oeq}} = \frac{1}{z_i} + \frac{1}{X_{mo}} \quad \text{where } X_{mo} = \frac{X_m}{-\frac{1}{A_V} + 1} \simeq X_m \tag{2.15.8}$$

where, z_o and z_{oeq} are the equivalent output resistance at transistor terminals and over-all equivalent output resistance at output terminals as shown in Fig. 2.15.4. Thus, the over-all equivalent output resistance at output terminals z_{oeq} is parallel combination of z_o and effective input Miller capacitance X_{mo} as shown in Fig. 2.15.4. Let us now analyze the effect of the parasitic capacitance on the transistor, having operation in high frequency region. As discussed earlier the parasitic capacitance C_{BC} (transistor parasitic capacitance between base and collector), C_{BE} (transistor parasitic capacitance between base and emitter), C_{CE} (transistor parasitic capacitance between collector and emitter), C_{wi} (parasitic capacitance between ground and input wiring circuit), and C_{wo} (parasitic capacitance between ground and output wiring circuit) has effect on the operation of transistor circuit. The parasitic capacitance C_{BC} connects the input and output terminal of the transistor circuit with the voltage gain of A_V. Thus, Miller capacitance can be applied to the transistor circuit and results in input and output Miller capacitance and is given by,

$$X_{mo} = \frac{1}{\omega C_{BC}}$$

$$X_{mi} - \frac{1}{\omega C_{BC}(1-A_V)} = \frac{1}{\omega C_{BCMi}}; C_{BCMi} = C_{BC}(1-A_V) \tag{2.15.9}$$

Now taking a simple case of voltage divider transistor circuit having input source v_i and is couple to the transistor circuit by an equivalent parasitic capacitor C_{ieq} as shown in Fig. 2.15.5. The ac equivalent of the given transistor circuit is given in Fig. 2.15.3. The input circuit with parasitic capacitances forms low pass filter as shown in Fig. 2.15.5. The actual input voltage V_i' given to the transistor circuit is given by,

$$V_i^' = \frac{V_i xc_{ieq}}{xc_{ieq} + z_i}$$

$$xc_{ieq} = \frac{1}{\omega_1 C_{ieq}}$$

$$z_i = R_2 \parallel R_1 \parallel \beta R_i$$

$$C_{ieq} = \omega C_{BC}(1-A_V) + C_{BE} + C_{wi} \tag{2.15.10}$$

where, V_i is input signal with angular frequency ω_i, R_1 and R_2 are voltage divider resistor as shown in Fig. 2.15.5. Thus, near zero frequency magnitude of the input signal to transistor circuit V_i' is unchanged, as impedance of the equivalent parasitic capacitor C_{ieq} is near to infinite. Now as the frequency input signal increases, the magnitude of the input signal to transistor circuit V_i' also decreases as shown in Fig. 2.15.1 (a) and (c). Thus, in high frequency range the transistor circuit behave as low pass filter and the amplitude of the signal decreases from maximum to zero value. The magnitude of the input signal can be normalized based on frequency of the input source as,

$$V_i^' = \frac{V_i xc_i}{\sqrt{(xc_i)^2 + (z_i)^2}}$$

$$V_i^' = \frac{V_i \frac{xc_i}{z_i}}{\sqrt{(1)^2 + (\frac{xc_i}{z_i})^2}}$$

$$V_i^' = \frac{V_i \frac{f_r}{f}}{\sqrt{(1)^2 + (\frac{f_r}{f})^2}} \text{ where } f_r = \frac{1}{2\pi z_i C_{ieq}} \tag{2.15.11}$$

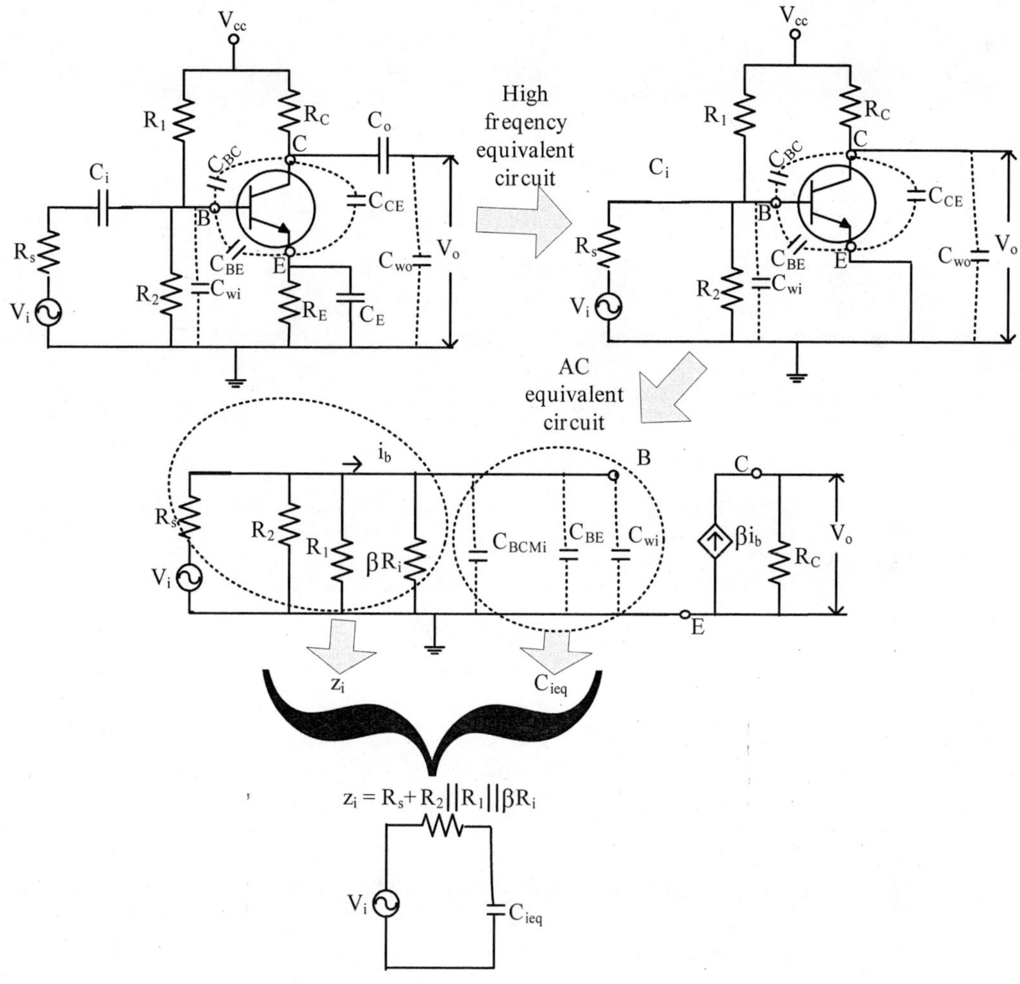

Fig. 2.14.5 Circuit schematic showing high frequency response analysis at input side for resistor divider transistor circuit.

The above expression clearly shows that for input source frequency $f << f_r$, there is no attenuation of the input signal and region of frequency operation is defined as medium frequency region. And for the frequency $f > f_r$, magnitude of the attenuates and diminishes to zero. The cut-OFF frequency f_r is the corner frequency and depends on the transistor circuit parameters.

UNIT SUMMARY

The operation of the bipolar junction transistor under both biased and unbiased working circumstances is covered in this basic unit on the transistor. The discussion of different transistor configurations, such as common base, common emitter, and common collector, followed. Once one has a solid understanding of these subjects, it is necessary to study the critical factor that determines how a bipolar junction transistor functions in an electrical circuit. In every electronic circuit, the bipolar junction transistor's biasing is a critical factor that determines how the transistor operates. The bipolar junction transistor's Q-point operating point is then covered. The next section discusses the several biasing circuits of the bipolar junction transistor. Following biasing, the BJT's function as an amplifier and switch is examined. This covers both closed- and open-switch BJT functioning. Also included are common base, common emitter, and common collector amplifier topologies along with their input and output characteristics. It is necessary to acquire the amplifiers' parameters, such as input impedance, voltage gain, current gain, and output admittance. It necessitates a brief signal analysis. The bipolar junction transistor's h-parameter model is used in this small signal analysis. The processes for using the small signal analysis to derive the amplifier parameters, such as input impedance, voltage gain, current gain, and output admittance, are described in depth. A reduced h-parameter model is also used for the tiny signal analysis in order to reduce the complexity of determining the amplifier parameters. The amplifier's high-frequency response is then shown in order to examine how the gain of the amplifier changes with frequency. Each of these ideas is necessary to comprehend how different electronic circuits work. The book's appendix contains a laboratory experiment pertaining to the input and output characteristics of a bipolar junction transistor in a common emitter configuration.

EXERCISES

Multiple Choice Questions

1.1 The relation between the doping concentration of emitter (E), base (B) and collector (C) in BJT is

a. B > C > E b. C > E > B

c. B > E > C d. E > C > B

1.2 When a pnp transistor is operated in saturated region, then its base-emitter junction and base-collector is

a. Forward Biased, Reverse Biased b. Reverse Biased, Reverse Biased

c. Forward Biased, Forward Biased d. Reverse Biased, Forward Biased

1.3 The BJT is

a. Current Controlled Device b. Voltage Controlled Device

1.4 The condition for choosing between the exact analysis and the simplified analysis is

a. $h_oR_L<0.1$ b. $h_fR_L<0.1$

c. $h_rR_L<0.1$ d. $h_iR_L<0.1$

Answers for Multiple Choice Questions			
2.1 **d**	2.2 **c**	2.3 **b**	2.4 **a**

Short Answer Type Questions

1.1 Explain the input and output characteristics of the transistor in common emitter configuration?

1.2 Explain the input and output characteristics of the transistor in common base configuration?

1.3 Explain the input and output characteristics of the transistor in common collector configuration?
1.4 Explain the h-parameter model of the transistor?
1.5 Explain the operation of BJT as a open switch?
1.6 Explain the operation of BJT as closed switch?

Long Answer Type Questions

1.1 Explain the construction and working of BJT?
1.2 Explain the frequency response RC coupled, direct coupled and transformer coupled amplifier?
1.3 Explain the Millers theorem?
1.4 Explain the low-frequency response and high-frequency response of CE amplifier?

KNOW MORE

For Additional Info Scan Me

REFERENCES

1. Robert Boylestad, Louis Nashelsky- Electronic Devices and Circuit Theory, Prentice Hall Upper Saddle River, New Jersey Columbus, Ohio
2. Donald L. Schilling, Charles Belove -Electronic Circuits: Discrete and Integrated, McGraw-Hill Book Company New York St. Louis San Francisco Toronto London Sydney.
3. Integrated Electronics Analog Digital Circuits, Jacob Millman and D. Halkias, McGraw Hill.
4. A. P. Godse & U. A. Bakshi. Electronic Devices and Circuits–I.

03 MOSFET

UNIT SPECIFICS

Through this unit we have discussed the following aspects:

- *Introduction to the MOSFET;*
- *MOSFET Transfer and Drain Characteristics;*
- *Various Configurations of the MOSFET;*
- *Analysis of MOSFET's in Ohmic Region;*
- *Biasing of the MOSFET's;*
- *Fixed Biased Circuit;*
- *Self Biased Circuit;*
- *Voltage Divider Circuit;*
- *Common Gate Circuit;*
- *Common Drain Circuit.*

RATIONALE

This fundamental unit on the MOSFET helps students to get a primary idea about the operation of the MOSFET under both unbiased and biased operating conditions. Next various configurations of the MOSFET like common gate, common source and common drain were discussed. After obtaining a good knowledge on these topics, next the crucial aspect that decides the operation of MOSFET in the electronic circuit needs to be studied. Biasing of the MOSFET is a crucial aspect that decides the operation of the MOSFET in any electronic circuit. Next, the operating point of the Q-point of the MOSFET is discussed. The operating point will ensure that the MOSFET always operates in stable and desired operating conditions. The various biasing circuits of the MOSFET are discussed next. After the biasing, the operation of MOSFET as a switch and amplifier is discussed. This includes the operation of MOSFET as closed switch and open switch. The parameters of the amplifiers like input impedance, voltage gain, current gain and output admittance need to be obtained. For which a small signal analysis is required. The detailed steps for obtaining the amplifier parameters like input impedance, voltage gain, current gain and output admittance using the small signal analysis is presented. All these concepts are essential to in understanding the operation of various electronic circuits. The laboratory experiment related to characteristics of the MOSFET is presented in the appendix of the book.

PRE-REQUISITES

Basic electrical circuit analysis

UNIT OUTCOMES

List of outcomes of this unit is as follows:

U3-O1: Understand the basics of MOSFET along with its operation

U3-O2: Realize various biasing circuits required the MOSFET along with the selection of the operating point

U3-O3: Realize various operations of the MOSFET in the electronic circuits

U3-O4: Determine the amplifier parameters by using the small signal analysis

U3-O5: Determine the design of common gate and common source configurations.

Unit-3 Outcomes	EXPECTED MAPPING WITH COURSE OUTCOMES (*1- Weak Correlation; 2- Medium correlation; 3- Strong Correlation*)				
	CO-1	CO-2	CO-3	CO-4	CO-5
U3-01	3	1	2	1	-
U3-02	3	1	2	1	-
U3-03	3	1	2	1	-
U3-04	3	1	3	1	-
U3-05	3	1	3	1	-

3.1 INTRODUCTION TO THE MOSFET'S

The Metal Oxide Semiconductor Field Effect Transistor (MOSFET) is the majority carrier device with three terminals. The three terminals of the MOSFET's are the gate, drain and source. MOSFETs are categorized into two types P and N-channel MOSFET based on the type of channel they used in its construction. In case of NPN or N-channel MOSFET, the N—Doped silicon (forming Drain and source terminal as shown in Fig. 3.1.1) with N-channel (connected to Gate terminal as shown in Fig. 3.1.1) is embedded within the P-doped silicon substrate as shown in Fig. 3.1.1. The same holds true for P-channel MOSFET where "N" is replaced by "P". Further, in an n-channel MOSFET (NMOS), the channel is formed by n-doped material (usually silicon), and a positive voltage applied to the gate relative to the source attracts electrons, creating a conducting channel between the source and drain. Thus, a small voltage at which the conducting channel is formed is called threshold gate source voltage. The gate source voltage controls the current flowing between drain and source terminals of the MOSFET. Based on the presence and absence of the channel between drain and source MOSFETs are categorized as– Enhancement and Depletion type. The structure of Enhancement and Depletion type MOSFET's are different as shown in Fig. 3.1.1. It can be noted the n-channel is present in the depletion type while the same is absent in the enhancement type between drain and source terminal. Further, the applied gate-source voltage V_{GS} is should be positive in case of an enhancement type, while for depletion type it resembles JFET which requires a negative voltage less than zero. The drain current is directly proportional to the difference between the applied gate source voltage and threshold gate source voltage for enhancement type of MOSFET's. The same is not applicable for Depletion type MOSFET's.

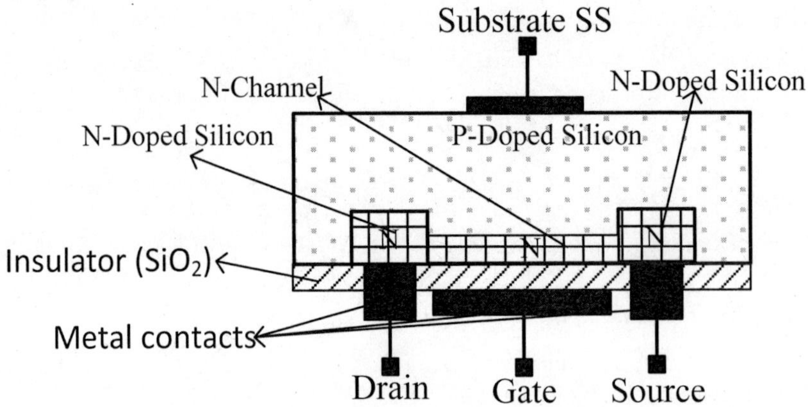

Fig. 3.1.1 Construction of N-channel MOSFET's.

In this chapter, we will discuss the depletion type MOSFET's, which is equally applicable to the JFETs. Further, same treatment is applicable to enhancement type MOSFETs and difference exist with respect to relation between the drain current and gate source voltage. Thus, with the inclusion of the required relationship between gate-source voltage V_{GS} and drain current I_D, the circuits having enhancement or depletion type MOSFET's can be analyzed as given in this chapter.

The depletion type MOSFET's is a three-terminal device named gate, drain and source. Further, an additional fourth terminal named substrate represented by 'SS' is also terminated as shown in Fig. 3.1.1. The additional terminal SS is shorted to the source terminal externally. Typically, the fourth terminal is internally shorted to source and MOSFET's device has three terminals. The MOSFET's are the majority carrier device i.e. conduction between drain and source is due to the electrons in the case of N-channel MOSFET's and holes in the case of P-channel MOSFET's. The MOSFET's are classified based on the channel between the drain and the source. If N-channel is employed between drain and source it comes under N-type MOSFET's. Similarly, if P-channel is employed between drain and source it comes under P-type MOSFET's. For the construction of the N-type MOSFET's slab silicon is doped with P-type material and is referred as substrate SS. In the given P-type slab, N-channel and N-doped regions are embedded as shown in Fig. 3.1.1. A layer of SiO_2 insulator is induced between the metal gate terminal and the N-channel apart from metal terminals of drain and source as can be seen in Fig. 3.1.1.

The MOSFET's are generally used for amplification and switching applications in electronic circuits. Further, based on the construction or type of channel added the categorization of MOSFET's can be done i.e. N-type and P-type. The drain and the source terminals are connected to the metallic contact to the n-doped region as shown. The N-doped region of the source and drain terminal is linked by N-channel in depletion N-type MOSFET's while the same is absent in enhancement N-type MOSFET's. The metal contact of the gate terminal is added over the thin layer of insulator SiO_2 as shown in Fig. 3.1.1. The thin layer of the insulator is kept over the N-channel to restrict the flow of electrons between the gate terminal and N-channel apart from adding a high input impedance characteristics. The symbols used for the P and N-type depletion MOSFET's are given in Fig. 3.1.2.

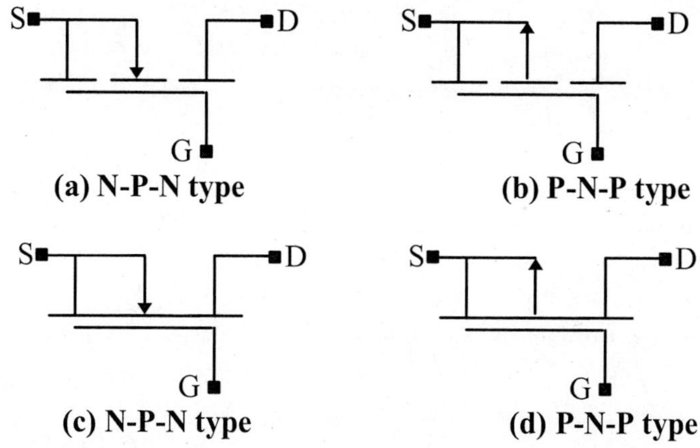

Fig. 3.1.2 Symbol of (a) N-P-N depletion type; (b) P-N-P depletion type; (c) N-P-N enhancement type, and (d) P-N- enhancement type MOSFET's.

3.2 DEPLETION MODE MOSFET TRANSFER AND DRAIN CHARACTERISTICS

The operation of the MOSFET is explained by considering an unbiased MOSFET which means no external dc voltage supply is connected to the MOSFET. Normally in the MOSFET, the N-channel between the drain and source restricts the flow of electrons between the drain and source terminal due to absence of charge carriers. Thus, ideally zero current flows between the drain and source terminals when no external dc voltage supply is connected or during zero biased conditions with. Thus, MOSFET's behave as an insulator with zero current through it apart from zero voltage across the drain and source.

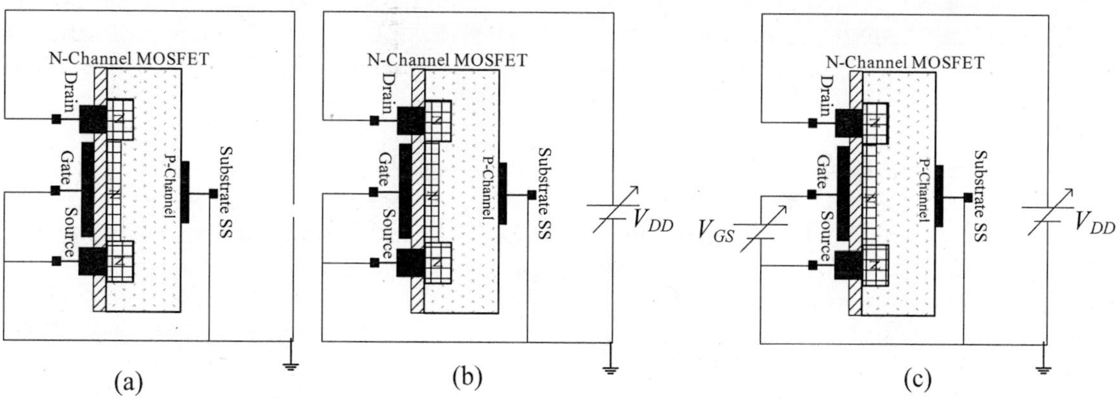

Fig. 3.2.1 Depletion mode MOSFET's circuit with (a) no biasing, (b) biasing with variable drain to source voltage V_{DD} and shorted gate to source terminal, and (c) biasing with variable drain to source voltage V_{DD} and gate to source voltage V_{GS} .

Now let us connect a variable DC supply voltage between drain to source terminals V_{DD}, which is initially kept zero and is gradually increased. Further, let the gate and source terminal be shorted as shown in Fig. 3.2.1(b). Now as the voltage V_{DD} is increased from zero to some small value same is reflected in the drain current I_D. The free electrons e- in the N-channel gets attracted towards the positive drain terminal as shown in Fig. 3.2.1(b). This results in the flow of the current from the drain to the source. The magnitude of the drain current I_D keeps on increasing as the drain voltage V_{DD} increases. The region of operation where the drain current I_D keeps on increasing as the drain voltage V_{DD} increases is termed an ohmic region similar to the active region in BJT. At particular drain voltage V_{DD}, the drain current I_D reaches its maximum saturation value I_{DSS}. Once the drain current I_D reaches its maximum saturation value I_{DSS}, there is no more increase in its value even with the increase in its drain voltage V_{DD}. The region of operation where there is no increase in the drain current I_D or it reaches its maximum saturation value I_{DSS} is termed as saturation region.

Now let us introduce a variable fix supply voltage across gate and source V_{GS} with magnitude as shown in Fig. 3.2.1(c) Taking $V_{GG} > 0$, now repeat the same process as discussed above for the fixed value of gate and source voltage V_{GS}. Now as the voltage V_{DD} is increased from zero to some small value same is reflected in the drain current I_D. The free electrons e- in the N-channel get attracted towards the positive drain terminal as shown in Fig. 3.2.1(c). This results in the flow of the current from the drain to the source. The magnitude of the drain current I_D keeps on increasing as the drain voltage V_{DD} increases. The region of operation where the drain current I_D keeps on increasing as the drain voltage V_{DD} increases is termed an ohmic region similar to active region in BJT. At particular drain voltage V_{DD}, the drain current I_D reaches its maximum saturation value I_{DSS}. The obtained value of the I_{DSS} is higher than what obtained for shorted gate to source terminal. Further, applied drain voltage V_{DD} corresponding to maximum saturation value I_{DSS} is also higher when compared with the short-circuit case as can be seen in Fig. 3.1.2. Once the drain current I_D reaches its maximum saturation value I_{DSS}, there is no more increase in its value even with the increase in its drain voltage V_{DD}. The region of operation where there is no increase in the drain current I_D or it reaches its maximum saturation value I_{DSS} is termed as saturation region.

Now taking $V_{GG} < 0$ or a negative small value (say (-1V)) repeat the same process as discussed above for the fixed value of gate and source voltage V_{GS}. Now as the voltage V_{DD} is increased from zero to some small value same is reflected in the drain current I_D. The free electrons e- in the N-channel get attracted towards the positive drain terminal as shown in Fig. 3.2.1. This results in the flow of the current from drain to source. The magnitude of the drain current I_D keeps on increasing as the drain voltage V_{DD} increases. The region of operation where the drain current I_D keeps on increasing as the drain voltage V_{DD} increases is termed as ohmic region. At particular drain voltage V_{DD}, the drain current I_D reaches its maximum saturation value I_{DSS}. The obtained value of the I_{DSS} is lower than what is obtained for shorted gate to the source terminal. Further, applied drain voltage V_{DD} corresponding to maximum saturation value I_{DSS} is also lower when compared with short-circuit case as can be seen in Fig. 3.2.2. Once the drain current I_D reaches its maximum saturation value I_{DSS}, there is no more increase in its value even with the increase in its drain voltage V_{DD}. The region of operation where there is no increase in the drain current I_D or it reaches its maximum saturation value I_{DSS} is termed as saturation region. Now if we still make gate and source voltage V_{GS} more negative (say -2V), the same characteristics repeat with the lower value of the maximum saturation value I_{DSS}. Thus, the maximum saturation value I_{DSS} with corresponding drain voltage V_{DD} keeps on decreasing as V_{GS} becomes more and more negative. This continues till the gate and source voltage V_{GS} reaches the threshold voltage of $-V_{tv}$. Once the gate and source voltage V_{GS} reaches the threshold voltage of $-V_{tv}$ there is no or zero flow of current between drain to source even with the increase of drain voltage V_{DD}. The given region of operation when there is no or zero flow of current between drain to source is termed as cut-OFF region. The three regions of operation – ohmic, saturation and cut-OFF region for the MOSFET's is shown in Fig. 3.2.2. Once completed, the obtained graph between drain current I_D and drain to source voltage (V_{DD}) can be plotted and is termed as drain characteristics as shown in Fig. 3.2.2. The drain characteristics when extended to get plot between gate and source voltage V_{GS} and drain current I_D gives the transfer characteristics as shown in Fig. 3.2.2.

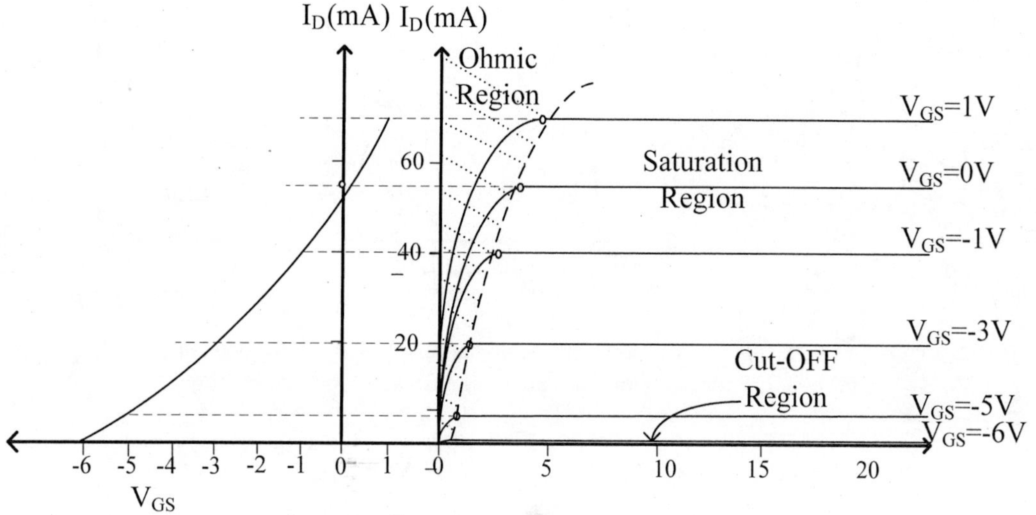

Fig. 3.2.2 Transfer and drain characteristics of the MOSFET.

To derive the transfer characteristics between gate to source voltage V_{GS} and drain current I_D apart from drain characteristics between drain to source voltage V_{DS} and drain current I_D it is required to have a variable DC voltage source between gate to source V_{GG} and drain to source terminals V_{DD}. For drain characteristics, initially gate to source terminal is shorted as shown in Fig 3.2.2.

The DC supply voltage between drain to source terminals V_{DD}, is initially kept zero and is gradually increased and readings are taken for the drain current I_D. and drain to source voltage V_{DS} for the given gate to source voltage V_{GS}. It can be noted that in ohmic region of operation, the MOSFET acts a voltage controlled variable resistor as depicted in Fig. 3.2.3(a). The resistance value between drain and source terminal is controlled by the voltage across gate and source V_{GS}. It can be noted from Fig. 3.2.2 drain characteristics that the slope of the drain characteristics line in ohmic region keeps on decreasing as the gate to source voltage V_{GS} becomes more and more negative. It becomes near to infinite for voltage lesser than (more negative) the threshold voltage V_{tv}. This property of the variable drain to source resistor is exploited in the circuit where gain control is required. The variable resistor between the drain and source terminal as the function of the gate to source voltage V_{GS} is given by,

$$r_D = \frac{r_o}{(1-\frac{|V_{GS}|}{V_{tv}})^2} \tag{3.2.1}$$

where, r_o is the resistance between drain and source terminal with gate to source voltage V_{GS} equal to zero, while , r_D is the resistance between drain and source terminal for the given gate to source voltage V_{GS}. It can be noted that in the given expression holds for gate to source voltage V_{GS} lying between $-V_{tv} \leq V_{GS} \leq 0$ apart from condition of MOSFETs drain to source voltage $V_{DS} \leq V_{tv}$. Further, in cut-OFF region where $V_{GS} \leq -V_{tv}$, the resistance r_D between drain and source terminal is infinite or near to the open circuit. Thus, MOSFET's show a very high resistance or open circuit between drain to source terminal in the cut-OFF region as depicted in Fig. 3.2.3 (b). Now for the region of operation in the saturation region, the device offers minimal resistance between drain to the source terminal. Thus, it acts as a switch with fix minimal resistance for the operation in saturation region as depicted in Fig. 3.2.3 (c).

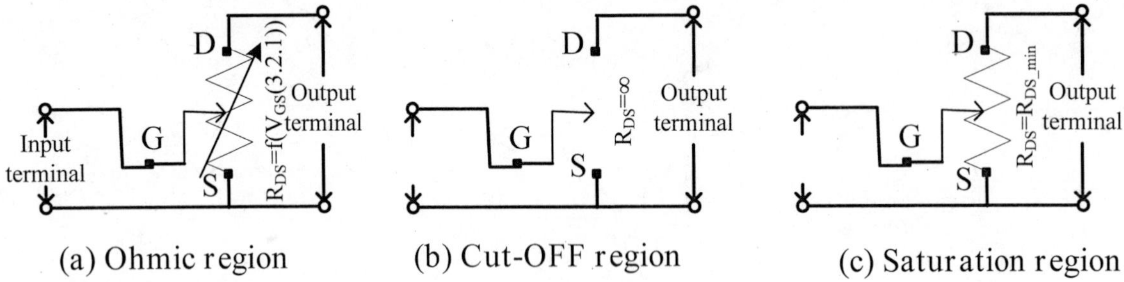

Fig. 3.2.3 MOSFET's resistance between drain and source terminal for, (a) ohmic region operating; (b) cut-OFF region operating, and (c) saturation region operating.

3.3 VARIOUS CONFIGURATIONS OF THE MOSFET

In general, most of electronic circuits are two-port networks which means there are two input terminals and two output terminals. The MOSFET is basically a three-terminal device. In order to construct an electronic circuit using the MOSFET, one terminal of the MOSFET needs to be common for the input and output terminals. This leads to a common gate, common source and common drain configuration as shown in Fig. 3.3.1. In the common gate configuration, the gate terminal is common to both the input and output terminals of the circuit. In the common source configuration, the source terminal is common to both the input and output terminals of the circuit. Similarly, in the common drain configuration, the drain terminal is common to both the input and output terminals of the circuit.

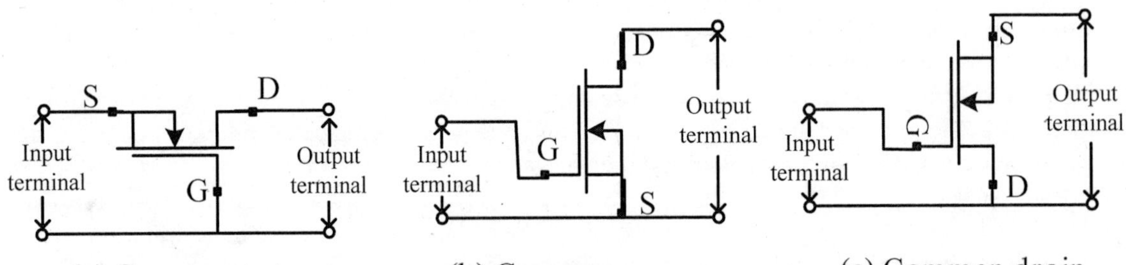

Fig. 3.3.1 Common gate, common source and common drain configurations of the MOSFET.

3.4 ANALYSIS OF MOSFET'S IN OHMIC REGION

MOSFET's devices are more popular compared to BJT due to high voltage gain and input impedance and low leakage current. Further, they are voltage-controlled devices which makes them more efficient when compared with BJT which are current controlled. In MOSFET'Ss the current in the drain terminal

I_D is controlled by the gate to source V_{GS}. The drain current is the function of the gate to source voltage V_{GS} and is given by,

$$I_D = I_{DSS}(1-\frac{V_{GS}}{V_{tv}})^2 \tag{3.4.1}$$

Where, I_{DSS} represents the saturation current for the gate to source voltage V_{GS} equal to zero; V_{tv} represents the peak threshold voltage of the MOSFET's. The above eq(3.4.1) is the non-linear equation. Thus, one needs to compute the gate to source voltage V_{GS} and solve the FET circuits. From eq(3.4.1) it dictates that there is the relation between the input gate to source voltage V_{GS} and output drain current I_D. Thus, if the derivative of the (3.4.1) is taken with respect to the input gate to source voltage V_{GS}, it simplifies to the linear equation and is given by,

$$\frac{dI_D}{d(V_{GS})} = \frac{d}{d(V_{GS})}(I_{DSS}(1-\frac{V_{GS}}{V_{tv}})^2)$$

$$\frac{dI_D}{d(V_{GS})} = 2(I_{DSS}(1-\frac{V_{GS}}{V_{tv}}))(\frac{d(-\frac{V_{GS}}{V_{tv}})}{d(V_{GS})}$$

$$\frac{dI_D}{d(V_{GS})} = 2(I_{DSS}(1-\frac{V_{GS}}{V_{tv}})) - \frac{1}{V_{tv}}$$

$$\frac{dI_D}{d(V_{GS})} = -\frac{2I_{DSS}}{V_{tv}}(1-\frac{V_{GS}}{V_{tv}}) \tag{3.4.2}$$

The (3.4.2) represents a linear equation of the straight line which relates the slope of (3.4.1) for the given the gate to source voltage V_{GS}. Further, LHS quantity in eq (3.4.2) i.e. represents the "transconductance" ($dI_D/d(V_{GS})$) and denoted by 'g_m'. The word "trans" refers to the relation between change in input and output, while conductance refers to the ratio of the current by voltage or change in output drain current to the change in input gate to source voltage V_{GS}. It can be noted that g_m is the function of the operating input gate to source voltage V_{GS}. Thus,

$$g_m = \frac{dI_D}{d(V_{GS})} = -\frac{2I_{DSS}}{V_{tv}}(1-\frac{V_{GS}}{V_{tv}}) \tag{3.4.3}$$

To ensure positive value of g_m eq (3.4.3) is rewritten as,

$$g_m = \frac{2I_{DSS}}{|V_{tv}|}(1-\frac{V_{GS}}{V_{tv}}) \tag{3.4.4}$$

Where $|V_{tv}|$ represents a absolute value or only positive value of the V_{tv} is considered. Now, for $V_{GS}= V_{tv}$, substituting in (3.4.4) we have, $g_m=0$. Similarly, for $V_{GS}= 0$, substituting in (3.4.4) we have, $g_m=2I_{DSS}/|V_{tv}|$, which is its maximum value and is represented by g_{mo}. Now, for $V_{GS}= V_{tv}/2$, substituting in (3.4.4) we have, $g_m= =I_{DSS}/|V_{tv}|= g_{mo}/2$. Thus, the relation between g_m and input gate to source voltage V_{GS} is a straight line and is given by,

$$g_m = \frac{2I_{DSS}}{|V_{tv}|}(1-\frac{V_{GS}}{V_{tv}}) = g_{mo}(1-\frac{V_{GS}}{V_{tv}})$$

$$\textit{where } g_{mo} = \frac{2I_{DSS}}{|V_{tv}|} \tag{3.4.5}$$

Thus, 3.4.5 actually represents a straight line equation, which simplifies the computation of the transconductance 'g_m'. Further, its value increases from zero at $V_{GS}= V_{tv}$ to the maximum of g_{mo} at $V_{GS}= 0$. Transconductance is actually the slope of the operating in the $V_{GS} - I_D$ characteristics. The slope for the $V_{GS} - I_D$ characteristics at the given operating point can be determined:

i. Graphically

For graphical computation, one needs to compute the change in the drain current to the given small change in gate to source voltage VGS about the operating point. Thus, transconductance 'gm' is given by,

$$g_m = \frac{\Delta I_D}{\Delta V_{GS}} \tag{3.4.6}$$

Thus, using the graph of $V_{GS} - I_D$ characteristics one can determine the change in input gate to source voltage ΔV_{GS} and its corresponding change in output drain current to ΔI_D to compute g_m.

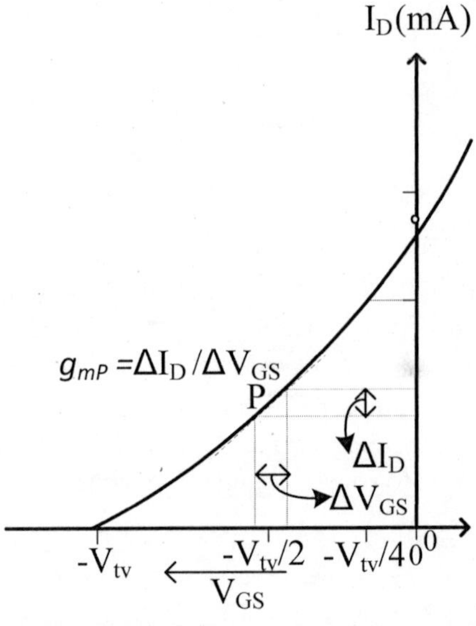

Fig. 3.4.1 Graphical computation of transconductance 'g_m' for the MOSFET

ii. Analytically.

For analytical one need to derive the expression for the derivative of the drain current I_D wrt gate to source voltage V_{GS} will result in,

$$g_m = \frac{2I_{DSS}}{|V_{tv}|}(1 - \frac{V_{GS}}{V_{tv}}) \tag{3.4.7}$$

The slope of the straight line is given by $2I_{DSS}/|V_{tv}|^2$. Its intercept on the y-axis (g_m) represents the maximum value of the transconductance obtained by putting $V_{GS}= 0$ in (3.4.7) and is given by,

$$g_m\Big|_{V_{GS}=0} = \frac{2I_{DSS}}{|V_{tv}|} = g_{mo} \tag{3.4.8}$$

At $V_{GS}= V_{tv}$; $g_m=0$

At $V_{GS}= V_{tv}/2$; $g_m=g_{mo}/2$

At $V_{GS}= V_{tv}/4$; $g_m=3g_{mo}/4$

At $V_{GS}= 0$; $g_m= g_{mo}$

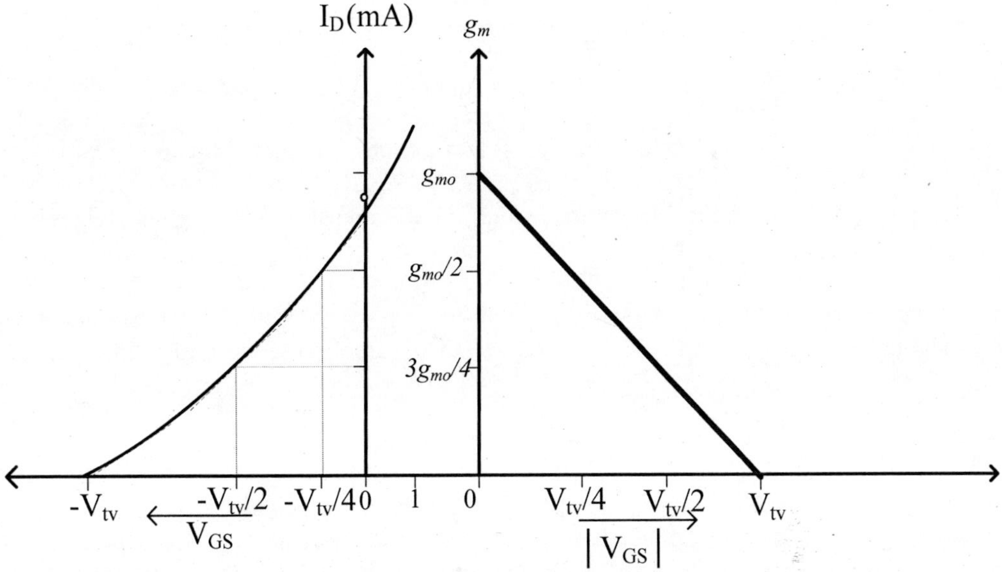

Fig. 3.4.2 Analytical computation of transconductance 'g_m' for the MOSFET and its variation with gate to source voltage V_{GS}

Till we have learnt about the expression of the transconductance 'g_m' in terms of V_{GS}. The magnitude of the transconductance 'g_m' can also be determined based on drain current I_D. From eq (3.4.1),

$$I_D = I_{DSS}(1-\frac{V_{GS}}{V_{tv}})^2$$

$$(\frac{I_D}{I_{DSS}})^{\frac{1}{2}} = (1-\frac{V_{GS}}{V_{tv}}) \qquad (3.4.9)$$

Now,

$$g_m = g_{mo}(1-\frac{V_{GS}}{V_{tv}})$$

$$\frac{g_m}{g_{mo}} = (1-\frac{V_{GS}}{V_{tv}})$$

$$\frac{g_m}{g_{mo}} = (\frac{I_D}{I_{DSS}})^{\frac{1}{2}} = (1-\frac{V_{GS}}{V_{tv}})$$

$$g_m = g_{mo}(\frac{I_D}{I_{DSS}})^{\frac{1}{2}} \qquad (3.4.10)$$

Thus, as the value of the drain current I_D increases, the corresponding value of the g_m also increases. The maximum value (saturated value) of the drain current is obtained when g_m is maximum at $V_{GS}=0$.

For, $I_D=I_{DSS}$; $g_m= g_{mo}$

At, $I_D=I_{DSS}/4$; $g_m= g_{mo}/2$

At, $I_D=I_{DSS}/2$; $g_m= g_{mo}/\sqrt{2}$

In other way if g_m is known then, I_D can be determined directly using,

$$I_D = I_{DSS}\left(\frac{g_m}{g_{mo}}\right)^2 \tag{3.4.11}$$

Thus, once drain current I_D is determined, it actually defines define the gain and operating point of the MOSFET circuit. Then, using the small signal model MOSFET one can determine the output of the MOSFET circuit. Now, as MOSFET devices have high input impedance device and ideally it is infinite between gate and source. This feature further helps in preventing the input signal distortion. Practically the impedance has the value more than 1000MΩ. While the output impedance between drain and source typically varies between 20 kΩ to 100 kΩ. In the datasheet typically the admittance parameter of Y_{ds} is provided having the value between 1/50 µS to 1/10 µS. The value of the drain to source impedance is given by,

$$z_{ds} = \frac{1}{Y_{ds}} \tag{3.4.12}$$

In the small ac equivalent circuit for the MOSFET, the output impedance represents the equivalent source to drain resistance of the MOSFET device. It also actually represents the slope of the output characteristics between the drain current I_D and the drain to source voltage V_{DS}. Near to zero slope of drain characteristics represents the infinite resistance. It can be observed that for the lower value of the input gate to source voltage near to threshold voltage V_{tv} the drain characteristics has zero slope or has infinite drain source resistance. Same is depicted in Fig. 3.2.3.

Further, it can be seen from the transfer output characteristics (plot of gate to source voltage versus drain current) shown in Fig. 3.4.2 the transconductance 'g_m' has the maximum value when the gate to source voltage is maximum or has zero value. The transconductance 'g_m' value decreases as the gate to source voltage decreases from zero to negative value of V_{tv}. The transconductance 'g_m' value equal to zero for the gate to source voltage of $-V_{tv}$. As the transconductance 'g_m' relates the gate to source voltage with the in the drain current and is the variable value which is the function of gate to source voltage. For the given value of the gate to source voltage the transconductance 'g_m' is a fix value and is given by ratio of drain current to gate to source voltage. In other words, it represents the slope of the transfer characteristics and for very small changes in gate to source voltage (ΔV_{GS} tends to zero), the slope transconductance 'g_m' is given by,

$$g_m = \frac{I_D}{V_{GS}} \tag{3.4.13}$$

Thus, for the small signal model,

$$I_D = g_m V_{GS} \tag{3.4.14}$$

For AC equivalent small signal model for the drain source region can be represented by a current source with magnitude given by transconductance 'g_m' times the gate to source voltage. Thus, MOSFET when working in ohmic region ideally have the fixed drain current given by (3.4.14). Thus, characteristics should be a straight line parallel to drain to source voltage axis. However, practically this is not true. The drain characteristics of the MOSFET device is having a slope which is represents the presence of resistance between drain to source terminal represented by 'r_d' as shown in Fig. 3.4.3. This resistance has high value and near to infinite for gate to source voltage 'V_{GS}' tending to negative of peak threshold voltage -V_{tv}. For this reason, the output characteristics of the MOSFET device is nearly parallel to voltage axis (drain to source voltage 'V_{DS}'). Thus, using the current gain and MOSFET drain to source resistance, the ac equivalent circuit for the three terminal MOSFET device can be drawn and is shown in Fig. 3.4.4. The given ac equivalent circuit can be used to replace the MOSFET terminals in the MOSFET network circuit.

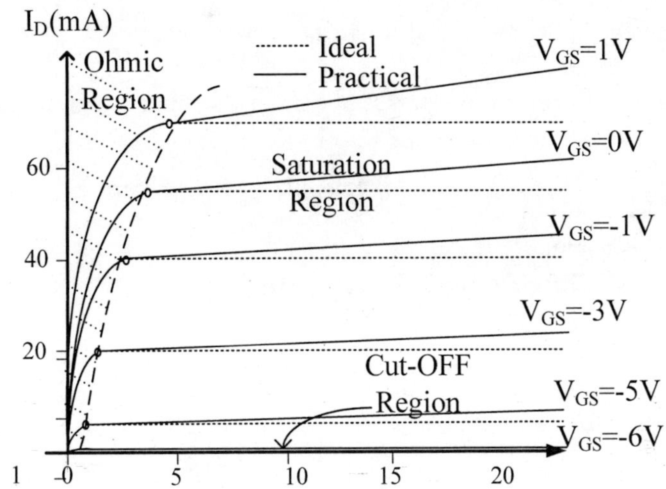

Fig. 3.4.3 Drain characteristics for the for ideal and practical MOSFET operating in saturation region.

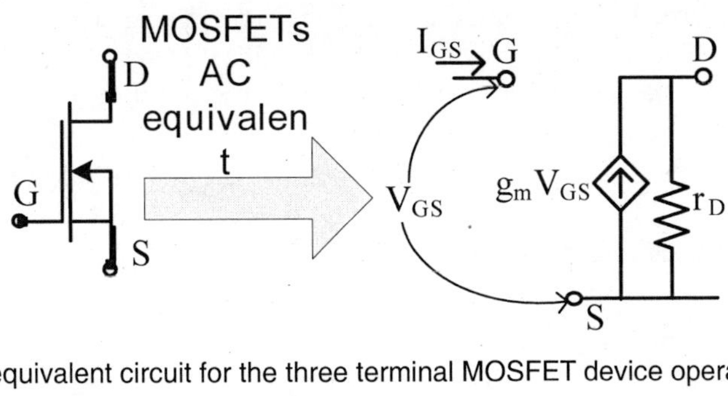

Fig. 3.4.4 AC equivalent circuit for the three terminal MOSFET device operating in ohmic region.

3.5 BIASING OF THE MOSFET'S

From above, it is clear that to solve the MOSFET problem, it is required to compute the gate to source voltage 'V_{GS}'. The gate to source voltage 'V_{GS}' can be determined by using DC analysis for computing operating point. Using DC analysis determine the gate to source voltage 'V_{GS}'. Once gate to source voltage 'V_{GS}' is computed, then determine the transconductance 'g_m'. Using the information of the transconductance 'g_m' and gate to source voltage 'V_{GS}' derive the ac equivalent circuit for the given MOSFET circuit.

Step involved in DC analysis of the MOSFET circuit is as follows:

i. Open circuit the coupling capacitors C_s, C_o, and C_E, if exist in the circuit.
ii. Short the input AC signal source.
iii. Solve the obtained MOSFET circuit for gate to source voltage 'V_{GS}'.
iv. As gate terminal is insulated by SiO_2, the gate current I_{GS} is considered to be zero.
v. Determine the transconductance 'g_m'.
vi. Determine the drain current I_D using transconductance 'g_m'.
vii. Using the drain current I_D determine, determine drain to source voltage V_{DS} by applying KVL in the outer loop having drain and source terminals.

Once the drain current I_D is determined, then the ac equivalent circuit for the given MOSFET network can be derived using following steps:

i. Short circuit all the coupling capacitors C_s, C_o, and C_E , if exist in the circuit.
ii. Short all the input DC sources defining the operating region for the MOSFET's.
iii. Replace the three terminal MOSFET device with its equivalent ac signal model as given in Fig. 3.4.4 (MOSFET ac equivalent).
iv. Solve the obtained circuit for the output voltage.

As in transistor circuits, MOSFET circuits are also categorized based on the DC biasing circuit as:

i. Fixed biased circuit
ii. Self biased circuit
iii. Voltage divider circuit
iv. Common gate circuit
v. Common drain circuit

3.6 FIXED BIASED CIRCUIT

The circuit diagram for the fixed biased circuit is given in Fig. 3.6.1. The given circuit has the fixed input gate voltage V_{GG} as given in Fig. 3.6.1. The fixed input gate voltage V_{GG} connected to the gate terminal "G" via gate resistor R_G. The gate resistor R_G is employed to limit the gate current and protect MOSFET device. To analyze the given circuit it is first necessary to determine the voltage across gate and source "V_{GS}". As MOSFET device has infinite input resistance, so current in the gate circuit I_{GS} is approximately equal to zero (≈ 0). As gate circuit I_{GS} is zero, the voltage across gate and source "V_{GS}" is equal to the DC supply voltage "V_{GG}" as depicted in Fig. 3.6.2.

$$-V_{GS} - V_{GG} = 0$$

$$V_{GS} = -V_{GG} \quad (3.6.1)$$

Once the voltage across gate and source "V_{GS}" is computed then the DC operating point of the given circuit can be computed. In other words, drain to source current "I_D" and transconductance 'g_m' can be computed using via a DC supply voltage "V_{GG}".

$$I_D = I_{DSS}(1 - \frac{V_{GG}}{V_{tv}})^2$$

$$g_m = \frac{I_D}{V_{GG}} \quad (3.6.2)$$

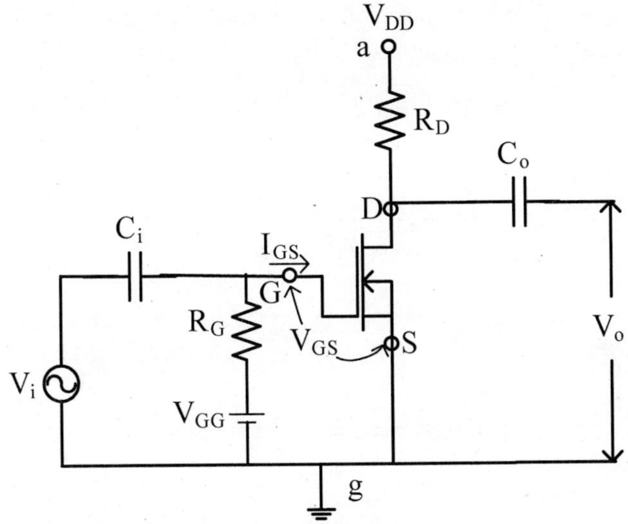

Fig. 3.6.1 The circuit diagram for the fixed biased circuit using depletion MOSFET.

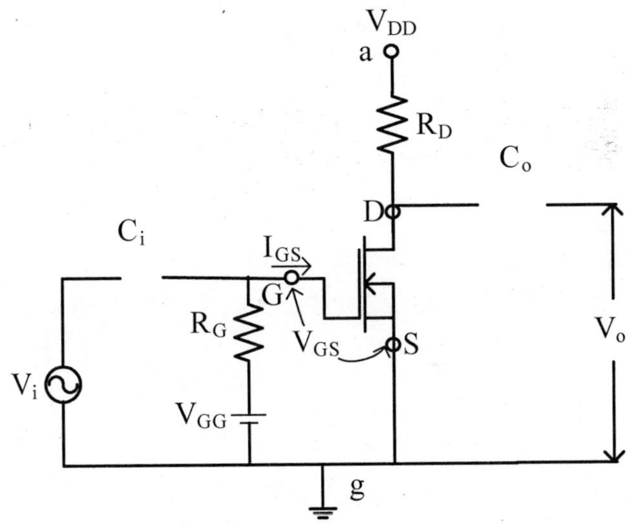

Fig. 3.6.2 The equivalent circuit diagram for DC analysis of the fixed biased circuit using DEPLETION MOSFET.

Now the other voltages can be determined by applying KVL at output circuit having loop at a-D-S-g-a.

$$V_{DD} - I_D R_D - V_{DS} = 0$$

$$V_{DS} = V_{DD} - I_D R_D \tag{3.6.3}$$

Thus, MOSFET operating drain to source voltage can be determined. Further, again using drain to source current "I_D" for the analysis of the AC equivalent circuit to determine the voltage gain for the input signal of the given MOSFET circuit. To derive the AC equivalent circuit one need to follow the steps given in above section. Using the AC equivalent circuit as shown in Fig. the equivalent input resistance is given by,

$$z_i = R_G \tag{3.6.4}$$

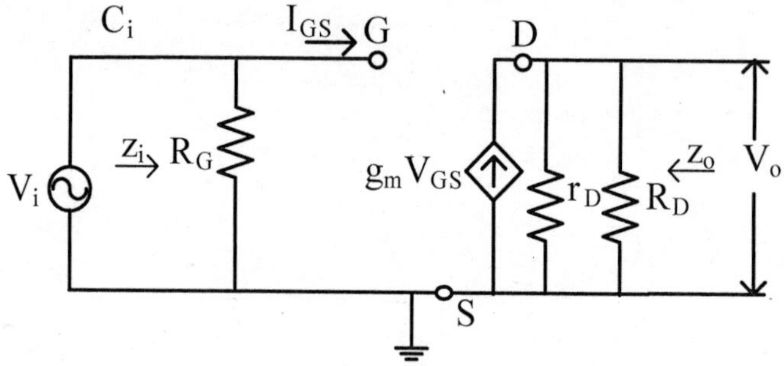

Fig. 3.6.3 The equivalent circuit diagram for AC analysis of the fixed biased circuit using depletion MOSFET.

For equivalent output resistance is obtained by shorting input source in the equivalent AC circuit. The obtained equivalent output resistance is given by,

$$z_o = r_D \parallel R_D \tag{3.6.5}$$

The voltage gain in the equivalent AC circuit is given by,

$$A_v = \frac{V_o}{V_i} \tag{3.6.6}$$

Where,

$$V_i = v_{Ac}$$

$$V_o = -g_m v_{Ac} z_o \tag{3.6.7}$$

Therefore,

$$A_v = \frac{V_o}{V_i} = \frac{-g_m v_{Ac} z_o V_o}{v_{Ac}}$$

$$A_v = -g_m z_o \tag{3.6.8}$$

The negative sign in the gain indicate output signal is 180° phase shifted with reference to the input signal.

3.7 SELF BIASED CIRCUIT

The circuit diagram for the self biased circuit is given in Fig. 3.7.1. As the name indicates the given circuit does not have any fixed input gate voltage V_{GG}. However, the drain terminal of the MOSFET circuit is connected to the ground via a resistor R_s as given in Fig. 3.7.1. To analyze the DC operating point of the given MOSFET circuit the given circuit is reduced to its DC equivalent using conditions given above. The DC equivalent circuit for the MOSFET circuit is given in Fig. 3.7.2. Using the DC equivalent circuit and applying KVL in the loop g-S-G-g we have,

$$-V_{GS} - I_D R_{So} - R_G I_{GS} = 0$$
$$V_{GS} = -I_D R_{So} - R_G I_{GS} \quad (I_{GS} = 0)$$
$$V_{GS} = -I_D R_{So} \tag{3.7.1}$$

Using Shockley equations we have,

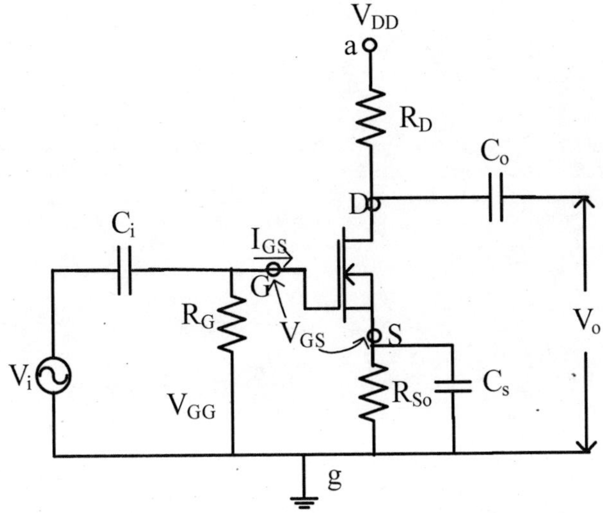

Fig. 3.7.1 The circuit diagram for the self-biased circuit using depletion MOSFET.

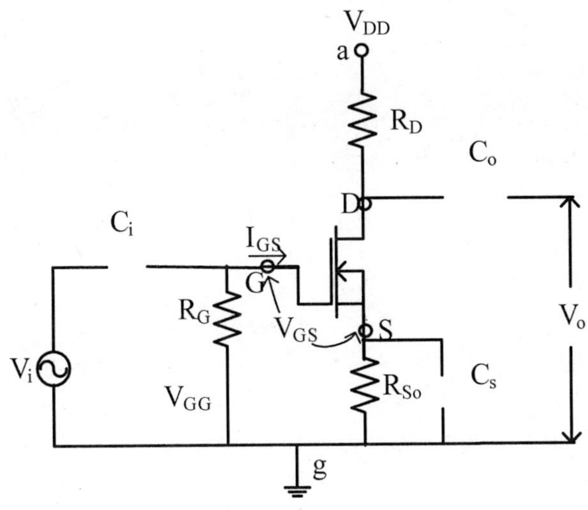

Fig. 3.7.2 The equivalent circuit diagram for DC analysis of the self biased circuit using depletion MOSFET.

$$I_D = I_{DSS}(1-\frac{V_{GS}}{V_{tv}})^2$$

$$I_D = I_{DSS}(1+\frac{I_D R_{So}}{V_{tv}})^2$$

$$\frac{I_D}{I_{DSS}} = (1+\frac{I_D R_{So}}{V_{tv}})^2$$

$$(\frac{I_D R_{So}}{V_{tv}})^2 + 2\frac{I_D R_{So}}{V_{tv}} - \frac{I_D}{I_{DSS}} + 1 = 0 \tag{3.7.2}$$

Solving the above quadratic equation (3.7.2) one can determine the drain current I_D and is given by,

$$(I_D)^2(\frac{R_{So}}{V_{tv}})^2 + I_D(\frac{2R_{So}}{V_{tv}} - \frac{1}{I_{DSS}}) + 1 = 0$$

$$I_D = \frac{-(\frac{2R_{So}}{V_{tv}} - \frac{1}{I_{DSS}}) \pm \sqrt{((\frac{2R_{So}}{V_{tv}} - \frac{1}{I_{DSS}})^2 - 4(\frac{R_{So}}{V_{tv}})^2}}{2(\frac{R_{So}}{V_{tv}})^2}. \tag{3.7.3}$$

Once the drain current I_D is computed one can determine the voltage across gate and source "V_{GS}" using,

$$V_{GS} = -I_D R_{So}$$

Further, the drain to source voltage across the MOSFET can be computed by applying KVL in the loop a-D-S-g-a we have,

$$V_{DD} - I_D R_D - I_D R_{So} - V_{DS} = 0$$

$$V_{DS} = V_{DD} - I_D(R_D + R_{So}) \tag{3.7.4}$$

Thus, MOSFET operating drain to source voltage can be determined. Further, again using drain to source current "I_{DS}" for the analysis of the AC equivalent circuit to determine the voltage gain for the input signal of the given MOSFET circuit. To derive the AC equivalent circuit one need to follow the steps given in above section. The equivalent input resistance is given by,

$$z_i = R_G \tag{3.7.5}$$

For equivalent output resistance is obtained by shorting input source in the equivalent ac circuit. The obtained equivalent output resistance is given by,

$$z_o = r_d \, || \, R_D \tag{3.7.6}$$

The voltage gain in the equivalent ac circuit is given by,

$$A_v = \frac{V_o}{V_i} \tag{3.7.7}$$

where,

$$V_i = v_{Ac}$$

$$V_o = -g_m v_{Ac} z_o$$

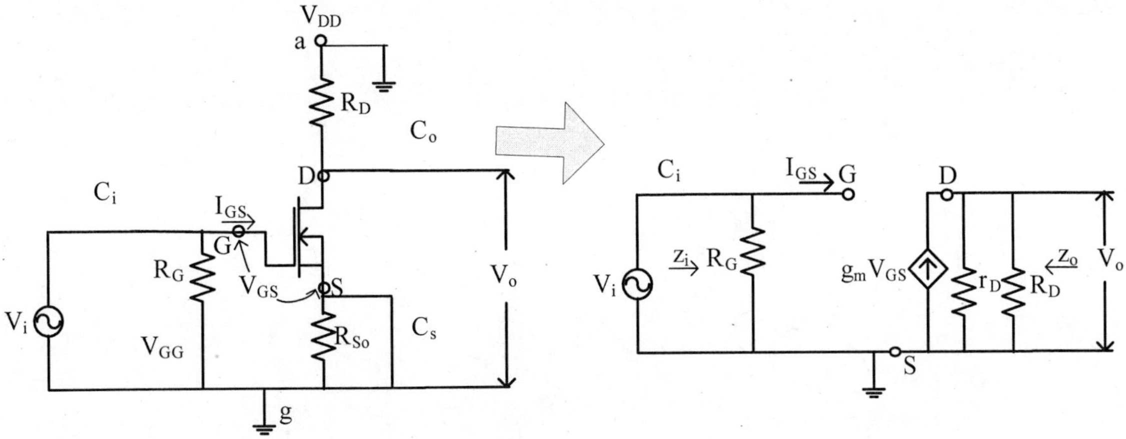

Fig. 3.7.3 The equivalent circuit diagram for AC analysis of the self biased circuit using depletion MOSFET.

Therefore,

$$A_v = \frac{V_o}{V_i} = \frac{-g_m v_{Ac} z_o V_o}{v_{Ac}}$$

$$A_v = -g_m z_o \tag{3.7.8}$$

The negative sign in the gain indicate output signal is 180° phase shifted with reference to the input signal.

3.8 VOLTAGE DIVIDER CIRCUIT

The circuit diagram for the voltage divider circuit is given in Fig. 3.8.1. The given circuit has two resistors R_1 and R_2 connected between V_{DD} and ground 'g' as shown in Fig 3.8.1. However, the drain terminal of the MOSFET circuit is connected to the ground via a resistor R_s as given in Fig. 3.8.1. To analyze the DC operating point of the given MOSFET circuit the given circuit is reduced to its DC equivalent using conditions given above. The DC equivalent circuit for the MOSFET circuit is given in Fig. 3.8.2. Using the DC equivalent circuit and applying KVL in the loop g-S-G-g we have,

$$-V_{GS} - I_D R_{So} + V_{R1} = 0$$

$$\text{where, } V_{R1} = \frac{V_{DD} R_1}{R_1 + R_2}$$

$$I_D = \frac{V_{R1} - V_{GS}}{R_{So}} \tag{3.8.1}$$

Using Shockley equations we have,

$$I_D = I_{DSS}(1 - \frac{V_{GS}}{V_{tv}})^2$$

$$\frac{V_{R1} - V_{GS}}{I_{DSS} R_{So}} = (1 - \frac{V_{GS}}{V_{tv}})^2$$

$$\frac{V_{R1}}{I_{DSS}R_{So}} - \frac{V_{GS}}{I_{DSS}R_{So}} = 1 - 2\frac{V_{GS}}{V_{tv}} + (\frac{V_{GS}}{V_{tv}})^2$$

$$(\frac{V_{GS}}{V_{tv}})^2 - 2\frac{V_{GS}}{V_{tv}} + \frac{V_{GS}}{I_{DSS}R_{So}} - \frac{V_{R1}}{I_{DSS}R_{So}} + 1 = 0 \quad (3.8.2)$$

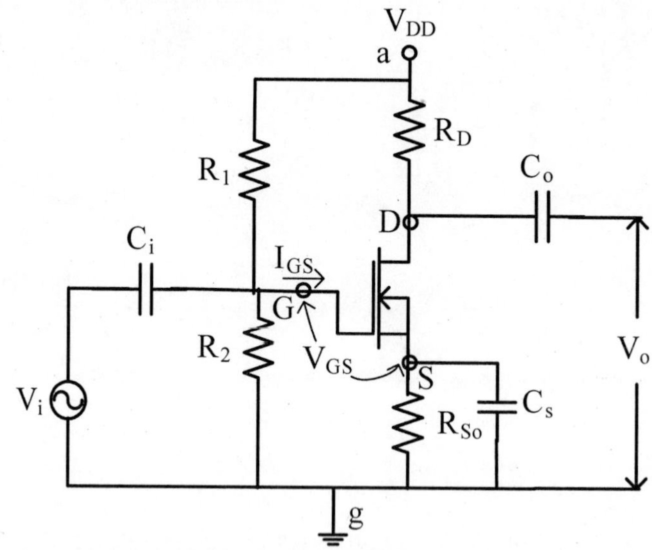

Fig. 3.8.1 The circuit diagram for the voltage divider biased circuit using depletion MOSFET.

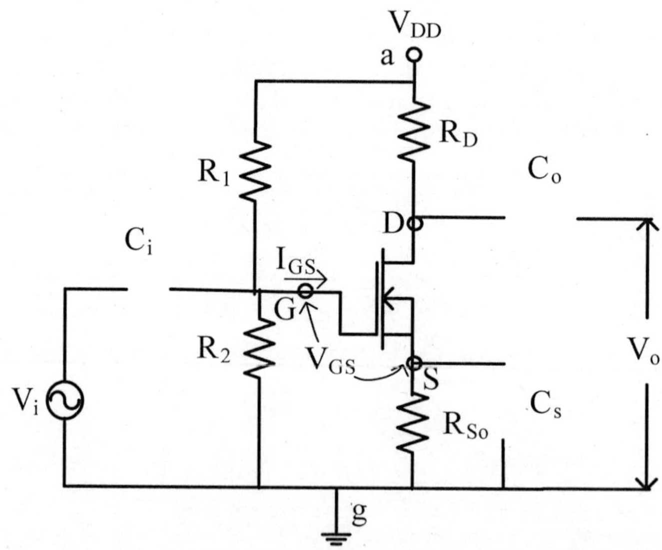

Fig. 3.8.2 The equivalent circuit diagram for DC analysis of the voltage divider biased circuit using depletion MOSFET.

Solving the above quadratic equation one can determine the drain current I_D and is given by,

$$(V_{GS})^2(\frac{1}{V_{tv}})^2 + V_{GS}(\frac{-2}{V_{tv}} + \frac{1}{I_{DSS}R_{So}}) - \frac{V_{R1}}{I_{DSS}R_{So}} + 1 = 0$$

$$V_{GS} = \frac{-(\frac{-2}{V_{tv}} + \frac{1}{I_{DSS}R_{So}}) \pm \sqrt{(\frac{-2}{V_{tv}} + \frac{1}{I_{DSS}R_{So}})^2 - 4(\frac{1}{V_{tv}})^2(-\frac{V_{R1}}{I_{DSS}R_{So}} + 1)}}{2(\frac{1}{V_{tv}})^2} \quad (3.8.3)$$

Once the voltage across gate and source "V_{GS}" is computed one can determine the drain current I_D using,

$$I_D = \frac{V_{R1} - V_{GS}}{R_{So}} \quad (3.8.4)$$

Further, the drain to source voltage across the MOSFET can be computed by applying KVL in the loop a-D-S-g-a we have,

$$V_{DD} - I_D R_D - I_D R_{So} - V_{DS} = 0$$

$$V_{DS} = V_{DD} - I_D(R_D + R_{So}) \quad (3.8.5)$$

Thus, MOSFET operating drain to source voltage can be determined. Further, again using drain to source current "I_{DS}" for the analysis of the AC equivalent circuit to determine the voltage gain for the input signal of the given MOSFET circuit. To derive the AC equivalent circuit, one need to follow the steps given in above section. The equivalent input resistance is given by,

$$z_i = R_1 \parallel R_2 \quad (3.8.6)$$

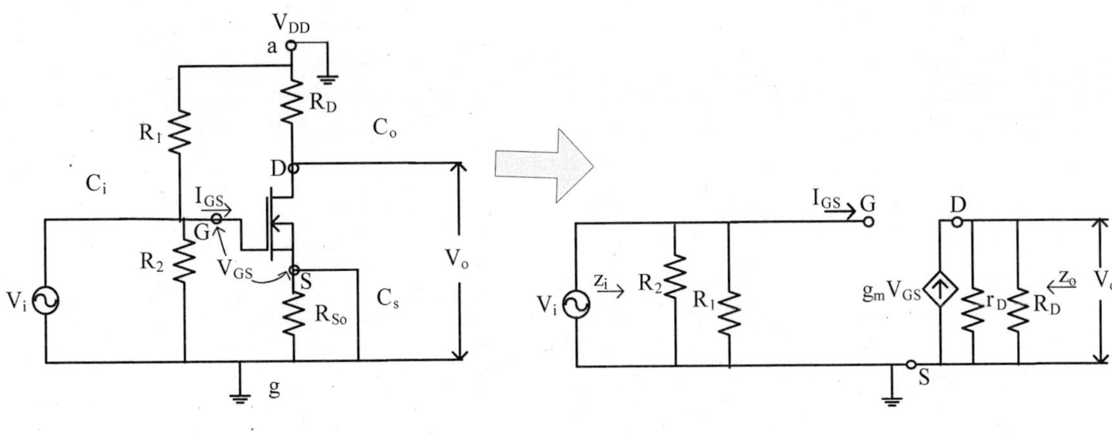

Fig. 3.8.3 The equivalent circuit diagram for AC analysis of the voltage divider biased circuit using depletion MOSFET.

For equivalent output resistance is obtained by shorting input source in the equivalent ac circuit. The obtained equivalent output resistance is given by,

$$z_o = r_D \parallel R_D \quad (3.8.7)$$

The voltage gain in the equivalent ac circuit is given by,

$$A_v = \frac{V_o}{V_i}$$

Where,

$$V_i = v_{Ac}$$
$$V_o = -g_m v_{Ac} z_o$$

Therefore,

$$A_v = \frac{V_o}{V_i} = \frac{-g_m v_{Ac} z_o V_o}{v_{Ac}}$$

$$A_v = -g_m z_o \quad (3.8.8)$$

The negative sign in the gain indicate output signal is 180° phase shifted with reference to the input signal.

3.9 COMMON GATE CIRCUIT

The circuit diagram for the common gate circuit is given in Fig. 3.9.1. The given circuit has input source connected to the source terminal via coupling capacitor as shown in Fig. 3.9.1. The gate terminal is shorted to the ground. The given circuit has zero phase displacement in the output signal. To analyze the DC operating point of the given MOSFET circuit the given circuit is reduced to its DC equivalent using conditions given above. The DC equivalent circuit for the MOSFET circuit is given in Fig. 3.9.2. Using the DC equivalent circuit and applying KVL in the loop g-S-G-g we have,

$$-V_{GS} - I_D R_{So} + V_{SS} = 0$$

$$I_D = \frac{V_{SS} - V_{GS}}{R_{So}} \quad (3.9.1)$$

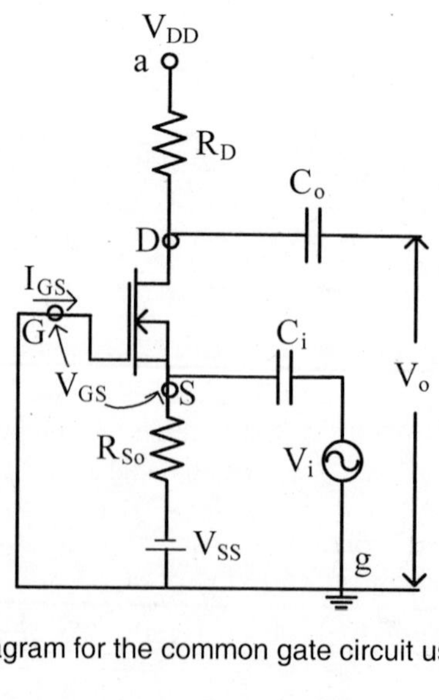

Fig. 3.9.1 The circuit diagram for the common gate circuit using depletion MOSFET.

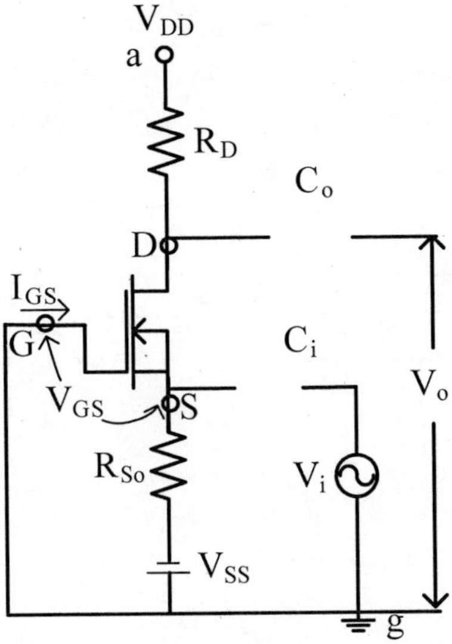

Fig. 3.9.2 The equivalent circuit diagram for DC analysis of the common gate circuit using depletion MOSFET.

Using Shockely equations we have,

$$I_D = I_{DSS}(1-\frac{V_{GS}}{V_{tv}})^2$$

$$\frac{V_{SS}-V_{GS}}{I_{DSS}R_{So}} = (1-\frac{V_{GS}}{V_{tv}})^2$$

$$\frac{V_{SS}}{I_{DSS}R_{So}} - \frac{V_{GS}}{I_{DSS}R_{So}} = 1-2\frac{V_{GS}}{V_{tv}}+(\frac{V_{GS}}{V_{tv}})^2$$

$$(\frac{V_{GS}}{V_{tv}})^2 - 2\frac{V_{GS}}{V_{tv}} + \frac{V_{GS}}{I_{DSS}R_{So}} - \frac{V_{SS}}{I_{DSS}R_{So}} + 1 = 0 \quad (3.9.2)$$

Solving the above quadratic equation one can determine the drain current I_D and is given by,

$$(V_{GS})^2(\frac{1}{V_{tv}})^2 + V_{GS}(\frac{-2}{V_{tv}}+\frac{1}{I_{DSS}R_{So}}) - \frac{V_{SS}}{I_{DSS}R_{So}} + 1 = 0$$

$$V_{GS} = \frac{-(\frac{-2}{V_{tv}}+\frac{1}{I_{DSS}R_S}) \pm \sqrt{(\frac{-2}{V_{tv}}+\frac{1}{I_{DSS}R_{So}})^2 - 4(\frac{1}{V_{tv}})^2(-\frac{V_{SS}}{I_{DSS}R_{So}}+1)}}{2(\frac{1}{V_{tv}})^2} \quad (3.9.3)$$

Once the voltage across gate and source "V_{GS}" is computed one can determine the drain current I_D using,

$$I_D = \frac{V_{SS}-V_{GS}}{R_S}$$

Further, the drain to source voltage across the MOSFET can be computed by applying KVL in the loop a-D-S-g-a we have,

$$V_{DD} - I_D R_D - I_D R_{So} - V_{DS} + V_{SS} = 0$$

$$V_{DS} = V_{DD} + V_{SS} - I_D (R_D + R_{So}) \tag{3.9.4}$$

Thus, MOSFET operating drain to source operating voltage can be determined. Further, again using drain to source current "I_{DS}" for the analysis of the AC equivalent circuit to determine the voltage gain for the input signal of the given MOSFET circuit. To derive the AC equivalent circuit one need to follow the steps given in above section leads to the equivalent circuit as shown in Fig. 3.9.3. The equivalent input resistance is computed using circuit shown in Fig. 3.9.4 and is given by,

$$z_i = \frac{V_i}{i_{in}} \tag{3.9.5}$$

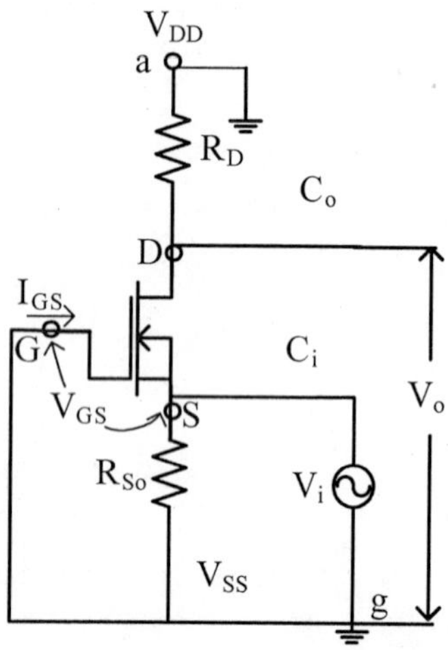

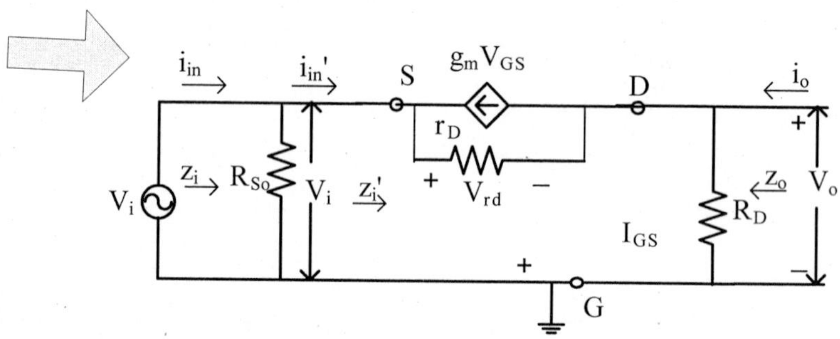

Fig. 3.9.3 The equivalent circuit diagram for AC analysis of the common gate circuit using depletion MOSFET.

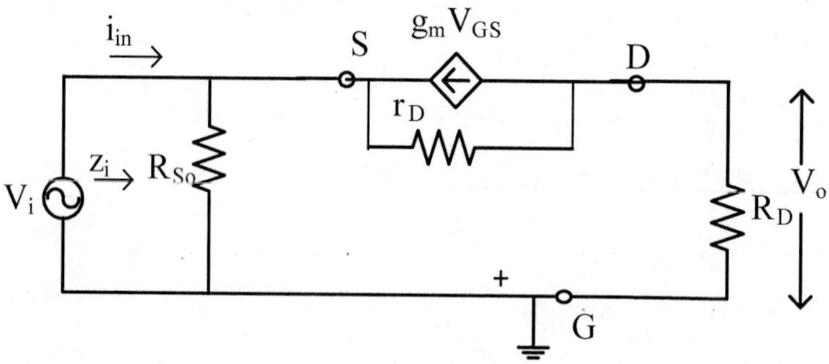

Fig. 3.9.4 The AC equivalent circuit diagram for determining input impedance zi for the common gate circuit using depletion MOSFET.

Applying KVL in loop G-S-D-G,

$$V_i - V_{rd} - V_o = 0$$

$$V_{rd} = V_i - V_o \tag{3.9.6}$$

Applying KCL at node 'S',

$$i_{in} - \frac{V_{rd}}{r_D} + g_m V_{GS} - \frac{V_i}{R_{So}} = 0$$

$$i_{in} - \frac{V_i - V_o}{r_D} + g_m V_{GS} - \frac{V_i}{R_{So}} = 0$$

$$i_{in} - \frac{V_i - V_o}{r_D} - g_m V_i - \frac{V_i}{R_{So}} = 0 \quad (V_{GS} = -V_i)$$

$$i_{in} = \frac{V_i - V_o}{r_D} + g_m V_i + \frac{V_i}{R_{So}} \tag{3.9.7}$$

Applying KCL at node 'D',

$$-\frac{V_o}{R_D} + g_m V_i - \frac{V_o - V_i}{r_D} = 0$$

$$g_m V_i + \frac{V_i}{r_D} = \frac{V_o}{R_D} + \frac{V_o}{r_D}$$

$$V_i(g_m + \frac{1}{r_D}) = V_o(\frac{1}{R_D} + \frac{1}{r_D}) \tag{3.9.8}$$

Substituting V_o from (3.9.8) to (3.9.7) we have,

$$V_o = \frac{V_i(g_m + \frac{1}{r_D})}{(\frac{1}{R_D} + \frac{1}{r_D})}$$

$$i_{in} = \frac{V_i - \dfrac{V_i(g_m + \frac{1}{r_D})}{(\frac{1}{R_D} + \frac{1}{r_D})}}{r_D} + g_m V_i + \frac{V_i}{R_{So}}$$

$$\frac{i_{in}}{V_i} = \frac{1 - \dfrac{1(g_m + \frac{1}{r_D})}{(\frac{1}{R_D} + \frac{1}{r_D})}}{r_D} + g_{mi} + \frac{1}{R_{So}}$$

$$\frac{i_{in}}{V_i} = \frac{(\frac{1}{R_D} + \frac{1}{r_D}) - (g_m + \frac{1}{r_D})}{(\frac{1}{R_D} + \frac{1}{r_D})r_D} + g_{mi} + \frac{1}{R_{So}}$$

$$\frac{i_{in}}{V_i} = \frac{(\frac{1}{R_D} - g_m)}{(\frac{1}{R_D} + \frac{1}{r_D})r_D} + g_m + \frac{1}{R_{So}}$$

$$\frac{i_{in}}{V_i} = \frac{(1 + r_D g_m)}{(R_D + r_D)} + \frac{1}{R_{So}}$$

$$\frac{i_{in}}{V_i} = \frac{(1 + r_D g_m)}{(R_D + r_D)} + \frac{1}{R_{So}} \simeq \frac{(r_D g_m)}{(r_D)} + \frac{1}{R_{So}} \simeq g_m + \frac{1}{R_{So}}$$

$$z_i = R_{So} \,||\, \frac{(R_D + r_D)}{(1 + r_D g_m)} \simeq R_{So} \,||\, \frac{1}{g_m} \tag{3.9.9}$$

For equivalent output resistance is obtained by shorting input source in the equivalent ac circuit as shown in Fig. 3.9.5. The obtained equivalent output resistance is given by,

$$z_o = r_D \,||\, R_D \tag{3.9.10}$$

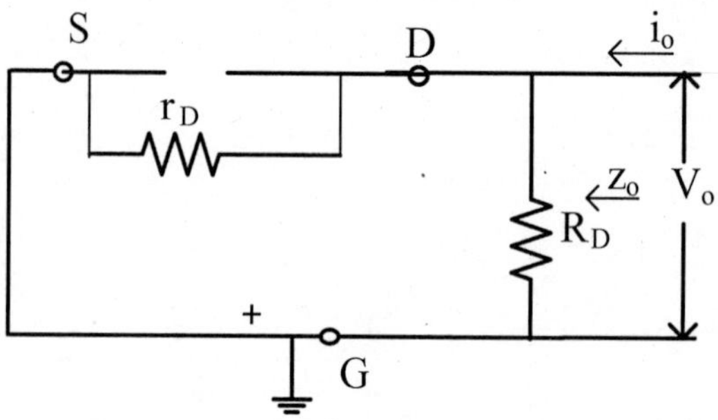

Fig. 3.9.5 The AC equivalent circuit diagram for determining output impedance zo for the common gate circuit using depletion MOSFET.

The voltage gain in the equivalent ac circuit is given by,

$$A_v = \frac{V_o}{V_i}$$

Where,

$$V_i = V_{GS}$$

$$V_o = I_D R_D$$

$$I_D - \frac{V_i - V_o}{r_D} - g_m V_i = 0$$

$$I_D = \frac{V_i - V_o}{r_D} + g_m V_i$$

$$V_o = (\frac{V_i - V_o}{r_D} + g_m V_i) R_D$$

$$\frac{V_o}{R_D} = (\frac{V_i - V_o}{r_D} + g_m V_i)$$

$$\frac{V_o}{R_D}+\frac{V_o}{r_D}=(\frac{V_i}{r_D}+g_mV_i)$$

$$V_o(\frac{1}{R_D}+\frac{1}{r_D})=V_i(\frac{1}{r_D}+g_m)$$

$$\frac{V_o}{V_i}=\frac{(\frac{1}{r_D}+g_m)}{(\frac{1}{R_D}+\frac{1}{r_D})}$$

$$A_v=\frac{V_o}{V_i}=\frac{R_D(1+g_mr_D)}{(r_D+R_D)} \tag{3.9.11}$$

Therefore,

$$A_v=\frac{R_D(1+g_mr_D)}{(r_D+R_D)}\approx\frac{R_Dg_mr_D}{(r_D)}$$

$$A_v\approx R_Dg_m \tag{3.9.12}$$

Thus, gain of the common gate circuit is the function of both trans-conductance and drain resistance. Further, output signal is in phase with the input signal.

3.10 COMMON DRAIN CIRCUIT

The circuit diagram for the common drain circuit is given in Fig. 3.10.1. The given circuit has input source connected to the gate terminal via coupling capacitor as shown in Fig. 3.10.1. The drain terminal is connected to dc supply directly. The given circuit has zero phase displacement in the output signal. The DC equivalent is same as self biased where output is shifted to the source terminal 'S' instead of drain terminal 'D' and drain terminal is directly connected to the source 'V_{DD}'. To analyze the DC operating point of the given MOSFET circuit the given circuit is reduced to its DC equivalent using conditions given above. The DC equivalent circuit for the MOSFET circuit is given in Fig. 3.10.2. Using the DC equivalent circuit and applying KVL in the loop g-S-G-g we have,

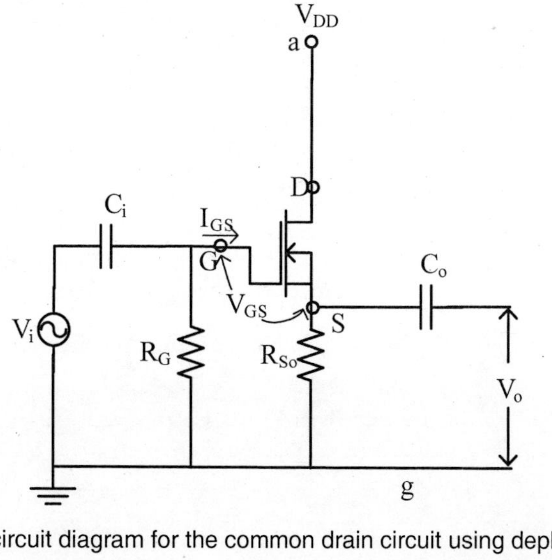

Fig. 3.10.1 The circuit diagram for the common drain circuit using depletion MOSFET.

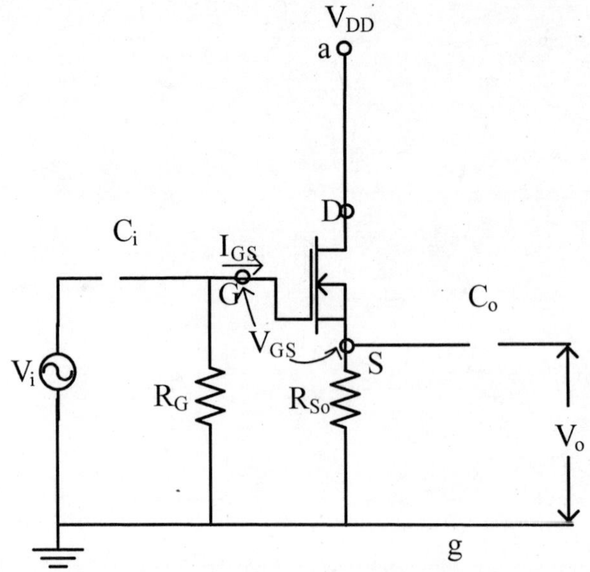

Fig. 3.10.2 The equivalent circuit diagram for DC analysis of the common drain circuit using depletion MOSFET.

$$V_{GS} = -I_D R_{So} - R_G I_{GS} \quad (I_{GS} = 0)$$

$$V_{GS} = -I_D R_{So}$$

$$I_D = \frac{-V_{GS}}{R_{So}} \tag{3.10.1}$$

Using Shockley equations we have,

$$I_D = I_{DSS}(1 - \frac{V_{GS}}{V_{tv}})^2$$

$$I_D = I_{DSS}(1 + \frac{I_D R_{So}}{V_{tv}})^2$$

$$\frac{I_D}{I_{DSS}} = (1 + \frac{I_D R_{So}}{V_{tv}})^2$$

$$(\frac{I_D R_{So}}{V_{tv}})^2 + 2\frac{I_D R_{So}}{V_{tv}} - \frac{I_D}{I_{DSS}} + 1 = 0 \tag{3.10.2}$$

Solving the above quadratic equation one can determine the drain current I_D and is given by,

$$(I_D)^2 (\frac{R_{So}}{V_{tv}})^2 + I_D(\frac{2R_{So}}{V_{tv}} - \frac{1}{I_{DSS}}) + 1 = 0$$

$$I_D = \frac{-(\frac{2R_{So}}{V_{tv}} - \frac{1}{I_{DSS}}) \pm \sqrt{((\frac{2R_{So}}{V_{tv}} - \frac{1}{I_{DSS}})^2 - 4(\frac{R_{So}}{V_{tv}})^2}}{2(\frac{R_{So}}{V_{tv}})^2} \tag{3.10.3}$$

Once the drain current I_D is computed one can determine the voltage across GATE and source "V_{GS}" using,

$$V_{GS} = -I_D R_{So} \tag{3.10.4}$$

Further, the drain to source voltage across the MOSFET can be computed by applying KVL in the loop a-D-S-g-a we have,

$$V_{DD} - I_D R_{So} - V_{DS} = 0$$

$$V_{DS} = V_{DD} - I_D (R_{So}) \tag{3.10.5}$$

Thus, MOSFET operating drain to source operating voltage can be determined. Further, again using drain to source current "I_D" for the analysis of the AC equivalent circuit to determine the voltage gain for the input signal of the given MOSFET circuit. To derive the AC equivalent circuit one need to follow the steps given in above section leads to the equivalent circuit as shown in Fig. 3.10.3. The equivalent input resistance is computed using circuit shown in Fig. 3.10.4 and is given by,

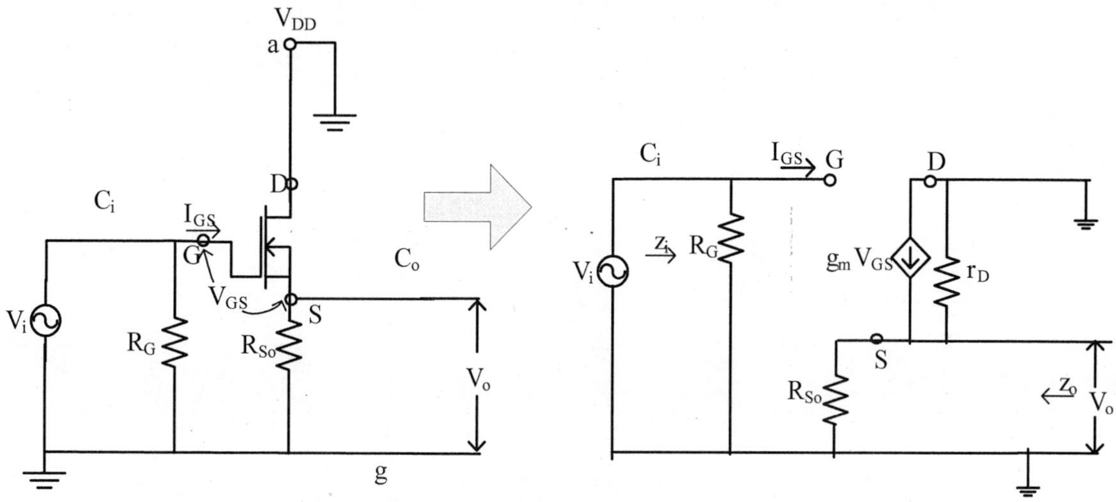

Fig. 3.10.3 The equivalent circuit diagram for AC analysis of the common drain circuit using depletion MOSFET.

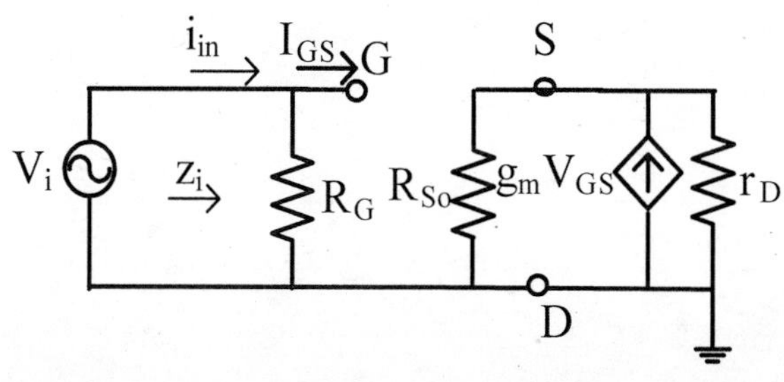

Fig. 3.10.4 The AC equivalent circuit diagram for determining input impedance z_i for the common drain circuit using depletion MOSFET.

$$z_i = \frac{V_i}{i_{in}}$$

$$z_i = R_G \tag{3.10.6}$$

For equivalent output resistance is obtained by shorting input source in the equivalent ac circuit as shown in Fig. 3.9.5. The obtained equivalent output resistance is obtained by applying KVL in loop G-S-D-G,

$$V_i - V_{GS} - V_o = 0$$

$$V_{GS} = V_i - V_o \tag{3.10.7}$$

Applying KCL at node 'S',

$$\frac{V_o}{r_D} - g_m V_{GS} + \frac{V_o}{R_{So}} - i_o = 0$$

$$i_o = \frac{V_o}{r_D} - g_m (V_i - V_o) + \frac{V_o}{R_{So}} \quad (V_i = 0)$$

$$i_o = \frac{V_o}{r_D} - g_m (-V_o) + \frac{V_o}{R_{So}}$$

$$\frac{i_o}{V_o} = \frac{1}{r_D} + g_m + \frac{1}{R_{So}}$$

$$\frac{i_o}{V_o} = \frac{1}{r_D} + \frac{1}{\frac{1}{g_m}} + \frac{1}{R_{So}}$$

$$\frac{1}{z_o} = \frac{1}{r_D} + \frac{1}{\frac{1}{g_m}} + \frac{1}{R_{So}}$$

$$z_o = r_D \,||\, \frac{1}{g_m} \,||\, R_{SO} \tag{3.10.8}$$

The voltage gain in the equivalent ac circuit is given by,

$$A_v = \frac{V_o}{V_i}$$

Where,

$V_i = i_{in} z_i$

$V_o = i_o\, z_o = g_m V_{GS}\, (r_D \,||\, R_{So}) \quad (V_{GS} = V_i - V_o)$

$V_o = g_m\, (V_i - V_o)(r_D \,||\, R_{So})$

$V_o\, (1 + g_m\, (r_D \,||\, R\_So)) = g_m\, V_i\, (r_D \,||\, R_{So})$

$$A_v = \frac{V_o}{V_i} = \frac{g_m(r_D \parallel R_{SO})}{(1 + g_m(r_D \parallel R_{SO}))} \tag{3.10.9}$$

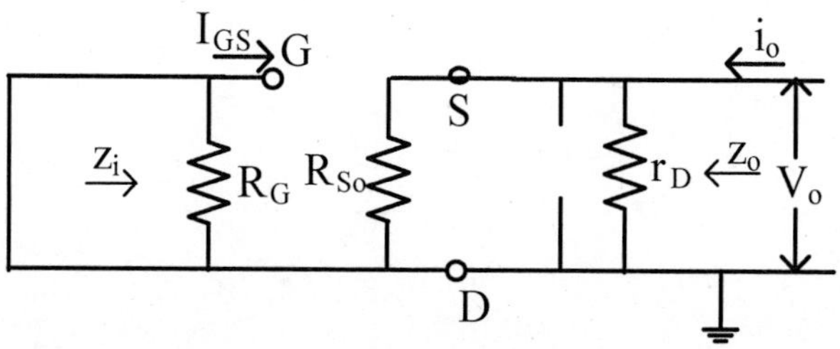

Fig. 3.10.5 The AC equivalent circuit diagram for determining output impedance z_o for the common drain circuit using depletion MOSFET.

Therefore,

$$A_v = \frac{V_o}{V_i} = \frac{g_m(r_D \parallel R_{SO})}{(1 + g_m(r_D \parallel R_{SO}))} \approx \frac{g_m R_{SO}}{(1 + g_m R_{SO})}$$

$$A_v = \frac{V_o}{V_i} \approx \frac{g_m R_{SO}}{(1 + g_m R_{SO})} \tag{3.10.10}$$

Thus, gain of the common gate circuit is the function of both trans-conductance and drain resistance. Further, output signal is in phase with the input signal.

Table I: Summary MOSFETs

S. No	Configuration	Diagram	Expression for Drain source voltage (V_{DS})	Expression for z_i, z_o, A_V
1	**Fixed Biased Circuit**	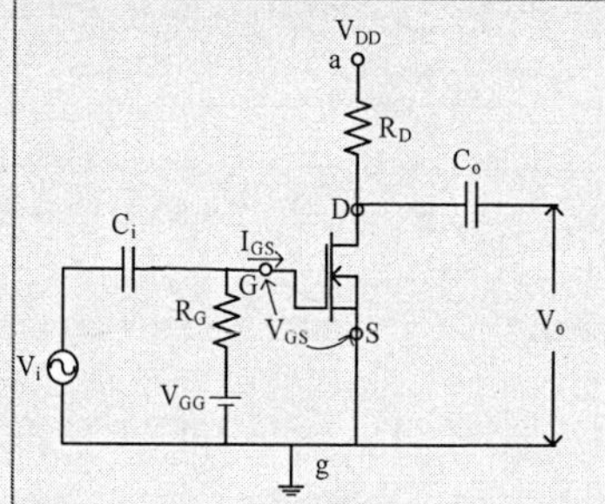	$V_{DS} = V_{DD} - I_D R_D$	$z_i = R_G$ $Z_o = r_D \parallel R_D$ $A_v = -g_m z_o$
2	**Self Biased Circuit**	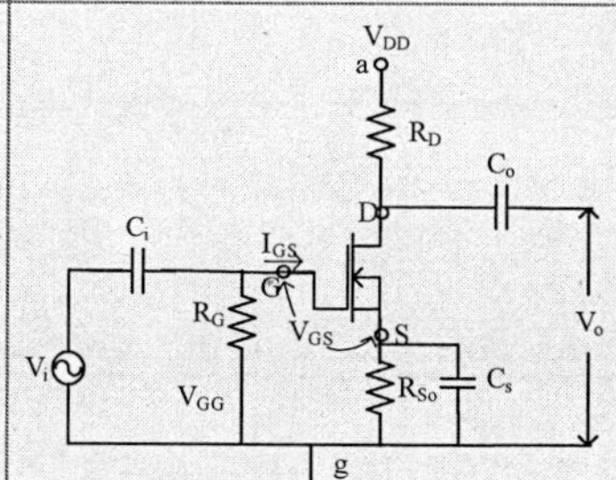	$V_{DS} = V_{DD}$ $- I_D(R_D + R_{So})$	$z_i = R_G$ $Z_o = r_D \parallel R_D$ $A_v = -g_m z_o$

3	**Voltage Divider Circuit**	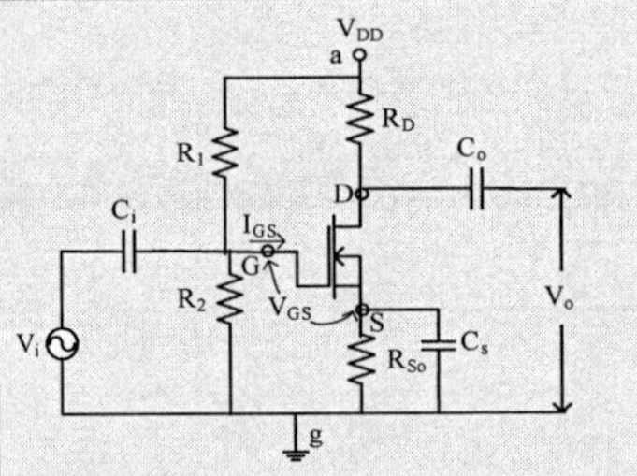	$V_{DS} = V_{DD} - I_D(R_D + R_{So})$	$Z_i = R_1 \parallel R_2$ $Z_o = r_D \parallel R_D$ $A_v = -g_m z_o$
4	**Common Gate Circuit**	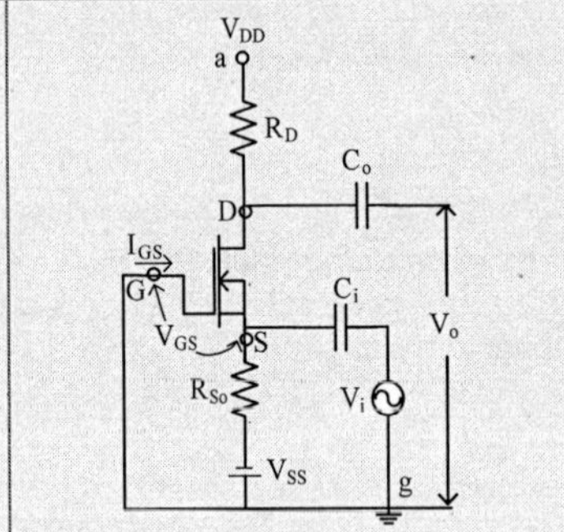	$V_{DS} = V_{DD} + V_{SS} - I_D(R_D + R_{So})$	$z_i = R_{So} \parallel \frac{(R_D + r_D)}{(1 + r_D g_m)}$ $z_o = r_D \parallel R_D$ $A_v = \frac{R_D(1 + g_m r_D)}{(r_D + R_D)}$
5	**Common Drain Circuit**	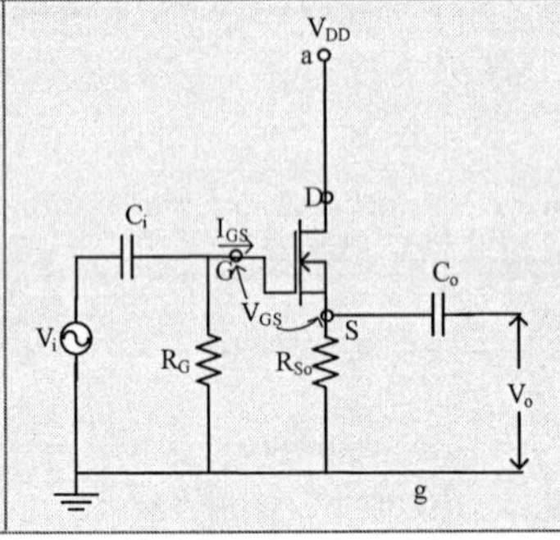	$V_{DS} = V_{DD} - I_D(R_{So})$	$z_i = R_G$ $z_o = r_D \parallel \frac{1}{g_m} \parallel R_{SO}$ $A_v = \frac{g_m(r_D \parallel R_{so})}{(1 + g_m(r_D \parallel R_{so}))}$

UNIT SUMMARY

The operation of the bipolar junction transistor under both biased and unbiased working circumstances is covered in this basic unit on the transistor. The discussion of different transistor configurations, such as common base, common emitter, and common collector, followed. Once one has a solid understanding of these subjects, it is necessary to study the critical factor that determines how a bipolar junction transistor functions in an electrical circuit. In every electronic circuit, the bipolar junction transistor's biasing is a critical factor that determines how the transistor operates. The bipolar junction transistor's Q-point operating point is then covered. The next section discusses the several biasing circuits of the bipolar junction transistor. Following biasing, the BJT's function as an amplifier and switch is examined. This covers both closed- and open-switch BJT functioning. Also included are common base, common emitter, and common collector amplifier topologies along with their input and output characteristics. It is necessary to acquire the amplifiers' parameters, such as input impedance, voltage gain, current gain, and output admittance. It necessitates a brief signal analysis. The bipolar junction transistor's h-parameter model is used in this small signal analysis. The processes for using the small signal analysis to derive the amplifier parameters, such as input impedance, voltage gain, current gain, and output admittance, are described in depth. A reduced h-parameter model is also used for the tiny signal analysis in order to reduce the complexity of determining the amplifier parameters. The amplifier's high-frequency response is then shown in order to examine how the gain of the amplifier changes with frequency. Each of these ideas is necessary to comprehend how different electronic circuits work. The book's appendix contains a laboratory experiment pertaining to the input and output characteristics of a bipolar junction transistor in a common emitter configuration.

EXERCISES

Multiple Choice Questions

3.1 What are the voltages across RD (3kΩ) and RS (0.82 kΩ)?

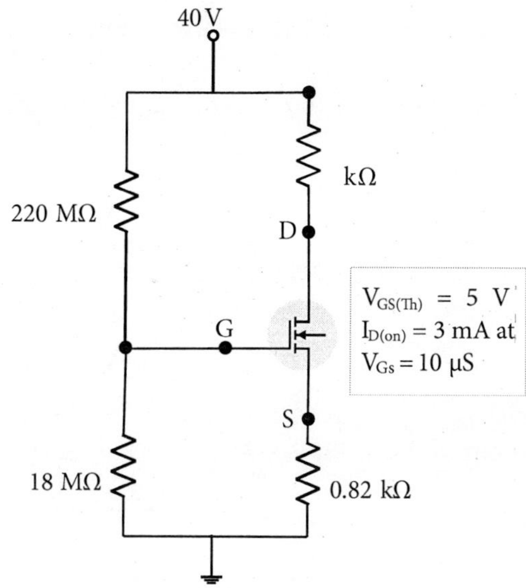

a. 0 V, 0 V
b. 5 V, 5 V
c. 10 V, 10 V
d. 20 V, 20 V

3.2 Calculate the value of RD.

a. 2 kΩ
b. 3 kΩ
c. 3.5 kΩ
d. 4.13 kΩ

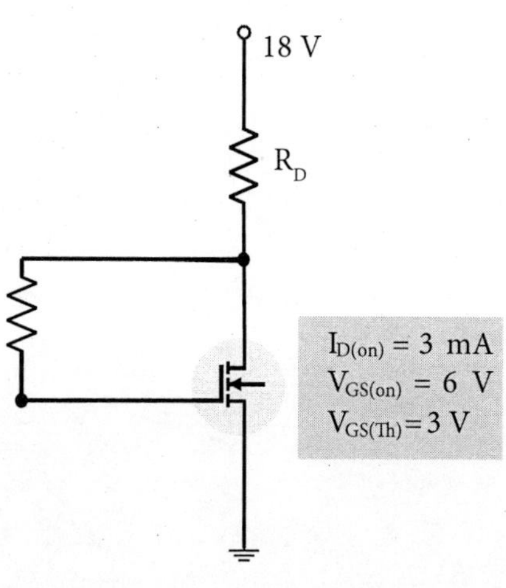

3.3 The slope of the dc load line in a self-bias configuration is controlled by _.

a. V_{DD} b. R_D

c. R_G d. R_S

3.4 The dc load line of common base FET configuration is controlled by ____.

a. V_{DD} and R_S b. R_D, R_S and V_{SS}

c. R_S and V_{SS} d. R_S

Answers for Multiple Choice Questions			
3.1 **a**	3.2 **d**	3.3 **d**	3.4 **d**

Short Answer Type Questions

3.1 Give the derivation of the Z_i, Z_o and voltage gain for the fixed biased JFET configuration?

3.2 Give the derivation of the Z_i, Z_o and voltage gain for the voltage divider JFET configuration?

3.3 Give the derivation of the Z_i, Z_o and A_v for self- biased common source JFET configuration with un-bypassed capacitor taking into account drain resistance r_d?

3.4 Give the derivation of the Z_i, Z_o and A_V for common gate FET configuration taking into account drain resistance r_d?

Long Answer Type Questions

3.1 Determine the following for the circuit in below figure (i) V_{GSQ} (ii) I_{DQ} (iii) V_{DS} (iv) V_D (v) V_G (vi) V_S

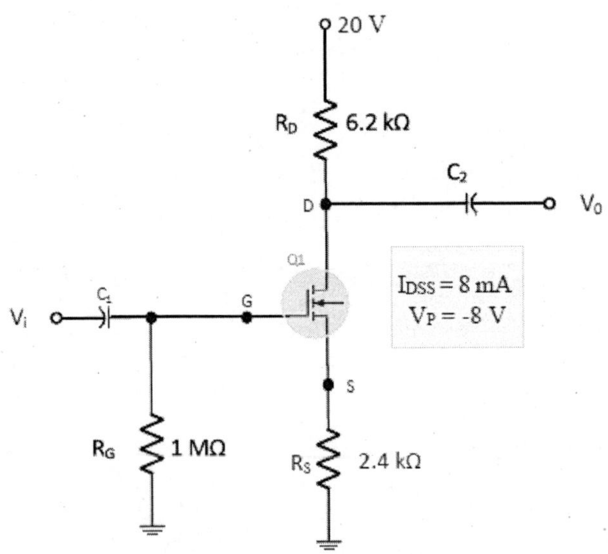

3.2 Determine the following for the circuit in the below figure (i)V_{GSQ} (ii) I_{DQ} (iii) V_{DS} (iv) V_D (v) V_G (vi) V_S

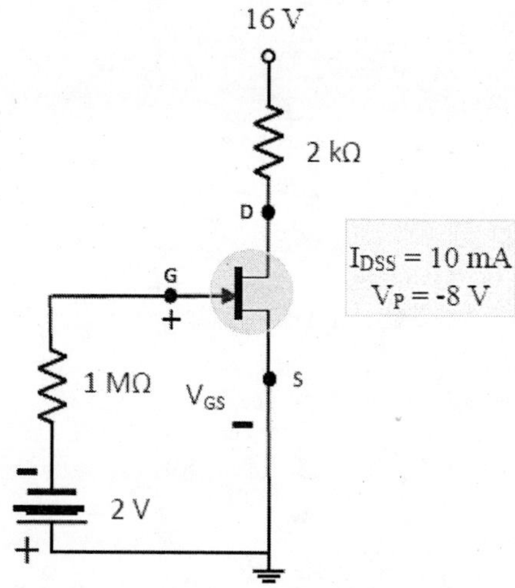

3.3 Derive the expressions for the input and output impedance for voltage shunt feedback. (FET – Amplifier)

3.4 Give the derivation of the Z_i, Z_o and A_v for self- biased common gate JFET configuration with bypassed capacitor

KNOW MORE

For Additional Info Scan Me

REFERENCES

1. Robert Boylestad, Louis Nashelsky- Electronic Devices and Circuit Theory, Prentice Hall Upper Saddle River, New Jersey Columbus, Ohio
2. Donald L. Schilling, Charles Belove -Electronic Circuits: Discrete and Integrated, McGraw-Hill Book Company New York St. Louis San Francisco Toronto London Sydney.
3. Integrated Electronics Analog Digital Circuits, Jacob Millman and D. Halkias, McGraw Hill.

04 POWER AMPLIFIER AND DIFFERENTIAL AMPLIFIER

UNIT SPECIFICS

Through this unit we have discussed the following aspects:

- *Introduction to Power Amplifier;*
- *Classification of Power Amplifiers;*
- *Differential Amplifiers.*

RATIONALE

This unit on the power amplifiers and differential amplifiers helps students to get a primary idea about the operation of the power and differential amplifiers. Firstly, the need for power amplifiers is explained. Next, the various classifications of the power amplifiers are discussed. This classification is required for the identification of the application based on various factors like efficiency, distortion etc. Next the need for differential amplifier along with the operation and analysis is explained.

PRE-REQUISITES

NIL

UNIT OUTCOMES

List of outcomes of this unit is as follows:

U4-O1: Understand the basics of power amplifiers

U4-O2: Understand various power amplifier configurations

U4-O3: Realize the various applications of the power amplifiers

U4-O4: Understand the basics of differential amplifiers

U4-O5: Realize the operation of the differential amplifier

Unit-4 Outcomes	EXPECTED MAPPING WITH COURSE OUTCOMES (*1- Weak Correlation; 2- Medium correlation; 3- Strong Correlation*)				
	CO-1	CO-2	CO-3	CO-4	CO-5
U4-01	1	-	1	3	-
U4-02	1	-	1	3	-
U4-03	1	-	1	3	-
U4-04	-	-	-	3	-
U4-05	1	-	1	3	-

4.1 INTRODUCTION TO POWER AMPLIFIERS

Power amplifiers play a crucial role in various electronic systems by amplifying low-power signals to higher power levels. They are widely used in audio systems, wireless communication systems, broadcasting equipment, and other applications. Linear power amplifiers are favoured when the goal is to amplify the input signal while maintaining its fidelity and minimizing distortion. They operate by varying the voltage or current of the input signal in proportion to the input signal itself. Selecting the appropriate power amplifier involves considering factors such as power requirements, load impedance, efficiency, distortion levels, and the intended application. Different amplifier topologies and designs exist to address specific needs, such as audio fidelity, efficiency, or linearity. The choice of amplifier depends on the specific requirements and constraints of the system or device being designed. The small signal analysis using h-parameters cannot be used for the power amplifiers due to the large signal nature at the input. The analysis of the power amplifiers is carried out by using the output characteristics and Q-point.

4.2 CLASSIFICATION OF POWER AMPLIFIERS

Based on the position of the Q-point, the power amplifiers are classified as:

4.2.1 Class A Power Amplifiers

The power amplifiers in which the Q-point lies exactly at the centre of the DC load line are called class A power amplifiers. Due to the location of the Q-point at the centre of the DC load line, the output signal of the amplifier will be available for the complete input signal as shown in Fig. 4.2.1.1. As a result, the output signal is available from 0^0 to 360^0. The output signal will have less distortion. The efficiency of the class A power amplifiers is very comparatively which is around 25% to 50%.

4.2.2 Class B Power Amplifiers

The power amplifiers in which the Q-point lies exactly on the X-axis of the output characteristics are called class B power amplifiers. Due to the location of the Q-point on the X-axis of the output characteristics, the output signal of the amplifier will be available for half cycle of the input signal as shown in Fig. 4.2.2.1. As a result, the output signal is available from 0^0 to 180^0. The output signal will have distortion higher than class A. The peak efficiency of the class B power amplifiers is higher than class A which is 78.5%. For obtaining two half cycles from the class B power amplifiers amplifier, two for each half cycle need to be used.

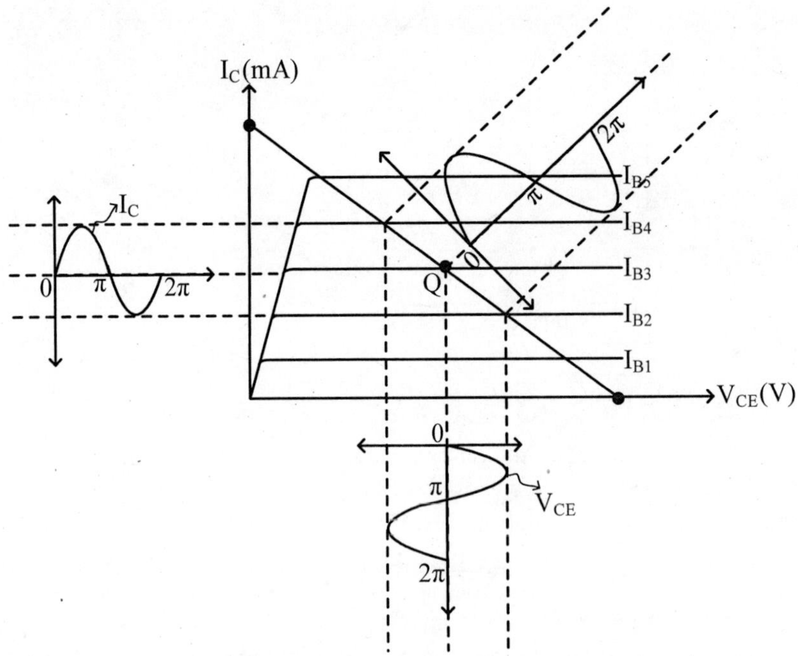

Fig. 4.2.1.1 Q-point for class A amplifier.

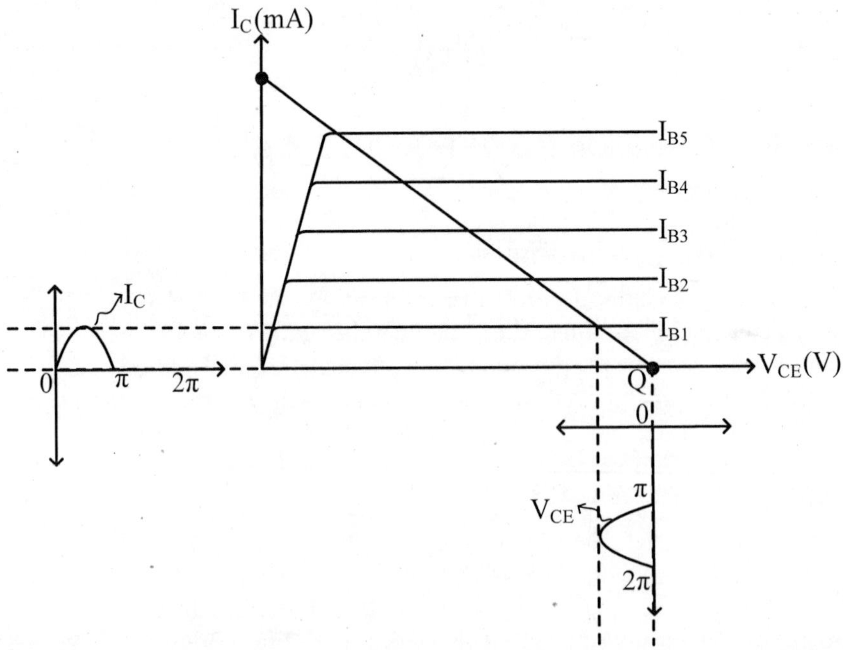

Fig. 4.2.2.1 Q-point for class B amplifier.

4.2.3 Class C Power Amplifiers

The power amplifiers in which the Q-point lies below the X-axis of the output characteristics are called class C power amplifiers. Due to the location of the Q-point below the X-axis of the output characteristics, the output signal of the amplifier will be available for less than half cycle of the input signal as shown in Fig. 4.2.3.1. As a result, the output signal is available below 180^0. The output signal will have distortion higher than class B. The efficiency of the class C power amplifiers is very high almost around 100%.

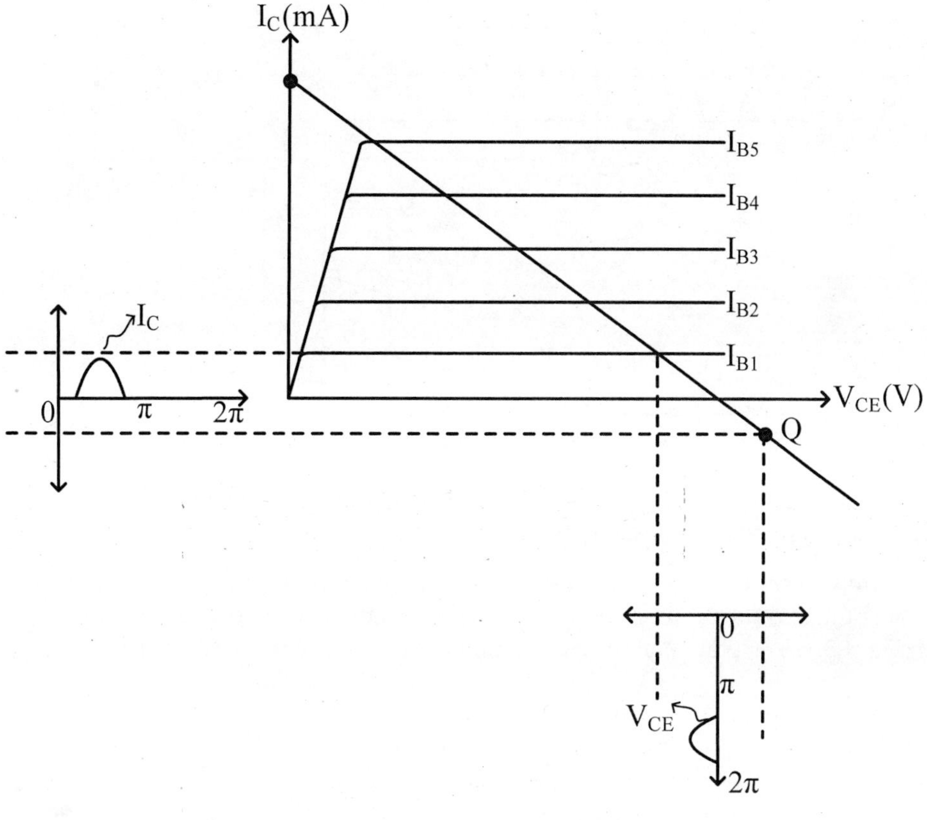

Fig. 4.2.3.1 Q-point for class C amplifier.

4.2.4 Class AB Power Amplifiers

The power amplifiers in which the Q-point is above the X-axis but below the center of the DC load line is called class AB power amplifiers. Due to the location of the Q-point above the X-axis but below the center of the DC load line, the output signal of the amplifier will be available for more than 180^0 but less than 360^0 of the input signal as shown in Fig. 4.2.4.1. The output signal will have distortion higher than class B but less than class A. The efficiency of the class AB power amplifiers is higher than class A but less than class B.

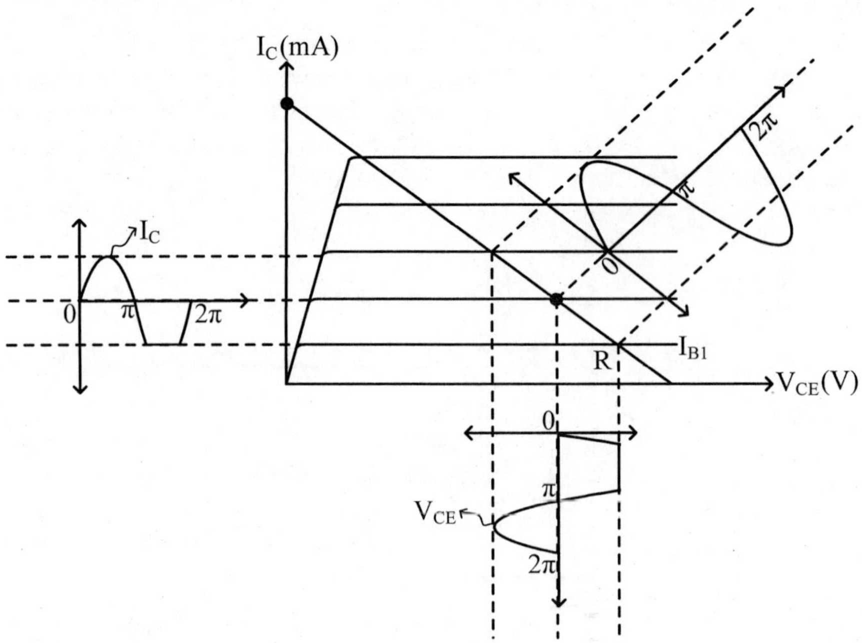

Fig. 4.2.4.1 Q-point for class AB amplifier.

4.3 DIFFERENTIAL AMPLIFIERS

Differential amplifier circuit is the basic unit for the operational amplifier. It consists of the two common emitter transistor circuits having a common emitter resistor shared between their emitter terminal as shown in Fig. 4.3.1. The other end of the common emitter resistor is connected to the negative supply $-V_{EE}$ as can be seen from Fig. 4.3.1. The circuit diagram for the differential amplifier is given in Fig. 4.3.1. There are two inputs and outputs terminal for the differential amplifier circuits as shown in Fig. 4.3.1. For DC analysis of the differential amplifier, the input AC signals are grounded and the operating voltage and current for the two transistors are computed as shown in Fig. 4.3.2. As the two transistors are identical, so below assumptions can be applied.

$$I_{B1} = I_{B2} = \frac{I_{C1}}{\beta} = \frac{I_{C2}}{\beta} = \frac{I_E}{2\beta} \quad (4.3.1)$$

Applying KVL at the outer loop of the transistor we have,

$$-V_{BE} - I_E R_E - (-V_{EE}) = 0$$

$$I_E = \frac{V_{EE} - V_{BE}}{R_E} = \frac{V_{EE} - 0.7}{R_E}$$

$$V_{cc} - I_{C1} R_c - V_{CE1} - I_E R_E - (-V_{EE}) = 0$$

$$V_{CE1} = V_{CE2} = V_{cc} - \frac{I_E}{2} R_c - I_E R_E + V_{EE} \quad (4.3.2)$$

Thus, the operating voltage (V_{CE1} and V_{CE2}) of both the transistor can be computed. Now, for AC analysis, it is required to have AC equivalent of the differential amplifier circuit. For getting AC equivalent circuit one need to ground the DC supply as shown in Fig. 4.3.3 and replace the transistor with its ac equivalent as shown in Fig. 4.3.3. Now for analysis ac equivalent circuit for transistor we

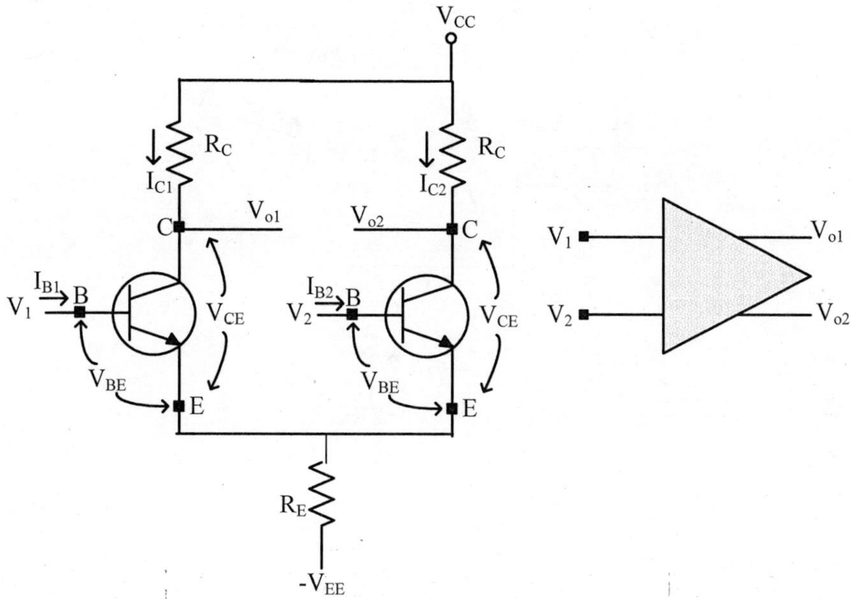

Fig. 4.3.1 Circuit schematic of differential amplifier using CE transistor.

assume,

$$V_1 - i_E R_E - i_{b1} R_i = 0$$
$$V_2 - i_E R_E - i_{b2} R_i = 0$$
$$V_1 - V_2 = (i_{b1} - i_{b2}) R_i \qquad (4.3.3)$$

Applying KVL at two outputs we have

$$v_{01} + i_{C1} R_C = 0$$
$$v_{02} + i_{C2} R_C = 0$$
$$v_{01} + i_{C1} R_C - (v_{02} + i_{C2} R_C) = 0$$
$$v_{01} - v_{02} + (i_{C1} - i_{C2}) R_C = 0 \qquad (4.3.4)$$

Applying KCL at node 'C1 ' and 'C2' we have

$$i_{C1} = \beta i_{b1}$$
$$i_{C2} = \beta i_{b2} \qquad (4.3.5)$$

Substituting (6.3.5) and (6.3.3) in (6.3.4) we have,

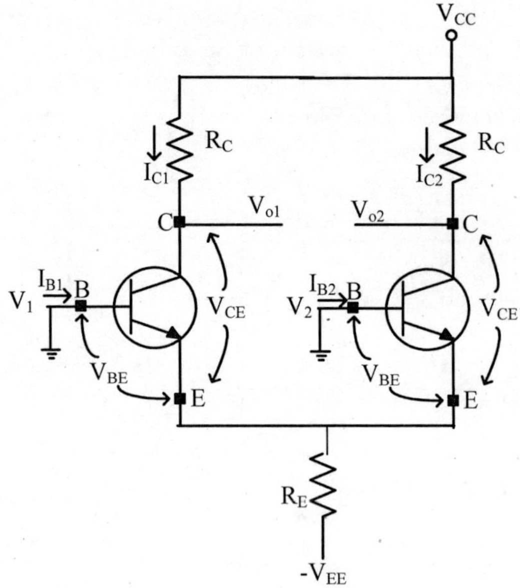

Fig. 4.3.2 Equivalent circuit schematic of differential amplifier for DC analysis.

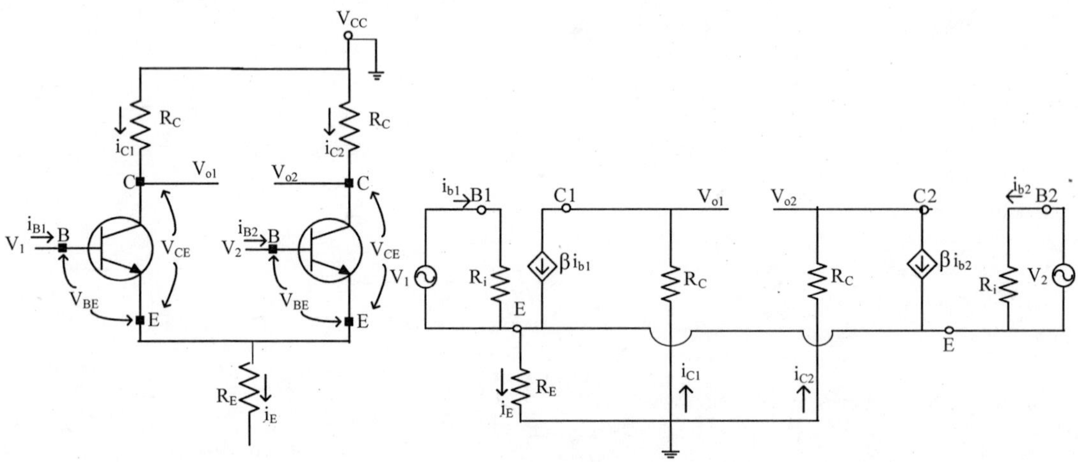

Fig. 4.3.3 Equivalent circuit schematic of differential amplifier for AC analysis.

$$v_{01} - v_{02} + (i_{C1} - i_{C2})R_C = 0$$

$$v_{01} - v_{02} + (\beta i_{b1} - \beta i_{b2})R_C = 0$$

$$v_{01} - v_{02} + (i_{b1} - i_{b2})\beta R_C = 0$$

$$(i_{b1} - i_{b2}) = \frac{V_1 - V_2}{R_i}$$

$$v_{01} - v_{02} + (\frac{V_1 - V_2}{R_i})\beta R_C = 0$$

$$v_{01} - v_{02} = -(\frac{\beta R_C}{R_i})(V_1 - V_2) \qquad (4.3.6)$$

Thus, difference of the output voltage is proportional to the difference in the input voltage. The emitter current can be derived by applying KCL at 'E' we have,

$$i_E - i_{b1} - \beta i_{b1} - i_{b2} - \beta i_{b2} = 0$$
$$i_E = (i_{b1} + i_{b2})(\beta + 1) \quad (4.3.7)$$

UNIT SUMMARY

This unit on power amplifiers and differential amplifiers have discussed the basic understanding of how these amplifiers work. The necessity of power amplifiers is first described. The several power amplifier classes are then covered. This classification is necessary in order to identify the application based on a number of variables, such as distortion and efficiency. The necessity of a differential amplifier is then discussed, along with its analysis and functioning.

EXERCISES

Multiple Choice Questions

4.1 In Class A power amplifier, the output current flows for

a. the full cycle of the input signal
b. half the cycle of the input signal
c. 3/4 of the cycle of the input signal
d. less than half the cycle of the input signal

4.2 Class power amplifier has the highest collector efficiency

a. C
b. A
c. B
d. None of these

4.3 In Class A power amplifier, the output current flows for

a. the full cycle of the input signal
b. half the cycle of the input signal
c. 3/4 of the cycle of the input signal
d. less than half the cycle of the input signal

4.4 The collector current in a Class-C amplifier

a. is not rich in harmonics
b. flows for 120° or less of the ac input cycle
c. in sinusoidal
d. flows for 240° or more of the ac input signal

Answers for Multiple Choice Questions			
4.1 **a**	4.2 **a**	4.3 **b**	4.4 **b**

Short Answer Type Questions

4.1 Explain in detail about class A power amplifiers?

4.2 Explain in detail about class B power amplifiers?

4.3 Explain in detail about class AB power amplifiers?

4.4 Explain in detail about class C power amplifiers?

Long Answer Type Questions

4.1 Explain in detail about various power amplifiers?

4.2 Explain in detail about the differential amplifiers?

KNOW MORE

For Additional Info Scan Me

REFERENCES

1. Robert Boylestad, Louis Nashelsky- Electronic Devices and Circuit Theory, Prentice Hall Upper Saddle River, New Jersey Columbus, Ohio
2. Donald L. Schilling, Charles Belove -Electronic Circuits: Discrete and Integrated, McGraw-Hill Book Company New York St. Louis San Francisco Toronto London Sydney.

05 OPERATIONAL AMPLIFIERS

UNIT SPECIFICS

Through this unit we have discussed the following aspects:

- *Introduction to op-amp;*
- *Basics of the op-amp;*
- *IC 741 op-amp;*
- *Internal structure of an op-amp;*
- *Ideal op-amp;*
- *Characteristics of ideal op-amp;*
- *Practical op-amp;*
- *Characteristics of practical op-amp*
- *Negative feedback in op-amp*
- *Procedure for analyzing the op-amp circuits.*

RATIONALE

This fundamental unit on op-amp helps students to get a primary idea about the basics of operational amplifiers. First, the history of the op-amp is explained. Next, the basics of the op-amp-like symbol, its terminals along the popular packages are explained. Thereafter, a popular op-amp IC 741 is introduced to the students. Next, the internal structure of the op-amp is discussed. The internal structure of the op-amp describes various stages in the op-amp which is responsible for the working of the op-amp. After obtaining a good knowledge on these topics, then the ideal op-amp along with its characteristics was discussed. Next, the practical op-amp along with its characteristics is explained. The limitations of the practical op-amp can be avoided by incorporating the negative feedback in the op-amp. So, the negative feedback in the op-amp is discussed. All these basic aspects are relevant to start designing the electronic circuits using op-amp. In addition to the above, the procedure for analyzing the op-amp circuits is also discussed. This analysis helps in the identifying the operation of various op-amp circuits. The laboratory experiment related to measurement of op-amp electrical parameters is presented in the appendix of the book

PRE-REQUISITES

NIL

UNIT OUTCOMES

List of outcomes of this unit is as follows:

U5-O1: Understand the basics of op-amps

U5-O2: Realize the internal structure of the op-amp

U5-O3: Understand the characteristics of the ideal op-amp versus the practical op-amp

U5-O4: Realize the necessity of the negative feedback in the practical op-amp

U5-O5: Realize the procedure for analyzing the op-amp circuits

Unit-5 Outcomes	EXPECTED MAPPING WITH COURSE OUTCOMES (1- *Weak Correlation; 2- Medium correlation; 3- Strong Correlation*)				
	CO-1	CO-2	CO-3	CO-4	CO-5
U5-O1	3	-	1	-	2
U5-O2	3	-	1	-	1
U5-O3	3	-	1	-	1
U5-O4	3	-	1	-	2
U5-O5	3	-	1	-	2

5.1 INTRODUCTION

Operational amplifiers, or op-amps, have found extensive applications in various fields, including communication systems and medical applications. The term "operational amplifier" originated from the early days when these devices were designed using vacuum tubes for performing mathematical operations. However, with the advancement of semiconductor technology, integrated circuit (IC) op-amps were developed. These IC op-amps were much more cost-effective and compact compared to their vacuum tube counterparts. The integration of multiple transistors and other components onto a single silicon chip allowed for the creation of op-amps that were inexpensive, small in size, and consumed less power. The development of IC op-amps revolutionized the field of linear integrated circuits, making them more accessible and widely used in various applications. These compact and economical devices played a crucial role in the proliferation of electronic systems, enabling advancements in areas such as telecommunications, medical devices, audio equipment, control systems, and more.

5.2 BASICS OF THE OP-AMP

5.2.1 Symbol of Op-amp

The symbol of the op-amp is shown in Fig. 5.2.1. The op-amp basically consists of one inverting terminal, one non-inverting terminal, one output terminal, one positive power supply terminal and one negative power supply terminal. The inverting input terminal is normally denoted by the notation (-). The non-inverting input terminal is denoted by the notation (+). The positive power supply terminal is denoted by the term V^+ and the negative power supply terminal is denoted by the term V^-.

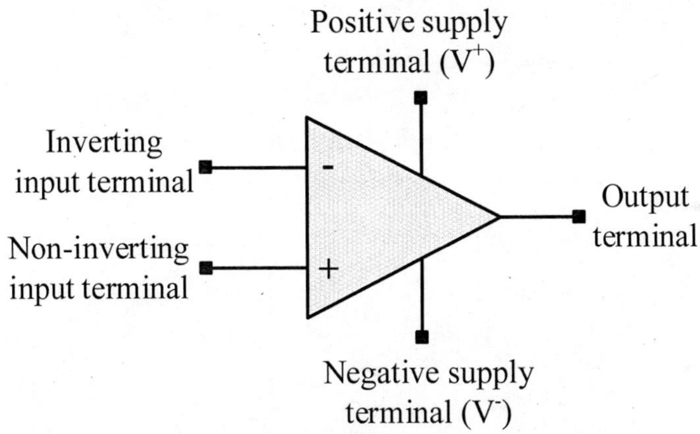

Fig. 5.2.1 Symbol of op-amp.

5.2.2 Popular Packages of Op-amp

The op-amps are basically available in these three packages

i. Metal can (TO) package

ii. Dual-in-line package (DIP)

iii. Flat package

Fig. 5.2.2.1 shows various popular packages of the op-amp. In the metal can (TO) package, the chip is

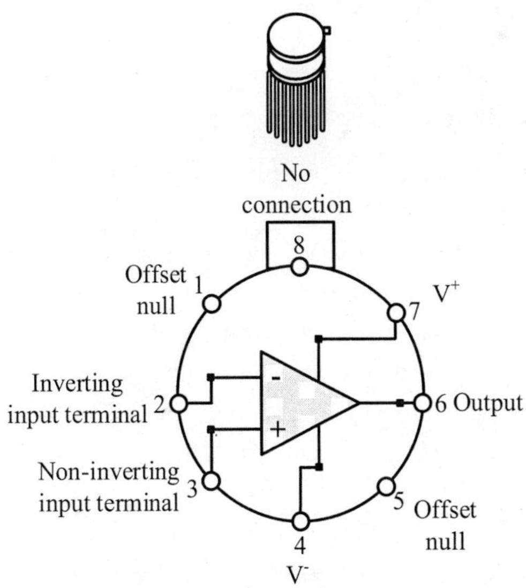

(a) 8-pin metal can package

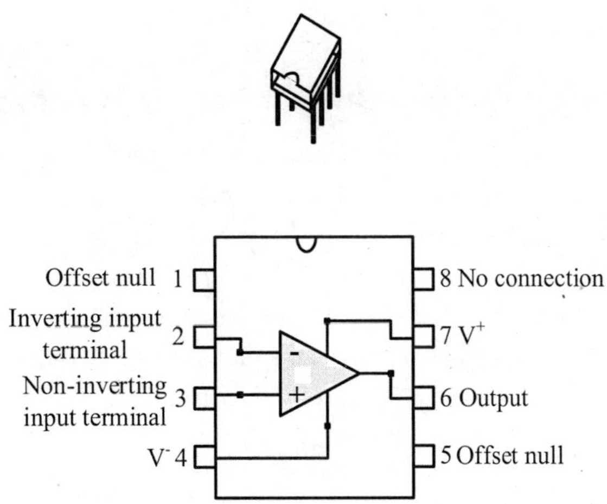

(b) 8-pin dual-in-line package

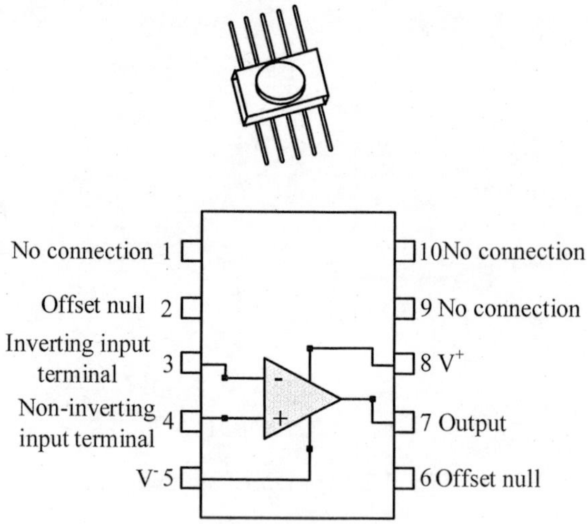

(c) 10 lead flat package

Fig. 5.2.2.1 Various packages of the op-amp.

enclosed in a metal or plastic case as shown in Fig. 5.2.2.1(a). The op-amps with metal can (TO) package are normally available 8, 10 and 13 pins. In the dual-in-line packages, the chip is enclosed in a plastic or ceramic case as shown in Fig. 5.2.2.1(b). The 8-pin dual-in-line package is more popular compared to the other packages. The flat package contains the chip enclosed in a rectangular ceramic case with the leads coming from either end as shown in Fig. 5.2.2.1(c). The flat package normally comes with 8, 10, 14 pins.

5.2.3 Power Supply Requirement for Op-amp

The op-amp normally requires two power supplies. The positive supply terminal of the op-amp is connected to the positive terminal of one power supply. Similarly, the negative supply terminal of the op-amp is connected to the negative terminal of the other supply. Fig. 5.2.3.1 shows the arrangement for the power supply of the op-amp. Normally the common point of the two power supplies must be grounded. Otherwise, double the input voltage will appear across the supply terminals of the op-amp which may damage the op-amp. The two power supplies required for the op-amp can be balanced or unbalanced. Fig. 5.2.3.2 shows the balanced and unbalanced power supplies required for the op-amp. But in practical applications, the balanced power supplies are used for the op-amp. The power supply range for the op-amp will be from ±5V to ±22V. The popular balanced power supplies for the op-amp are ±15V. Instead of two independent power supplies, the balanced power supply for the op-amp can be obtained by using the circuit shown in Fig. 5.2.3.3. The circuit shown in Fig. 5.2.3.3(a) consists of two resistors of equal value and two capacitors of equal value. The value of the resistor will be generally in kΩ so that no current is drawn from the supply. The two capacitors whose value is normally between 0.01 to 10μF will provide the decoupling from the source. The circuit shown in Fig. 5.2.3.3(b) uses the zener diodes for obtaining the balanced power supply for the op-amp. The value of the series resistance

Rs will be chosen in such a way that the Zener diode will operate in the breakdown mode. Sometimes, the reversal of power supplies will damage the op-amp. In order to avoid damage of the op-amp, the protection mechanism needs to be included in the power supply arrangement. The circuit shown in Fig. 5.2.3.3(c) have the protection mechanism in the power supply arrangement of the op-amp. The diodes D_1 and D_2 are used to protect the op-amp due to the reversal of the power supply.

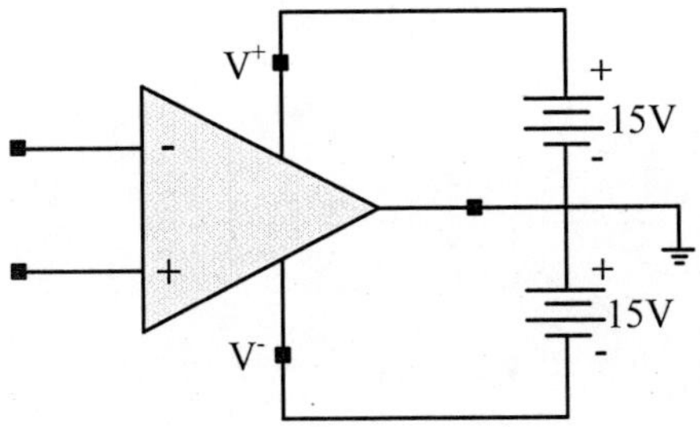

Fig. 5.2.3.1 Power supply connections of op-amp.

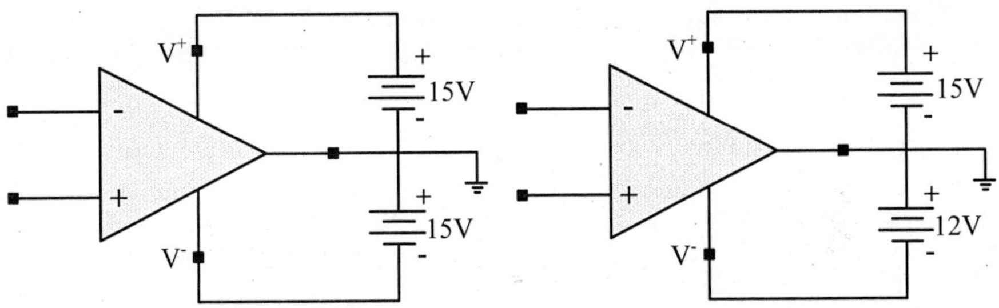

(a) Balanced power supply for op-amp

(b) Unbalanced power supply for op-amp

Fig. 5.2.3.2 Balanced and unbalanced power supply connections of the op-amp.

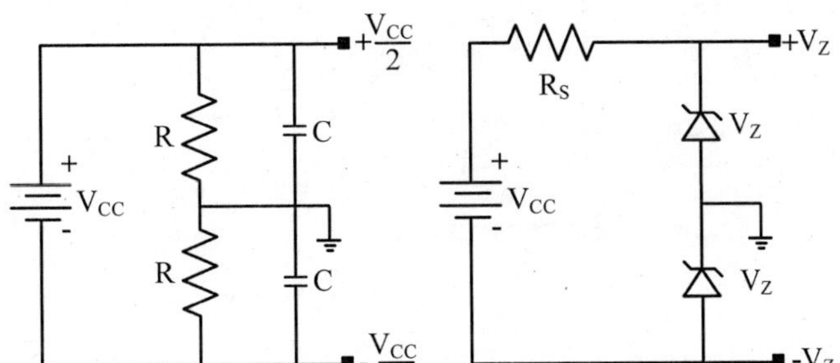

(a)

(b)

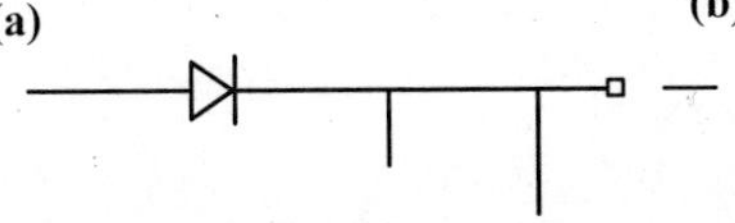

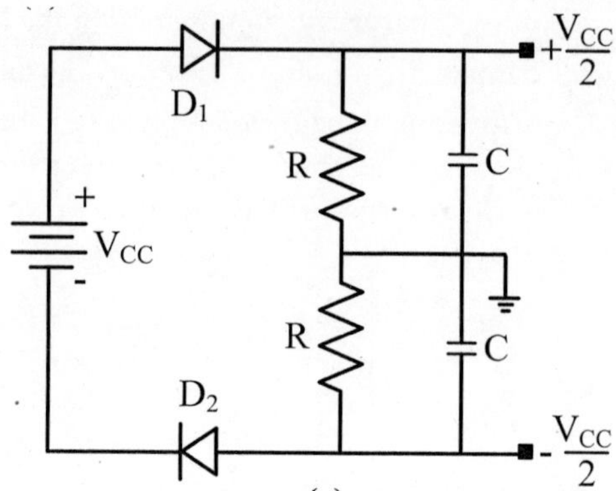

(c)

Fig. 5.2.3.3 Circuits for obtaining the balanced power supply for the op-amp from the single power supply.

5.3 IC 741 OP-AMP

The IC 741 is a more popularly available op-amp. Different manufacturers like Fairchild Semiconductors, Motorola, and National Semiconductor manufacture the IC 741 by adding some prefixes to 741 like μA, MC, and LM. The IC 741 op-amp is normally an 8-pin IC and is available in the dual-in-line package. The pin configuration of IC 741 is shown in Fig. 5.3.1. The pins 2 and 3 are inverting and non-inverting input terminals of the op-amp. The pin 6 is the output pin of the op-amp. The pins 4 and 7 are used for the negative power supply and positive power supply terminals of the op-amp. The pins 1 and 5 are used to nullify the offset voltage. So, these are called offset null pins. The pin 8 does not have any connection.

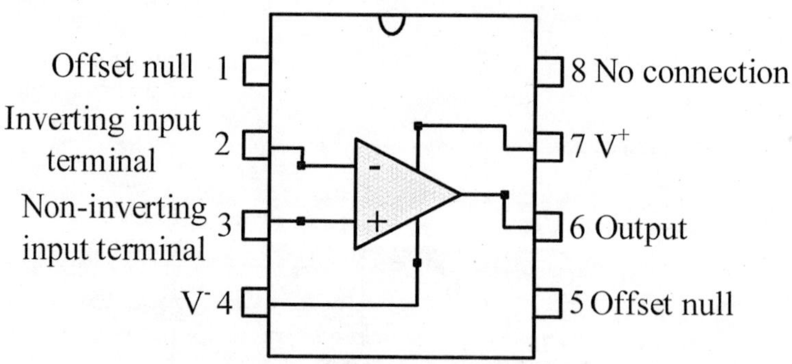

Fig. 5.3.1 Pin configuration of IC 741 op-amp.

5.4 INTERNAL STRUCTURE OF AN OP-AMP

The op-amp is basically a multistage amplifier that is obtained by cascading four blocks as shown in Fig. 5.4.1. The four cascaded blocks are:

i. Input stage
ii. Intermediate stage
iii. Level-shifting stage
iv. Output stage.

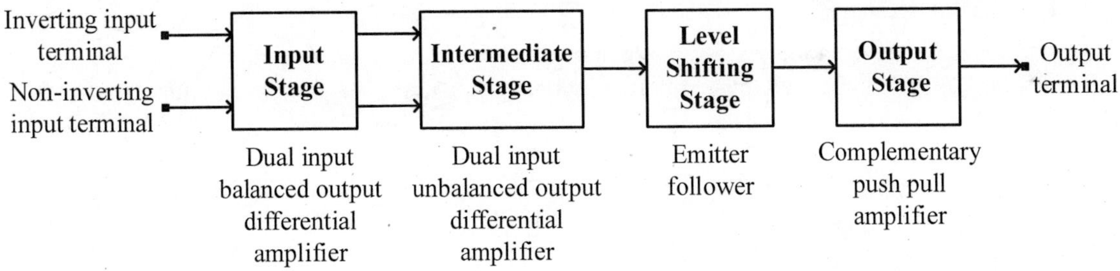

Fig. 5.4.1 Block diagram of op-amp.

5.4.1 Input Stage

The dual-input balanced output differential amplifier is commonly used as the input stage in op-amps for several reasons, including its ability to provide high gain and high input resistance. The input stage of an op-amp is responsible for amplifying the small input signals and providing the majority of the overall gain of the amplifier. By using a dual-input differential amplifier as the input stage, the op-amp can achieve a high voltage gain, typically in the range of tens of thousands to hundreds of thousands. Additionally, the dual-input differential amplifier offers a high input resistance, which is desirable to minimize the loading effect on the signal source. High input resistance means that the input stage draws very little current from the source, ensuring that the signal source is not significantly affected or distorted by the connection to the op-amp input. Overall, the dual-input balanced output differential amplifier meets the requirements of high gain and high input resistance, making it an ideal choice for the input stage of op-amps.

5.4.2 Intermediate stage

In order to have a high gain in op-amp, an additional stage i.e., the intermediate stage needs to be cascaded with the input stage. Since the intermediate stage has two input terminals and one output terminal, a dual input unbalanced differential amplifier is normally used as the intermediate stage in the op-amp.

5.4.3 Level-shifting Stage

The level-shifting stage is used in op-amp to eliminate the DC offset at the output of the intermediate stage and provide the desired output level. The level-shifting stage is responsible for shifting the DC operating point of the output signal to the desired level. One commonly used configuration for the level-shifting stage is the emitter follower, also known as a common collector amplifier. The emitter follower provides a high input resistance, allowing it to effectively isolate the previous stages and avoid loading effects. By using an emitter follower as the level-shifting stage, the output of the intermediate stage is connected to the emitter of the transistor. The transistor acts as a buffer, providing a low output resistance while maintaining a high input resistance. This configuration allows the output of the intermediate stage to be coupled to the load or subsequent stages without significantly affecting the signal. Furthermore, the emitter follower stage also helps to eliminate the DC offset present at the output of the intermediate stage. The transistor's biasing arrangement and voltage divider network in the emitter follower configuration ensures that the output voltage is shifted to the desired level, typically ground or another reference voltage, effectively cancelling out any DC offset.

5.4.4 Output Stage

The output stage in an op-amp is a complementary push-pull amplifier. The purpose of output stage is to handle high output voltage swings, and high current capability of op-amp. In addition to the above, the output stage also needs to have low output resistance so that the complete output voltage will appear across the load.

5.5 IDEAL OP-AMP

The ideal op-amp resembles same like a differential amplifier. If V_1 and V_2 are two inputs to the ideal op-amp, then the output voltage V_o is given by

$$V_O = A_d V_d + A_c V_c \quad (5.5.1)$$

where A_d is the differential gain

A_c is the common mode gain

V_d is the differential voltage which is the difference between two input voltages V_1 and V_2.

V_c is the common mode voltage which is the average of two input voltages V_1 and V_2.

In the case of the ideal op-amp, the common mode gain is zero. Therefore, the expression for the output voltage V_o of the ideal op-amp is given by

$$V_O = A_d V_d = A_d (V_1 - V_2) \quad (5.5.2)$$

The common mode rejection ratio (CMRR) of the op-amp is the ratio of the differential gain A_d to the common mode gain A_c. In the case of the ideal op-amp, the common mode gain A_c is zero so the CMRR is infinite.

Using the expression (5.5.2), the equivalent circuit of the op-amp can be obtained as shown in Fig. 5.5.1. In the case of the ideal op-amp, the input resistance R_{in} is equal to infinity and the output resistance R_O is equal to zero. Due to infinite input resistance, the op-amp will not draw any input current. Therefore, the input currents I_1 and I_2 of the op-amp are zero. The open loop voltage gain of the ideal op-amp (A_{OL}) is infinite so for a finite value of output voltage, the difference of the input voltages V_d is zero. If a graph is plotted between the difference of the input voltages V_d versus the output voltage of op-amp V_O, then the transfer curve of the op-amp can be obtained. The transfer curve of the ideal op-amp is shown in Fig. 5.5.2.

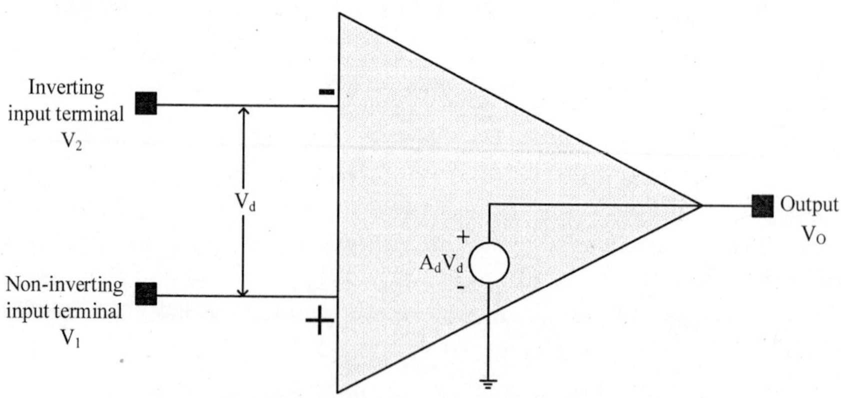

Fig. 5.5.1 Equivalent circuit of the ideal op-amp.

5.6 CHARACTERISTICS OF IDEAL OP-AMP

The characteristics of ideal op-amp are:

i. The open loop gain of the ideal op-amp (A_{OL}) is infinite.
ii. The input resistance of the ideal op-amp (R_{in}) is infinite.
iii. The output resistance of the ideal op-amp (R_O) is zero.
iv. The ideal op-amp will have zero output voltage for the zero differential voltage. There is no input offset voltage (V_{ios}) in the output of the ideal op-amp.

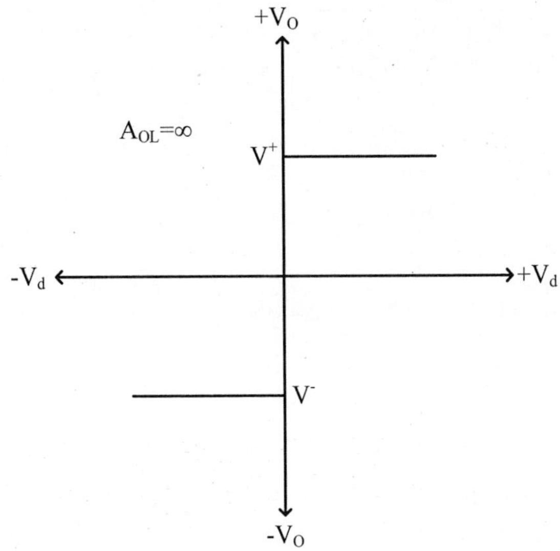

Fig. 5.6.1 Transfer curve of the ideal op-amp.

v. The ideal op-amp has infinite bandwidth. The ideal op-amp can work with the input signals that have a frequency between 0 to infinity.
vi. The ideal op-amp is having infinite common-mode rejection ratio (CMRR) because of the zero common-mode gain.
vii. The ideal op-amp is having an infinite slew rate (S). The output of the op-amp will change immediately with the change in the input voltage. The slew rate of the op-amp can be calculated by

$$\text{Slew rate } S = \left.\frac{dV_O}{dt}\right|_{\text{maximum}} = \infty \tag{5.6.1}$$

viii. The ideal op-amp is having zero value of power supply rejection ratio (PSSR). This means the variation in the input offset voltage V_{ios} due to the variation in one power supply while keeping the other power supply constant is zero. Therefore, the power supply sensitivity of ideal op-amp is zero. The PSSR can be calculated by s

$$PSSR = \left.\frac{\Delta V_{ios}}{\Delta V^+}\right|_{V^- = \text{Constant}} = \left.\frac{\Delta V_{ios}}{\Delta V^-}\right|_{V^+ = \text{Constant}} = 0 \tag{5.6.2}$$

ix. The temperature will not change the characteristics of the ideal op-amp.

5.7 PRACTICAL OP-AMP

In the case of the practical op-amp, the open loop voltage gain (A_{OL}) is finite. Due to this finite value of the A_{OL}, the output voltage will vary linearly for a small range of differential voltage V_d. To explain this, let us consider op-amp which is having a power supply of ±15V. For IC 741 op-amp, the open loop gain A_{OL} is 200000. Therefore, using the expression (5.5.2), the range of differential voltage will be

$$V_d = \frac{V_O}{A_d} = \frac{\pm 15}{200000} = \pm 75\mu V \tag{5.7.1}$$

If the differential voltage V_d is in between ±75μV, then the output voltage will vary linearly with the differential voltage V_d. If the differential voltage V_d is greater than or equal to +75μV, then the output voltage V_O will saturate at +15V. Similarly, if the differential voltage V_d is less than or equal to -75μV, then the output voltage V_O will saturate at -15V. Due to this, the transfer curve of the practical op-amp will deviate from the ideal op-amp. The transfer curve of the practical op-amp is shown in Fig. 5.7.1. The equivalent circuit of the practical op-amp is shown in Fig. 5.7.2. In the practical op-amp, the open loop voltage gain (A_{OL}) is finite and there exist a high value of input resistance R_{in} and low value of output resistance R_O.

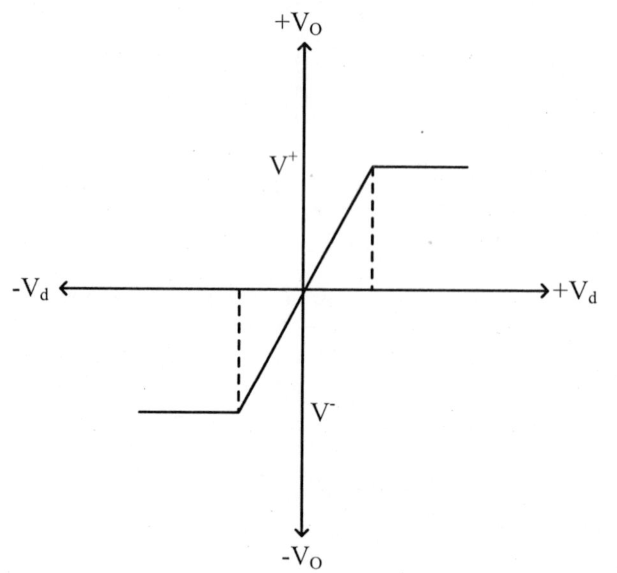

Fig. 5.7.1 Transfer curve of the practical op-amp.

5.8 CHARACTERISTICS OF PRACTICAL OP-AMP

The characteristics of the practical op-amp are:

i. The open loop gain of the practical op-amp (A_{OL}) is finite. Its value is in several thousands.
ii. The input resistance of the ideal op-amp (R_{in}) is finite. Its value is typically greater than 1MΩ.
iii. The output resistance of the ideal op-amp (R_O) is not zero. Its value is typically around 1 or 2 ohms.
iv. In the practical op-amp, there exist some output voltage for the zero differential voltage V_d. Therefore, in order to make the output voltage zero, a small input voltage of order millivolts

needs to be applied at the input terminals such that the output voltage is zero. This voltage is called input offset voltage.

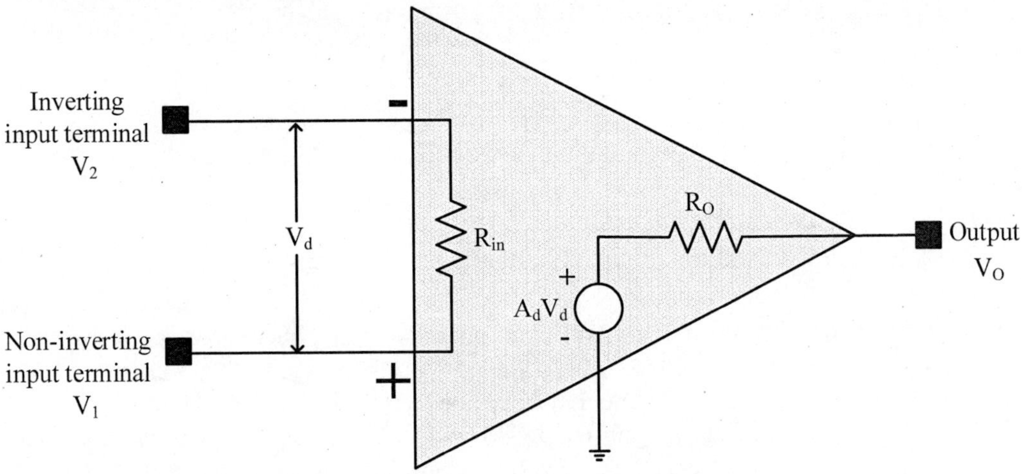

Fig. 5.8.1 Equivalent circuit of the practical op-amp.

v. The bandwidth of the practical op-amp is very narrow. So, the practical op-amp in the open loop can handle low-frequency input signals.

vi. In the practical op-amp, there exists some input bias current. If I_{b1} and I_{b2} are the bias currents flowing the input terminals of the op-amp as shown in Fig. 5.8.1. Then the input bias current I_b is nothing but the average of the two input currents. In Fig. 5.8.1, both the op-amp terminals are grounded. The power supply terminals of the op-amp are necessary to bias the op-amp. Therefore, the input bias current I_b is given by

$$I_b = \frac{I_{b1} + I_{b2}}{2} \tag{5.8.1}$$

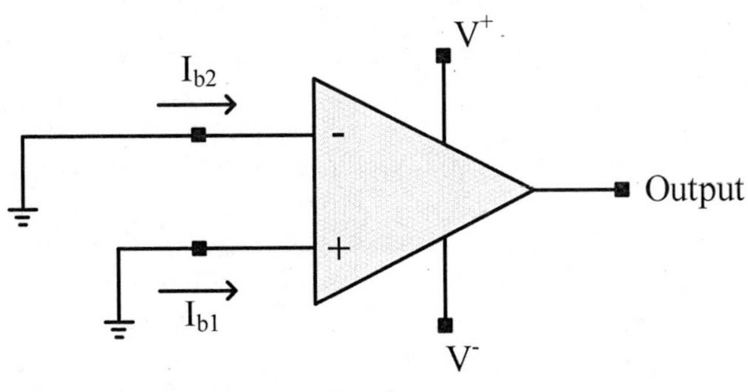

Fig. 5.8.2 Op-amp input bias currents.

The input bias currents I_{b1} and I_{b2} are the base bias currents in the input stage of the op-amp. The magnitude of these currents is not the same because of the nonidealities of the transistors in the differential amplifier. The magnitude of the input bias current is very small typically in the order of nano amperes.

vii. In the practical op-amp, apart from the input bias current, the input offset current also exists. The input offset current is the current which is the algebraic difference of two input bias currents I_{b1} and I_{b2}. Therefore, the input bias current is equal to

$$I_b = |I_{b1} - I_{b2}| \tag{5.8.2}$$

viii. The slew rate is defined as the maximum rate of change in the output voltage per time. The unit of the slew rate is volt per microsecond. Slew rate is a measure of how quickly an op-amp can change its output voltage in response to a step input signal.

$$\text{Slew rate } S = \left.\frac{dV_O}{dt}\right|_{\text{maximum}} \tag{5.8.3}$$

ix. The gain-bandwidth product (GBW) of an operational amplifier (op-amp) is the frequency at which the open-loop voltage gain of the op-amp is unity. In other words, when the op-amp's gain is set to one (0 dB), the frequency at which the output signal magnitude equals the input signal magnitude is the gain-bandwidth product. For example, if an op-amp has a gain-bandwidth product of 1 MHz, it means that when the op-amp is configured for unity gain, the bandwidth of the op-amp is 1 MHz. Beyond this frequency, the gain is less than unity. The gain-bandwidth product is an important figure of merit for op-amps as it helps to determine the maximum usable bandwidth in different applications. It is worth noting that the actual open-loop gain of the op-amp decreases as the frequency increases beyond the gain-bandwidth product.

5.9 NEGATIVE FEEDBACK IN OP-AMP

Normally, the op-amp have high open loop gain. Due to this high voltage gain, the linear region in the output of op-amp will obtain for the small range of differential voltage V_d and will saturate to the supply voltage if the differential voltage V_d is out of this range as shown in Fig. 5.7.1. This limits the usage of open loop op-amp in practical applications. In addition to this, the bandwidth of the practical op-amp is very less. Due to this, the op-amp will work only with input signals of a particular band of low frequency.

For the usage of op-amp in practical applications, the voltage gain of op-amp needs to be reduced. In addition to this, the bandwidth of the practical op-amp also needs to be increased. For obtaining these two objectives, negative feedback is introduced in the op-amp. Fig. 5.9.1 shows the implementation of negative feedback in the op-amp. V_S is the input signal applied at the non-inverting input terminal of the op-amp. The load resistance R_L is connected at the output terminal and V_O is the voltage across the load resistance. In order to have negative feedback the resistors R_F and R_1 are used. The resistor R_F is connected between the negative input terminal and the output terminal of the op-amp. The resistor R_1 is connected between the negative input terminal of the op-amp and the ground. By making this arrangement, a part of the output voltage is fed back to the input in the opposite phase. Let A_{OL} is the open loop gain of an op-amp, A is the gain of an op-amp with feedback, V_1 be the voltage at the non-inverting terminal, V_2 be the voltage at inverting terminal and V_d be the differential voltage. The gain of the op-amp with feedback is given by

$$A = \frac{V_O}{V_S} \tag{5.9.1}$$

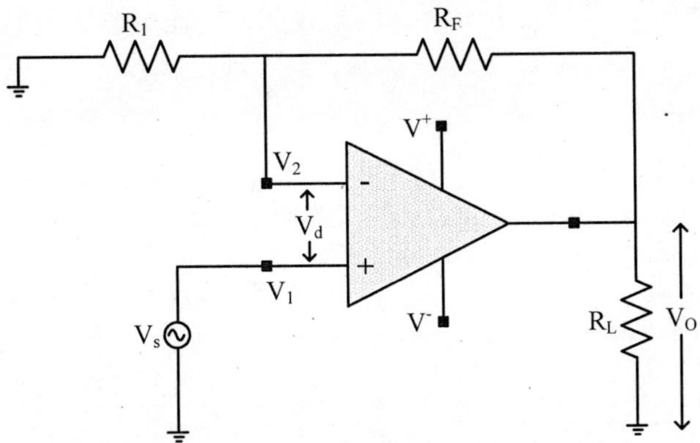

Fig. 5.9.1 Op-amp with negative feedback.

and the output voltage V_O is again given by

$$V_O = A_{OL} V_d = A_{OL}(V_1 - V_2) \tag{5.9.2}$$

Substituting (5.9.2) in (5.9.1) gives

$$A = \frac{A_{OL}(V_1 - V_2)}{V_S} \tag{5.9.3}$$

The expressions for the voltages at the noninverting input terminal V_1 and the inverting input terminal V_2 is given by

$$V_1 = V_S$$

$$V_2 = \frac{R_1 V_O}{R_1 + R_F} \tag{5.9.4}$$

substituting (5.9.4) in (5.9.3) gives

$$A = \frac{A_{OL}\left(V_S - \frac{R_1 V_O}{R_1 + R_F}\right)}{V_S} \tag{5.9.5}$$

simplifying (5.9.5) gives

$$A = A_{OL}\left(1 - \frac{AR_1}{R_1 + R_F}\right) = A_{OL} - \frac{A_{OL}AR_1}{R_1 + R_F}$$

$$\Rightarrow A\left(1 + \frac{A_{OL}R_1}{R_1 + R_F}\right) = A_{OL}$$

$$\Rightarrow A = \frac{A_{OL}}{\left(1 + \frac{A_{OL}R_1}{R_1 + R_F}\right)} = \frac{A_{OL}}{(1 + A_{OL}\beta)} \tag{5.9.6}$$

where β is called the feedback factor and its value is equal to

$$\beta = \frac{R_1}{R_1 + R_F} \tag{5.9.7}$$

Therefore, the voltage gain op-amp with the negative feedback is given by

$$A = \frac{A_{OL}}{(1 + A_{OL}\beta)} \tag{5.9.8}$$

From the equation (5.9.8), it can be observed that the op-amp with negative feedback have reduced gain compared to the open loop gain of the op-amp. This results in the increase in the range of differential voltage V_d making it suitable for practical applications.

Let f_{OL} be the cut-off frequency of the open loop op-amp and f be the cutoff frequency of op-amp with the negative feedback. The gain bandwidth product of the open loop op-amp and the op-amp with negative feedback will be equal. Therefore

$$A_{OL} f_{OL} = Af \tag{5.9.9}$$

substituting (5.9.8) in (5.9.9) gives

$$A_{OL} f_{OL} = \frac{A_{OL}}{(1 + A_{OL}\beta)} f$$

$$\Rightarrow f = f_{OL}(1 + A_{OL}\beta) \tag{5.9.10}$$

From (5.9.10), it can be observed that the cut-off frequency of op-amp with negative feedback is high compared to the cutoff frequency of open loop op-amp, Therefore, the op-amp with negative feedback can work with the input signals of wide range of frequencies.

Let R_{in} be the input resistance of the op-amp in an open loop and R_{inf} be the input resistance of the op-amp with negative feedback. If I_{in} is the input current then the input resistances R_{in} and R_{inf} of the op-amp in an open loop and with open with negative feedback are given by

$$R_{in} = \frac{V_d}{I_{in}} \text{ and } R_{inf} = \frac{V_S}{I_{in}} \tag{5.9.11}$$

The R_{inf} can be written as

$$R_{inf} = \frac{V_S}{\frac{V_d}{R}} = R\frac{V_S}{V_d} \tag{5.9.12}$$

substituting (5.9.1) and (5.9.2) in (5.9.12), we get

$$R_{inf} = R\frac{V_S}{V_d} = R\frac{\frac{V_O}{A}}{\frac{V_O}{A_{OL}}} = R\frac{A_{OL}}{A} \tag{5.9.13}$$

substituting (5.9.8) in (5.9.13), we get

$$R_{inf} = R\frac{A_{OL}}{\frac{A_{OL}}{(1 + A_{OL}\beta)}} = R(1 + A_{OL}\beta) \tag{5.9.14}$$

From (5.9.14), it can be observed that the input resistance of op-amp with negative feedback is high compared to the input resistance of open loop op-amp.

Similarly, if R_O be the output resistance of the op-amp in an open loop and R_{Of} be the output resistance of the op-amp with negative feedback. Then the expression for the output resistance of the op-amp with negative feedback is given by

$$R_{Of} = \frac{R_O}{1 + A_{OL}\beta} \tag{5.9.15}$$

5.10 PROCEDURE FOR ANALYSING THE OP-AMP CIRCUITS

The op-amp circuits are analysed by considering the following assumptions:

i. The input current drawn by the op-amp is zero.

ii. Virtual ground

Due to the very high value of the input resistance, the input current drawn op-amp is very low and can be neglected. This is the first assumption used for solving the op-amp circuit. The second assumption is the concept for virtual ground. As said earlier, in the open loop configuration of the op-amp, the differential voltage V_d across the input terminals of the op-amp is very small and can be considered as zero. Therefore,

$$V_d = V_1 - V_2 = 0 \tag{5.10.1}$$

therefore

$$V_1 = V_2 \tag{5.10.2}$$

So, using (5.10.2) if the noninverting input terminal of the op-amp is grounded, then the inverting input terminal of the op-amp is also zero. Though there is no physical connection, a virtual connection between the noninverting and inverting input terminal because of which the inverting input terminal of the op-amp is grounded. Using these two assumptions, any op-amp circuit can be analysed.

UNIT SUMMARY

The chapter started with the history of the op-amp. Next, the basics of the op-amp-like symbol, its terminals along the popular packages are explained. Thereafter, a popular op-amp IC 741 is introduced. Next, the internal structure of the op-amp is discussed. The internal structure of the op-amp describes various stages in the op-amp which is responsible for the working of the op-amp. After obtaining a good knowledge on these topics, then the ideal op-amp along with its characteristics was discussed. Next, the practical op-amp along with its characteristics is explained. By adding negative feedback to the op-amp, the practical op-amp's limits can be circumvented. So, the negative feedback in the op-amp is discussed. All of these fundamental elements are necessary to begin creating electronic circuits with an op-amp. Along with the aforementioned, the process for examining the op-amp circuits is also covered. This analysis aids in determining how different op-amp circuits function. The book's appendix contains the laboratory experiment pertaining to the measuring of op-amp electrical parameters.

EXERCISES

Multiple Choice Questions

5.1 For an ideal op–amp, which of the following is correct?

a. Zero input impedance and infinite bandwidth

b. Infinite input impedance and infinite bandwidth

c. Infinite input impedance and zero bandwidth
d. Zero input impedance and zero bandwidth

5.2 In order for an output to swing above and below a zero reference, the op-amp circuit requires
a. Zero offset
b. A resistive feedback network
c. A wide bandwidth
d. A negative and positive supply

5.3 In an op-amp, the input impedance is and the output impedance is .
a. high; high
b. low; low
c. low; high
d. high; low

5.4 CMRR for an op-amp should be
a. As large as possible
b. close to zero
c. close to unity
d. As small as possible

Answers for Multiple Choice Questions			
5.1 **b**	5.2 **d**	5.3 **d**	5.4 **a**

Short Answer Type Questions

5.1 Explain the pin diagram of IC 741?
5.2 Explain the power supply requirements of op-amp?
5.3 Explain the negative feedback in the op-amp?

Long Answer Type Questions

5.1 Explain in detail the block diagram of op-amp?
5.2 Explain in detail the ideal characteristics of op-amp?
5.3 Explain in detail the practical characteristics of an op-amp?
5.4 Explain in detail the ideal and practical voltage transfer curve of op-amp?

KNOW MORE

For Additional Info Scan Me

REFERENCES

1. *Ramakanth A. Gayakwad, "OP-amps and Linear Integrated Circuits", PHI, Fourth Edition, 2010.*
2. *Adel S. Sedra, Kenneth C. Smith, Arun N. Chandorkar, "Microelectronic Circuits: Theory and Applications", Oxford University Press, Seventh Edition, (1 June 2017).*
3. *U. A. Bakshi and A. P. Godse. "Operational Amplifiers and Linear ICs".*

06 LINEAR APPLICATIONS OF OPERATIONAL AMPLIFIERS

UNIT SPECIFICS

Through this unit we have discussed the following aspects:

- *Assumptions for op-amp Analysis;*
- *Op-amp as inverting amplifier;*
- *Non-inverting amplifier;*
- *Voltage follower (unity follower);*
- *Summing amplifier circuit;*
- *Integrator circuit;*
- *Differentiator circuit;*
- *Instrumentation amplifier;*
- *Op-amp as low pass filter;*
- *Op-amp as high pass filter;*
- *Op-amp as band pass filter;*
- *Proportional controller;*
- *Proportional integral controller;*
- *Proportional integral derivative controller;*
- *Analog to digital converter;*
- *Voltage regulator;*
- *Phase shift oscillator;*
- *Wien bridge oscillator;*
- *Lag lead compensator.*

RATIONALE

This chapter deals with the application of the operational amplifier in various signals processing circuits. The chapter starts with the basic inverting and the non-inverting operational amplifiers which represents the fundamental circuits for the op-amp. Further, details of the ideal operational amplifier circuit are also given. The given ideal operational amplifier circuit is again used in the analysis of the op-amps application. The chapter gives the details of various op-amp based electronic circuits like adder, integrator, differentiator, filter circuits, analogue to digital converter, instrumentation amplifier circuit etc. The student will not only learn how to analyze the op-amps circuits but can design new applications of op-amps with the given details in the chapter. Using the given approach for analysis of op-amps circuit student will be able investigate new op-amps circuits. Further, additional experiments based on the op-amps are added in the appendix to enhance the practical knowledge of the students..

PRE-REQUISITES

Nil

UNIT OUTCOMES

List of outcomes of this unit is as follows:

U6-O1: Understand and design the op-amps as inverting and non-inverting amplifier.

U6-O2: Understand and design the op-amps for summer, integrator and differentiator circuit.

U6-O3: Understand and design the op-amps for implementation of filter circuit application.

U6-O4: Understand and design the op-amps for implementation of PID controller circuit.

U6-O5: Understand and design the op-amps for implementation ADC, compensator and instrumentation amplifier circuit.

Unit-6 Outcomes	EXPECTED MAPPING WITH COURSE OUTCOMES (*1- Weak Correlation; 2- Medium correlation; 3- Strong Correlation*)				
	CO-1	CO-2	CO-3	CO-4	CO-5
U6-01	3	1	3	-	3
U6-02	3	3	1	-	3
U6-03	3	2	1	-	3
U6-04	3	2	1	-	3
U6-05	3	2	1	-	3

6.1 ASSUMPTIONS FOR OP-AMP ANALYSIS

Operational amplifier (op-amp) typically employed in most of the signal conditioning circuits. For analysis of the complex circuits having op-amp is typically done considering the ideal their characteristics. Ideal op-amp have infinite input resistance between its input terminals, zero output resistance and has infinite gain as shown in Fig. 6.1.1(a). However, practical op-amp have high input resistance (in MΩ) between its input terminals, zero output resistance and has finite gain as shown in Fig. 6.1.1(b). To analyse basic op-amp circuits both ideal and practical circuit equivalent can be used. Typically, ideal equivalent circuit is employed as it simplifies the analysis of the op-amp circuit. While doing the analysis of the op-amp circuits below are assumptions taken:

i. Open loop gain of the op-amp circuit in open loop is infinite.
ii. The resistance R_i between the input terminal is infinite.
iii. The resistance Ro at the output terminal is zero.
iv. Current in the input terminals of the op-amp is zero.
v. The resistance R_i between the input terminal is infinite.
vi. As the currents in the input terminals are zero, so potential of input terminals of the op-amp is equal as shown in Fig. 6.1.2.

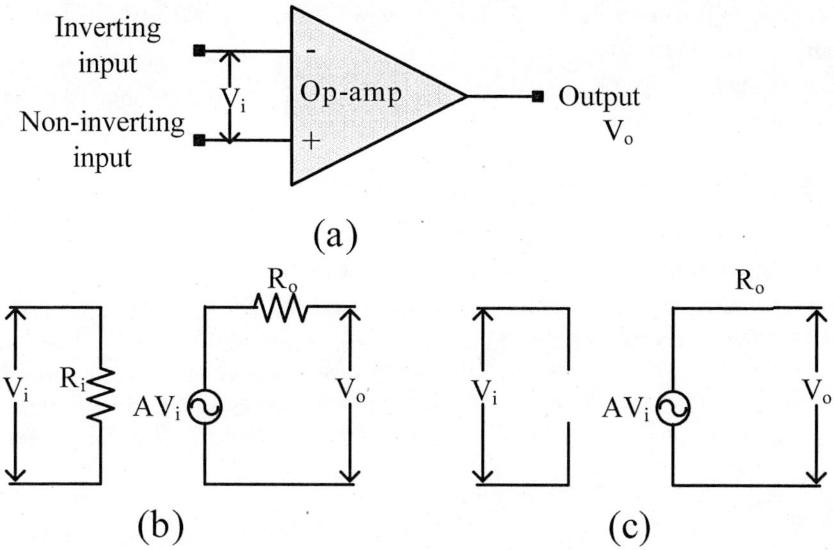

Fig. 6.1.1 Op-amp (a) symbol, (b) practical equivalent circuit schematic, and (c) ideal equivalent circuit schematic

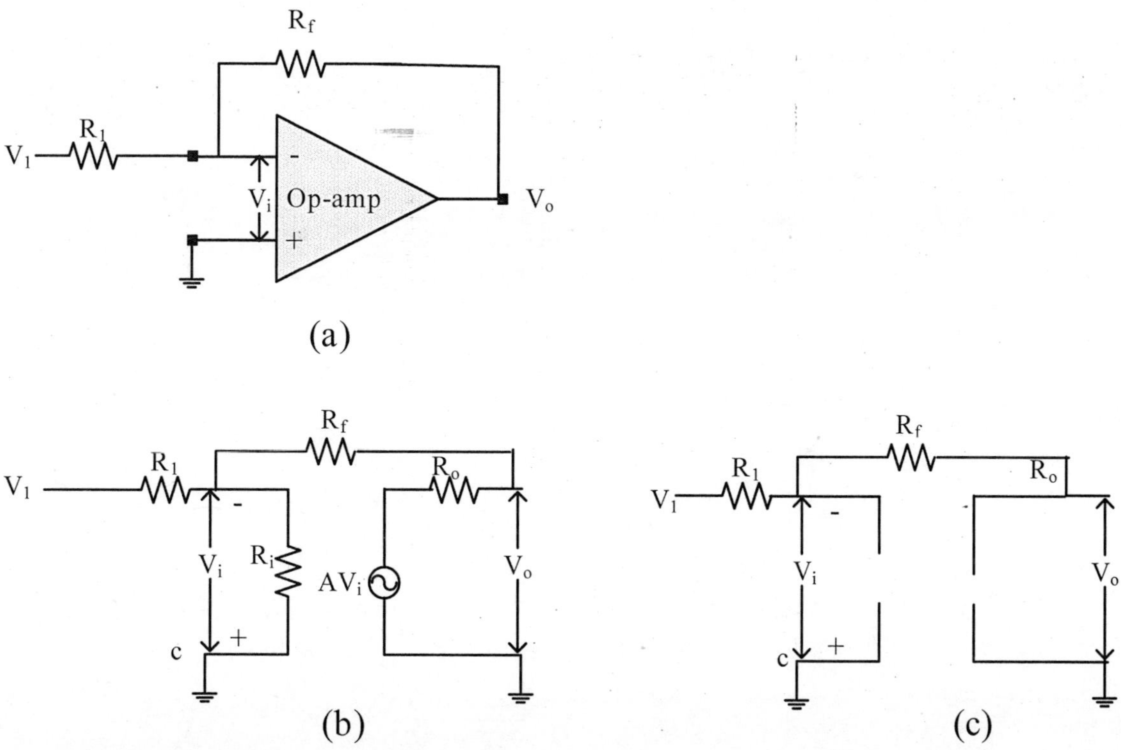

Fig. 6.1.2 Inverting op-amp (a) circuit schematic, (b) practical equivalent circuit schematic for inverting op-amp, and (c) ideal equivalent circuit schematic for inverting op-amp.

Typically, op-amp circuits can be classified into two broad categories based on the input applied at the inverting or non-inverting terminal:

i. Inverting amplifier
ii. Non-inverting amplifier

6.2 OP-AMP AS INVERTING AMPLIFIER

The circuit configuration of the inverting amplifier is shown in Fig. 6.2.1. It consists of two resistors at the inverting terminal feedback and input connected to output and input source respectively as shown in Fig. 6.2.1. The non-inverting terminal connected to ground. Thus the, input signal is applied at the inverting terminal via resistor R_1 with non-inverting terminal grounded as shown Fig. 6.2.1 The output of the op-amp is amplified input signal which is 180° out of phase with input. The amplification in the input signal can be computed using the KCL at node 'a' as can be seen in Fig. 6.2.1. Using the concept of virtual ground or equating the potential both the input terminals and applying KCL at node 'a' we have,

$$\frac{-V_1}{R_1} + \frac{0 - Vo}{R_f} = 0$$

$$\frac{V_1}{R_1} = \frac{-Vo}{R_f}$$

$$\frac{Vo}{V_1} = \frac{-R_f}{R_1} \tag{6.2.1}$$

It can be noted that there is a negative sign in the expression (6.2.1) which supports the fact that the output voltage is 180° phase shifted from the input voltage.

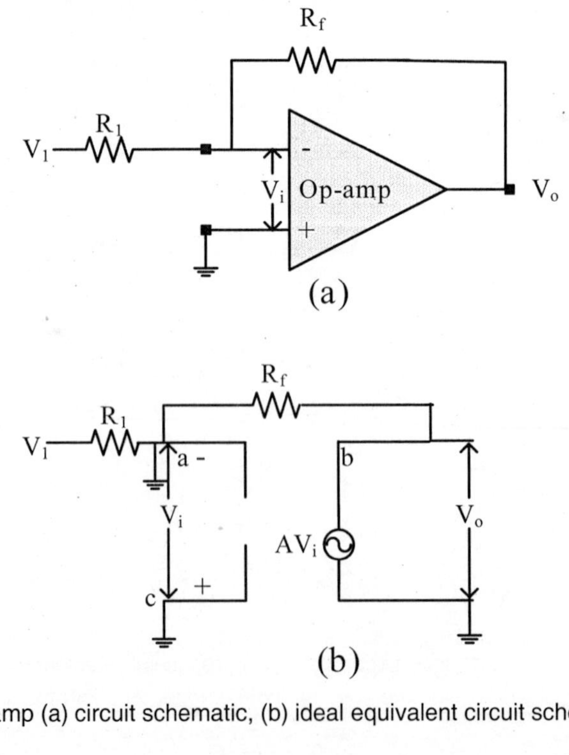

Fig. 6.2.1 Inverting op-amp (a) circuit schematic, (b) ideal equivalent circuit schematic for inverting op-amp.

6.3 NON-INVERTING AMPLIFIER

The circuit configuration of the inverting amplifier is shown in Fig. 6.3.1. It consists of two resistors at the inverting terminal feedback and an input connected to output and ground respectively as shown in Fig. 6.3.1. The non-inverting terminal is connected to input source. Thus, the input signal is applied at the non-inverting terminal as shown Fig. 6.3.1. The output of the op-amp is an amplified input signal which is in phase with input. The amplification in the input signal can be computed using the KCL at node 'a' as can be seen in Fig. 6.3.1. Using the concept of virtual ground or equating the potential both the input terminals and applying KCL at node 'a' we have,

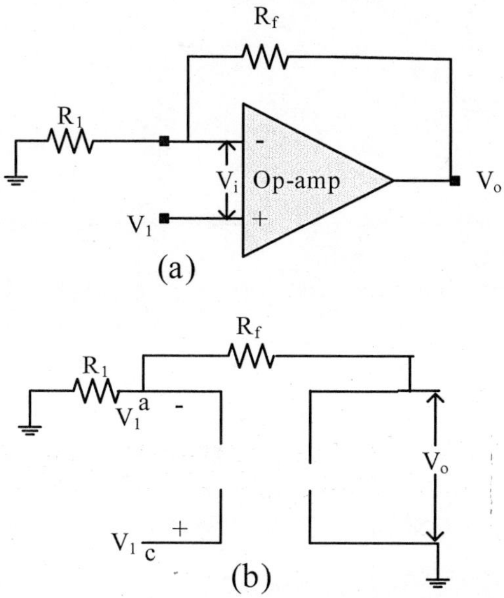

Fig. 6.3.1 Non-inverting op-amp (a) circuit schematic, (b) ideal equivalent circuit schematic for non-inverting op-amp.

$$\frac{V_1}{R_1} + \frac{V_1 - Vo}{R_f} = 0$$

$$V_1\left(\frac{1}{R_1} + \frac{1}{R_f}\right) = \frac{Vo}{R_f}$$

$$\frac{Vo}{V_1} = \left(1 + \frac{R_f}{R_1}\right) \tag{6.3.1}$$

It can be noted that there is no negative sign in the expression (6.3.1) which supports the fact that the output voltage is in exact phase with the input voltage.

6.4 VOLTAGE FOLLOWER (UNITY FOLLOWER)

The circuit schematic of the voltage or unity follower is shown in Fig. 6.4.1. The inverting terminal connected to output, while the non-inverting terminal is connected to input source. The output of the op-amp is input signal which is in phase with it. The gain in the input signal can be computed using the

KCL at node 'a' as can be seen in Fig. 6.4.1. Using the concept of virtual ground or equating the potential both the input terminals and applying KVL in loop a- b-g, we have,

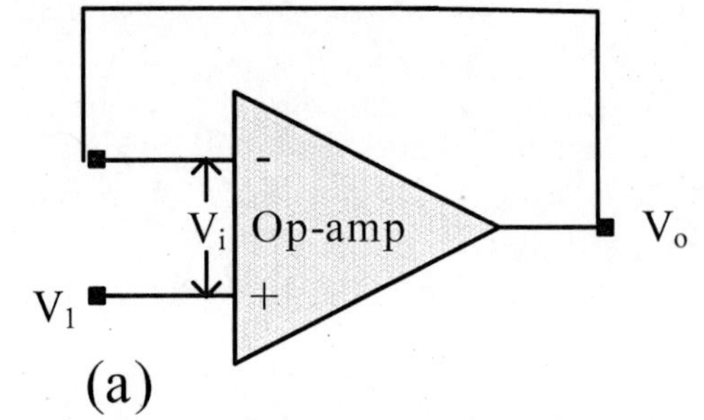

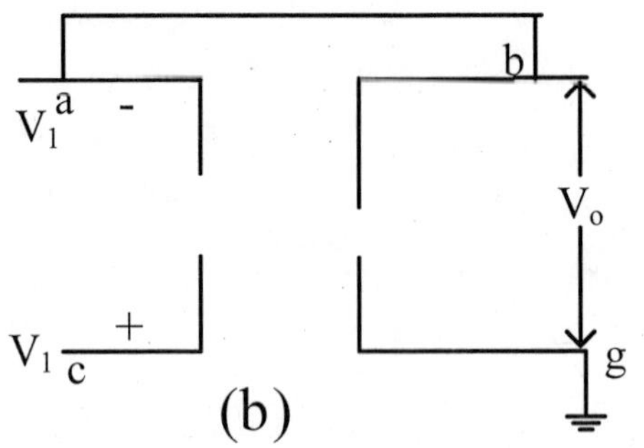

Fig. 6.4.1 Voltage follower using op-amp (a) circuit schematic, (b) ideal equivalent circuit schematic.

$$V_1 - Vo = 0$$

$$V_1 = Vo \tag{6.4.1}$$

Thus, given circuit gives output which is same as an input. The given circuit is typically used as a buffer circuit which has high input impedance low output resistance.

6.5 SUMMING AMPLIFIER CIRCUIT

The circuit schematic for the summing amplifier is shown in Fig. 6.5.1. It consists of input resistors with input voltage signal and feedback resistor as shown in Fig. 6.5.1. The non-inverting terminal connected to ground. Thus the, input signals to be summed are applied at the inverting terminal via three resistors R_1, R_2, R_3 as shown Fig. 6.5.1. The output of the op-amp is sum of the input signals amplified by ratio of feedback resistor and input resistor. The amplification in the input signal can be computed using the KCL at node 'a' as can be seen in Fig. 6.5.1. Using the concept of virtual ground or equating the potential both the input terminals and applying KCL at node 'a' we have,

$$\frac{-V_1}{R_1} + \frac{-V_2}{R_2} + \frac{-V_3}{R_3} + \frac{0 - Vo}{R_f} = 0$$

$$\frac{-V_1}{R_1} + \frac{-V_2}{R_2} + \frac{-V_3}{R_3} = \frac{-Vo}{R_f}$$

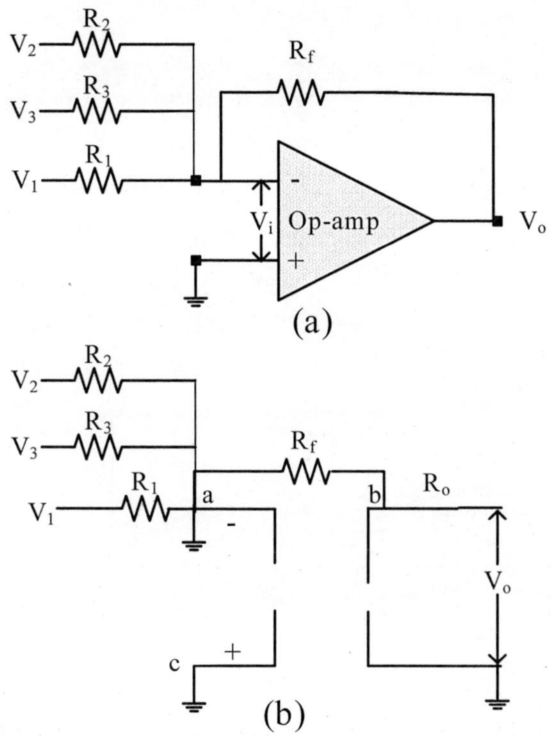

Fig. 6.5.1 Summer circuits using op-amp (a) circuit schematic, (b) ideal equivalent circuit schematic.

$$Vo = -R_f\left(\frac{-R_f}{R_1}V_1 + \frac{-R_f}{R_2}V_2 + \frac{-R_f}{R_3}V_3\right) \quad \text{if } R_1 = R_2 = R_3 = R$$

$$Vo = \frac{-R_f}{R}(V_1 + V_2 + V_3) \tag{6.5.1}$$

6.6 INTEGRATOR CIRCUIT

The circuit configuration of the inverting amplifier is shown in Fig. 6.6.1. It consists of a resistor and the feedback capacitor connected at the inverting terminal as shown in Fig. 6.6.1. The other end of the resistor and the feedback capacitor connected to input signal and output respectively. The input signal is applied at the inverting terminal via resistor R_1 with non-inverting terminal grounded as shown Fig. 6.6.1. The output of the op-amp is integrated input signal which is 180° out of phase with input with amplification defined by the input resistor and feedback capacitor value. The amplification in the input signal can be computed using the KCL at node 'a' as can be seen in Fig. 6.6.1. Using the concept of virtual ground or equating the potential both the input terminals and applying KCL at node 'a' using Laplace domain representation we have,

$$\frac{-V_1}{R_1} + \frac{0 - Vo}{\frac{1}{sC_f}} = 0$$

$$\frac{V_1}{R_1} = \frac{-Vo}{\dfrac{1}{sC_f}}$$

$$Vo = -\frac{V_1}{sC_f R_1} \tag{6.6.1}$$

In the time domain the expression of the output voltage is given by,

$$Vo = \frac{-1}{C_f R_1}\int V_1 dt \tag{6.6.2}$$

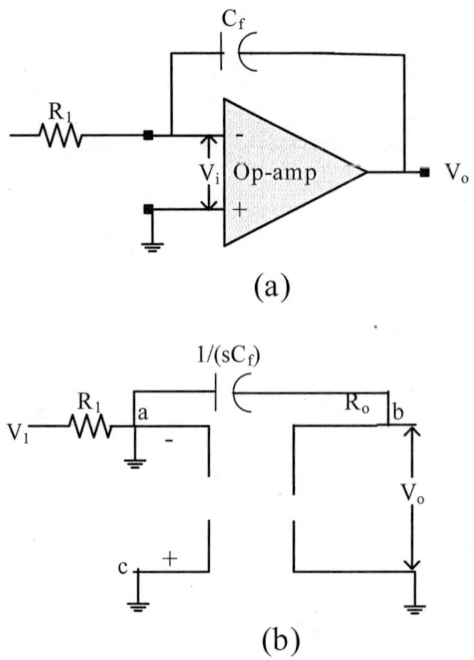

Fig. 6.6.1 Integrator circuit using op-amp (a) circuit schematic, (b) ideal equivalent circuit schematic.

6.7 DIFFERENTIATOR CIRCUIT

The circuit configuration of the inverting amplifier is shown in Fig. 6.7.1. It consists of a capacitor C_1 and the feedback resistor R_f connected at the inverting terminal as shown in Fig. 6.7.1. The other end of the capacitor C_1 and the feedback resistor R_f is connected to input signal and output respectively. The input signal is applied at the inverting terminal via capacitor C_1 with non-inverting terminal grounded as shown Fig. 6.7.1. The output of the op-amp is differentiated input signal which is 180° out of phase with input with amplification defined by the capacitor C_1 and the feedback resistor R_f value. The amplification in the input signal can be computed using the KCL at node 'a' as can be seen in Fig. 6.7.1. Using the concept of virtual ground or equating the potential both the input terminals and applying KCL at node 'a' we have,

$$\frac{-V_1}{\frac{1}{sC_1}}+\frac{0-Vo}{R_f}=0$$

$$\frac{V_1}{\frac{1}{sC_1}}=\frac{-Vo}{R_f}$$

$$Vo=-V_1(sC_1R_f) \tag{6.7.1}$$

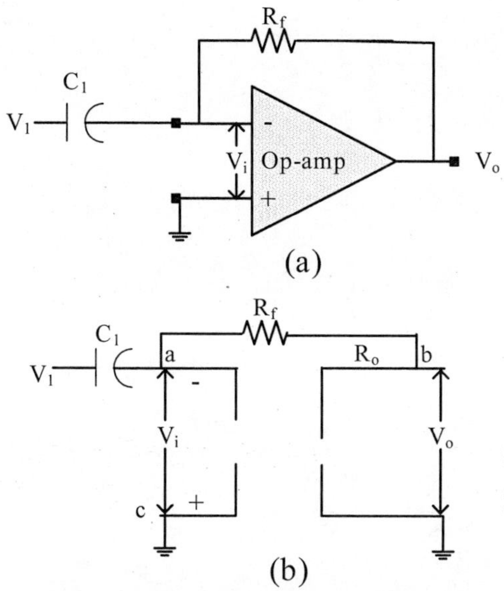

Fig. 6.7.1 Differentiator circuit using op-amp (a) circuit schematic, (b) ideal equivalent circuit schematic.

In time domain the expression of the output voltage is given by,

$$Vo=-(C_1R_f)\frac{\partial V_1}{dt} \tag{6.7.2}$$

6.8 INSTRUMENTATION AMPLIFIER

The instrumentation amplifier circuit given in Fig. 6.8.1 is typically used in the multimeter for the measurement of the voltage between two points. Using the op-amp equivalent circuit and applying KCL we have node co,

$$V_{o2}=2V_v$$

Applying KCL we have node a2,

$$\frac{V_2-V_{o2}}{R_1}+\frac{V_2-V_1}{R_a}=0$$

$$\frac{V_2}{R_1}+\frac{V_2-V_1}{R_a}=\frac{V_{o2}}{R_1} \tag{6.8.1}$$

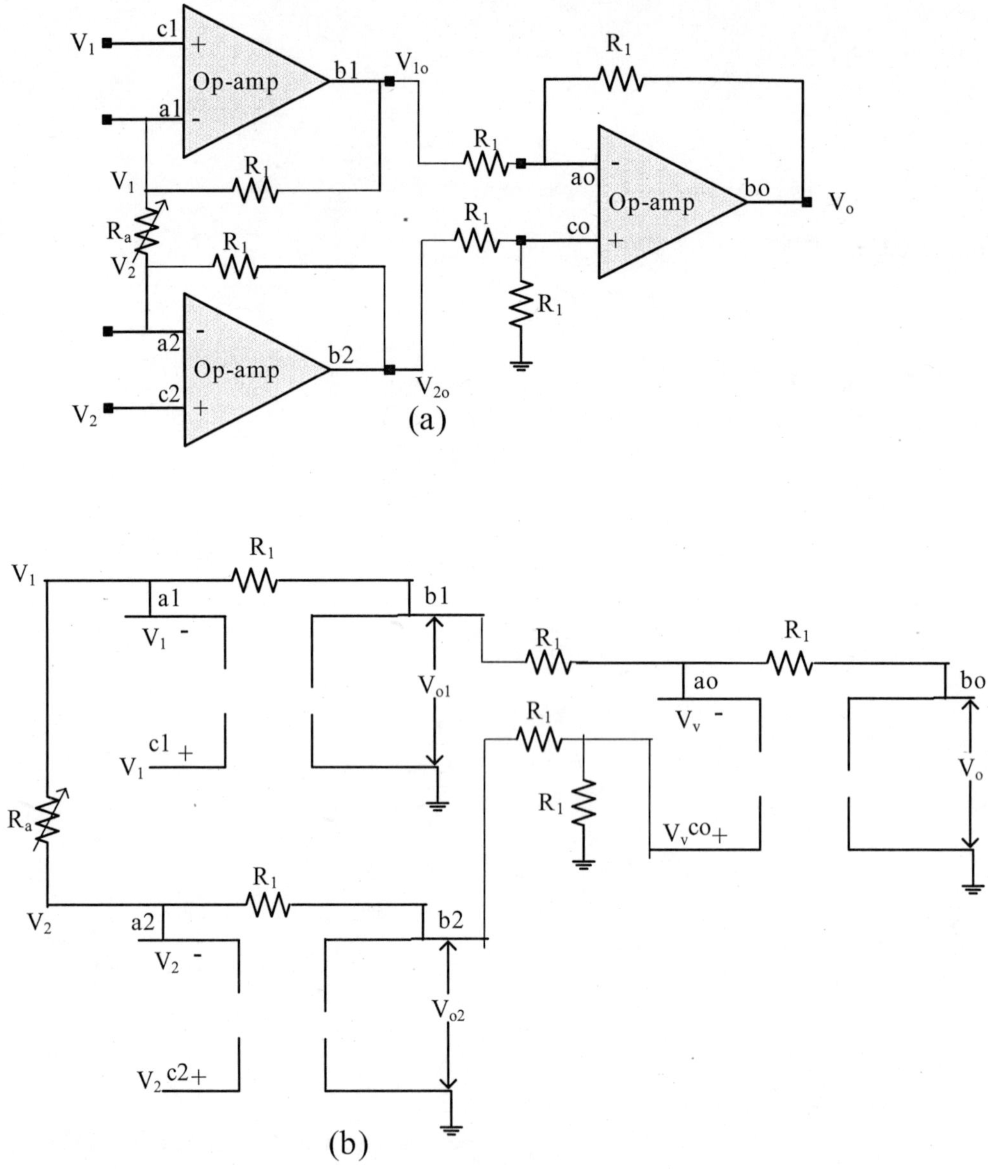

Fig. 6.8.1 Instrumentation amplifier circuit using op-amp (a) circuit schematic, (b) ideal equivalent circuit schematic.

Applying KCL we have node a1,

$$\frac{V_1 - V_{o1}}{R_1} + \frac{V_1 - V_2}{R_a} = 0$$

$$\frac{V_1}{R_1} + \frac{V_1 - V_2}{R_a} = \frac{V_{o1}}{R_1} \qquad (6.8.2)$$

Applying KCL we have node ao,

$$\frac{V_v - V_{o1}}{R_1} + \frac{V_v - V_o}{R_1} = 0$$

$$2V_v - V_{o1} = V_o$$

$$\frac{V_{o2}}{R_1} - \frac{V_{o1}}{R_1} = \frac{V_o}{R_1} \tag{6.8.3}$$

Substituting V_{o1} and V_{o2} we have,

$$\frac{V_2}{R_1} + \frac{V_2 - V_1}{R_a} - \left(\frac{V_1}{R_1} + \frac{V_1 - V_2}{R_a}\right) = \frac{V_o}{R_1}$$

$$\frac{V_2 - V_1}{R_1} + \frac{(2(V_2 - V_1)}{R_a} = \frac{V_o}{R_1}$$

$$V_o = (V_2 - V_1)(1 + \frac{2R_1}{R_a}) \tag{6.8.4}$$

Thus, output voltage V_o is proportional to difference of voltage between input terminals. Or,

$$V_o = K(V_2 - V_1) \text{ where } K=(1 + \frac{2R_1}{R_a})$$

6.9 OP-AMP AS LOW PASS FILTER

The first order low pass filter using op-amp is given in Fig. 6.9.1. The input signal is passed through the low pass RC filter. The RC low pass filter passes the signal as it is with frequency less than the cut-OFF frequency. While the signal with frequency greater than cut-OFF frequency are attenuated with -20db/decade as shown in Fig. 6.9.2. The signal 'V_{1f}' amplitude at the non-inverting terminal and cut-OFF frequency of the low pass filter can be computed as,

$$V_{1f} = \frac{V_1(\frac{1}{sC})}{\frac{1}{sC} + R}$$

$$V_{1f} = \frac{V_1}{1 + sCR} \tag{6.9.1}$$

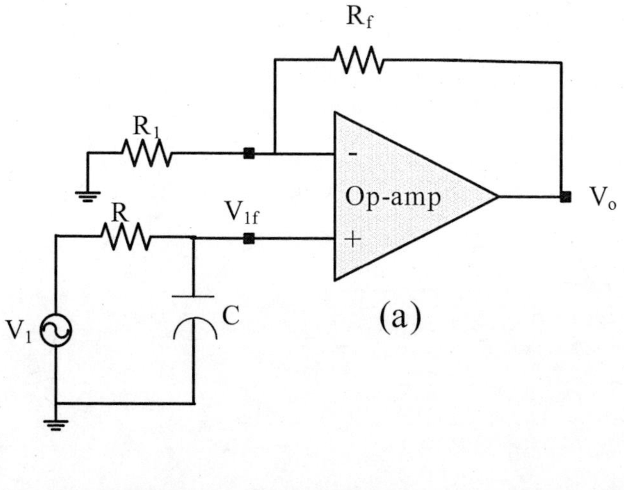

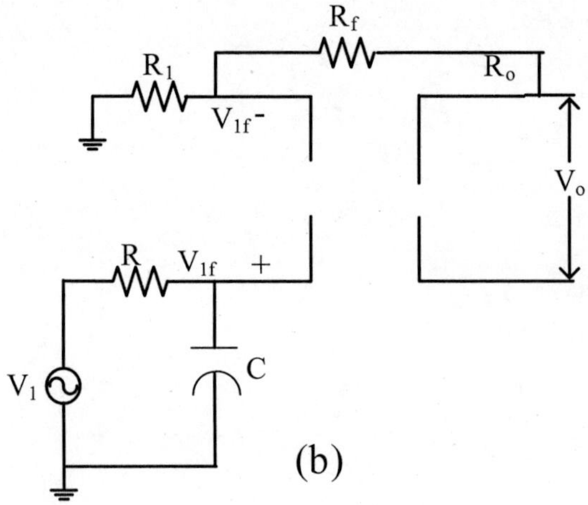

Fig. 6.9.1 Active low pass filter circuit using op-amp (a) circuit schematic, (b) ideal equivalent circuit schematic.

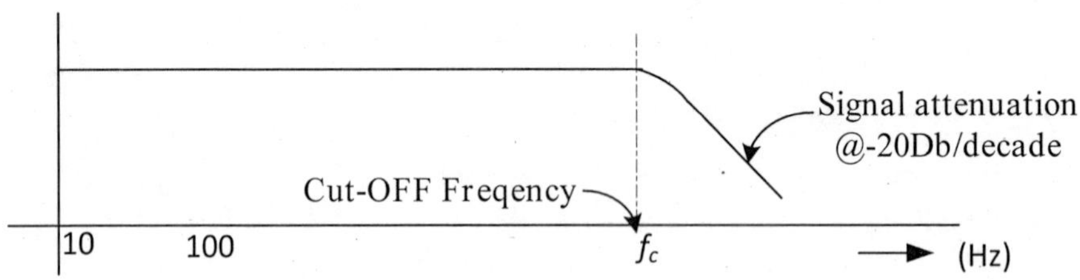

Fig. 6.9.2 Plot of frequency versus gain for low pass filter circuit.

The cut-OFF frequency f_c is computed by,

$$1 + sCR = 0$$

$$s = \left| \frac{-1}{CR} \right|$$

$$f_c = \frac{1}{CR} \tag{6.9.2}$$

Magnitude of the is defined by the,

$$V_{1f} = \frac{V_1}{1 + sCR}$$

$$V_{1f} = \frac{V_1}{\sqrt{1 + (sCR)^2}}$$

$$V_{1f} = \frac{V_1}{\sqrt{1 + \left(\frac{s}{\frac{1}{CR}}\right)^2}} = \frac{V_1}{\sqrt{1 + \left(\frac{f}{f_c}\right)^2}} \tag{6.9.3}$$

For input signal frequency f much less than f_c ($f<<f_c$), the non-inverting terminal voltage V_{1f} is given by,

$$V_{1f} = \left| \frac{V_1}{\sqrt{1+(\frac{f}{f_c})^2}} \right|$$

$$\frac{f}{f_c} << 1$$

$$V_{1f} = |V_1| \tag{6.9.4}$$

Thus, the non-inverting terminal voltage V_{1f} remains same as input signal V_1 for the frequency of the signal. For input signal frequency f much greater than f_c ($f>>f_c$), the non-inverting terminal voltage V_{1f} is given by,

$$V_{1f} = \left| \frac{V_1}{\sqrt{1+(\frac{f}{f_c})^2}} \right|$$

$$\frac{f}{f_c} >> 1$$

$$V_{1f} = \left| \frac{V_1}{\frac{f}{f_c}} \right| = \left| \frac{f_c V_1}{f} \right| \tag{6.9.5}$$

Thus, the non-inverting terminal voltage V_{1f} attenuates for the frequency of the signal greater than cut-OFF frequency with -20db/decade . The signal at non-inverting terminal voltage V_{1f} get amplified by the gain given by,

$$\frac{V_1}{R_1} + \frac{V_1 - Vo}{R_f} = 0$$

$$V_1(\frac{1}{R_1} + \frac{1}{R_f}) = \frac{Vo}{R_f}$$

$$\frac{Vo}{V_1} = (1 + \frac{R_f}{R_1}) \tag{6.9.6}$$

6.10 OP-AMP AS HIGH PASS FILTER

The first order high pass filter using op-amp is given in Fig. 6.10.1. The input signal is passed through the high pass RC filter. The RC high pass filter passes the signal as it is with frequency greater than the cut-OFF frequency. While the signal with frequency less than cut-OFF frequency are attenuated with -20db/decade as shown in Fig. 6.10.2. The signal 'V_{1f}' amplitude at the non-inverting terminal and cut-OFF frequency of the low pass filter can be computed as,

$$V_{1f} = \frac{V_1(R)}{\frac{1}{sC} + R}$$

$$V_{1f} = \frac{sCRV_1}{1 + sCR} \qquad (6.10.1)$$

The cut-off frequency f_c is computed by,

$$1 + sCR = 0$$

$$s = \left|\frac{-1}{CR}\right|.$$

$$f_c = \frac{1}{2\pi CR} \qquad (6.10.2)$$

Magnitude of the is defined by the,

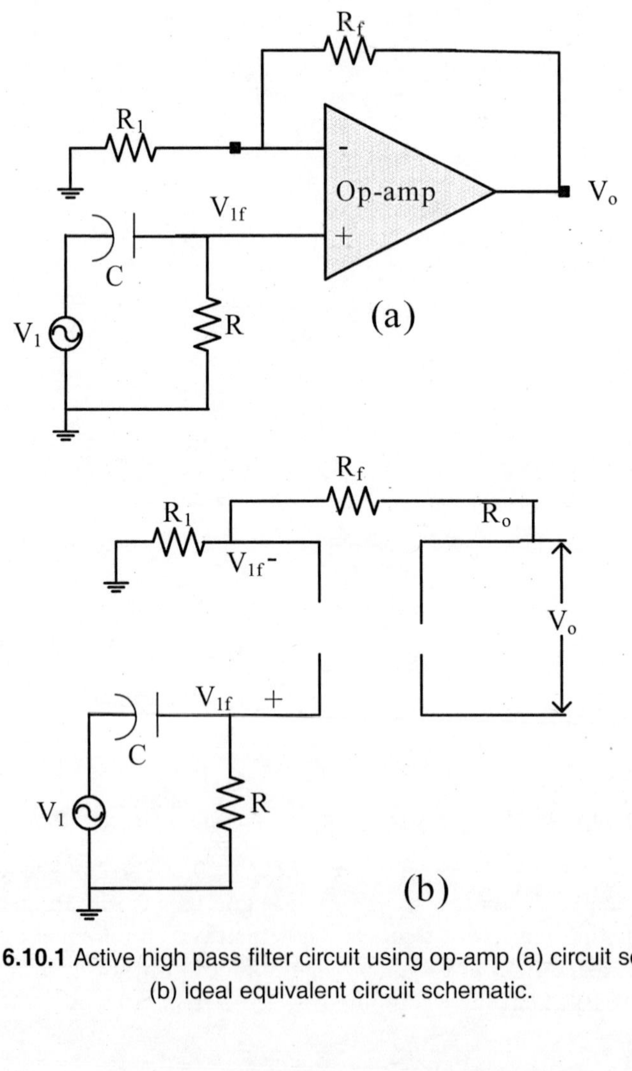

Fig. 6.10.1 Active high pass filter circuit using op-amp (a) circuit schematic, (b) ideal equivalent circuit schematic.

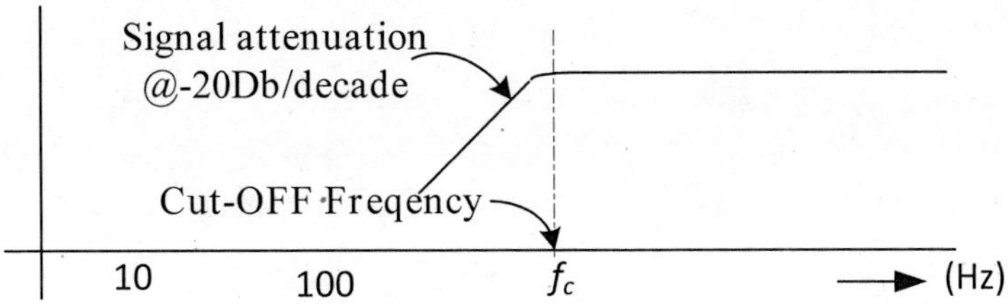

Fig. 6.10.2 Plot of frequency versus gain for high pass filter circuit.

$$V_{1f} = \frac{sCRV_1}{1+sCR}$$

$$V_{1f} = \frac{|V_1|}{\sqrt{\frac{1}{(sCR)^2}+1}}$$

$$V_{1f} = \frac{V_1}{\sqrt{1+(\frac{CR}{s})^2}} = \frac{V_1}{\sqrt{1+(\frac{f_c}{f})^2}} \tag{6.10.3}$$

For input signal frequency f much less than f_c ($f<<f_c$), the non-inverting terminal voltage V_{lf} is given by,

$$V_{1f} = \left|\frac{V_1}{\sqrt{1+(\frac{f_c}{f})^2}}\right| \tag{6.10.6}$$

$$\frac{f}{f_c} << 1$$

$$V_{1f} = \left|\frac{V_1}{\frac{f_c}{f}}\right| = \left|\frac{fV_1}{f_c}\right| \tag{6.10.4}$$

Thus, the non-inverting terminal voltage V_{lf} attenuates for the frequency of the signal greater than cut-OFF frequency with -20db/decade . For input signal frequency f much greater than f_c ($f>>f_c$), the non-inverting terminal voltage V_{lf} is given by,

$$V_{1f} = \left|\frac{V_1}{\sqrt{1+(\frac{f_c}{f})^2}}\right|$$

$$\frac{f}{f_c} >> 1$$

$$V_{1f} = \left|\frac{V_1}{1}\right| = |V_1| \tag{6.10.5}$$

Thus, the non-inverting terminal voltage V_{If} remains same as input signal V_1 for the frequency of the signal. The signal at non-inverting terminal voltage V_{If} get amplified by the gain given by,

$$\frac{V_1}{R_1} + \frac{V_1 - Vo}{R_f} = 0$$

$$V_1\left(\frac{1}{R_1} + \frac{1}{R_f}\right) = \frac{Vo}{R_f}$$

$$\frac{Vo}{V_1} = \left(1 + \frac{R_f}{R_1}\right)$$

6.11 OP-AMP AS BAND PASS FILTER

It is combination of low pass and high pass filter. The output of the high pass filter becomes the input for the low pass filter. The high pass filter is cascaded with low pass filter. The lower cut OFF frequency of given band pass filter is defined by the high pass filter. While the higher cut-OFF frequency of given band pass filter is defined by the low pass filter. The same is depicted in Fig. 6.11.1. Thus, the cut-OFF frequency of the high pass filter is lower than the cut-OFF frequency of the low pass filter. The design of the band pass filter is actually the design of design of high and low pass filter with cut-OFF frequency f_{cH} and f_{cL} respectively. The required gain in the input signal can be of adjusted by proper design of input and feedback resistor in the inverting terminal. The same is depicted in Fig. 6.11.2. The design for RC components for low and high pass filter is given by,

$$f_{cH} = \frac{1}{2\pi C_H R_H}$$

$$f_{cL} = \frac{1}{2\pi C_L R_L} \tag{6.11.1}$$

The over-all gain A of the band-pass filter with cascaded op-amp is given by,

$$A = A_H A_L \tag{6.11.2}$$

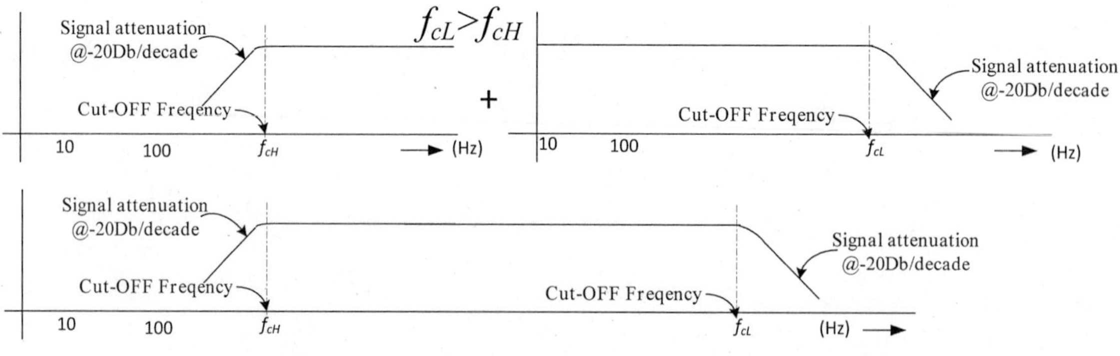

Fig. 6.11.1 Plot of frequency versus gain for band pass filter circuit.

Where, A_H and A_L are the defined gains of the high and low pass filter respectively. The chosen design values of the input and feedback resistors at inverting terminal of the op-amp for low (R_L,R_{fL}) and high(R_H,R_{fH}) pass filter are computed considering,

$$A_H = (1 + \frac{R_{fH}}{R_{1H}})$$

$$A_L = (1 + \frac{R_{fL}}{R_{1L}}) \tag{6.11.3}$$

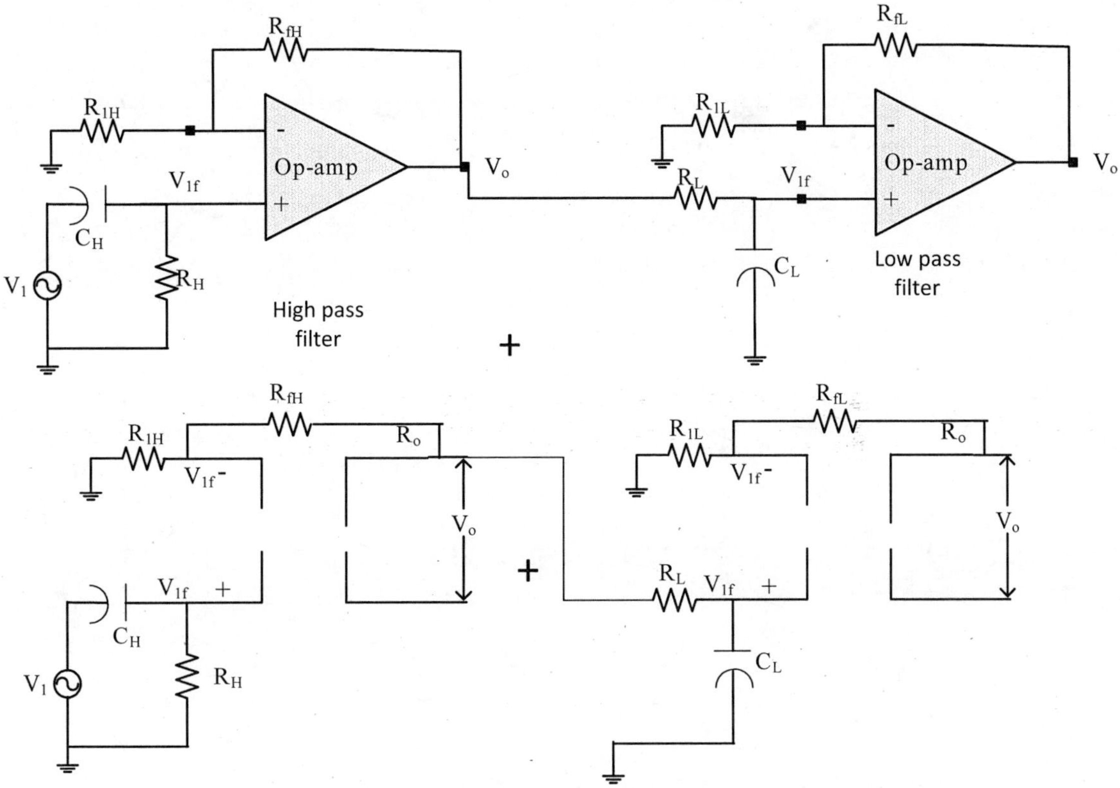

Fig. 6.11.2 Active band pass filter circuit using op-amp (a) circuit schematic, (b) ideal equivalent circuit schematic.

6.12 PROPORTIONAL CONTROLLER

The proportional controller gives output which proportional to its input signal. The op-amp having input at inverting terminal gives the output which proportional to its input. However, obtained output is 180 degrees out of phase with the input. The obtained output when again given inverting amplifier with unity gain forms the proportional controller as shown in Fig. 6.12.1. The required proportional gain 'k' can be obtained using proper design value of input and feedback resistors at inverting terminal of the op-amp and is given by,

$$k = \left| \frac{R_f}{R_1} \right| \tag{6.12.1}$$

The output of the P controller circuit is given by,

$$Vo = kV_1 \tag{6.12.2}$$

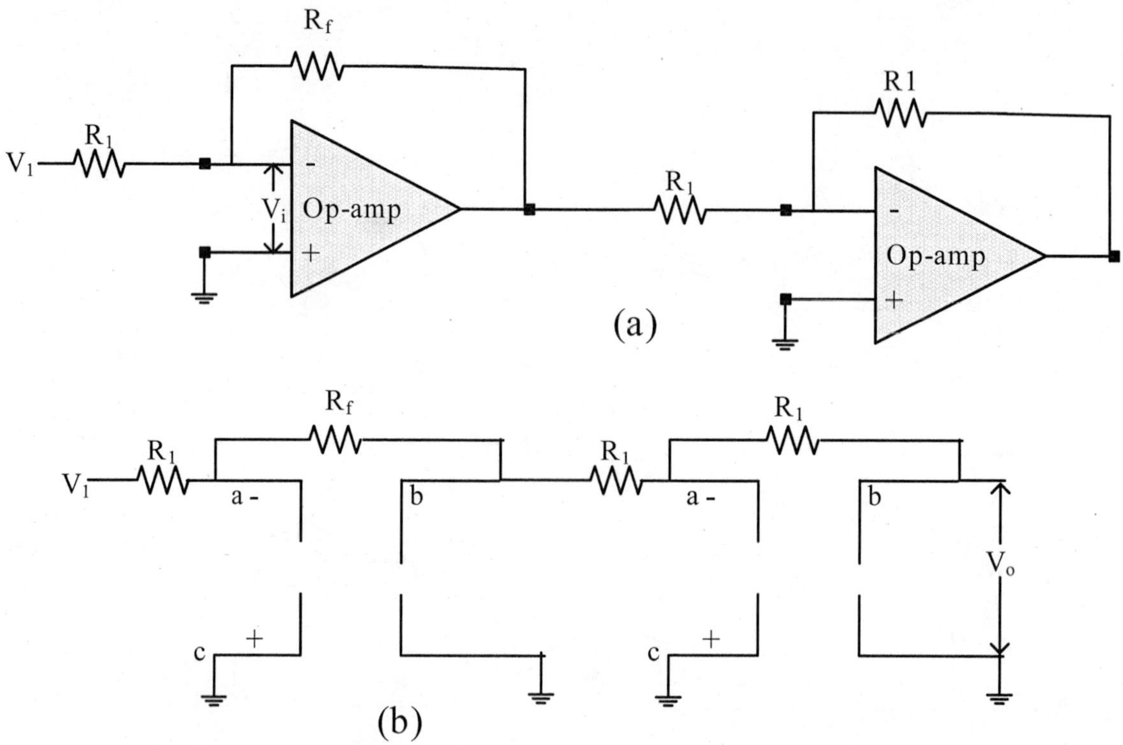

Fig. 6.12.1 Proportional controller circuit using op-amp (a) circuit schematic, (b) ideal equivalent circuit schematic.

6.13 PROPORTIONAL INTEGRAL CONTROLLER

The proportional integral (PI) controller gives output which is sum of both proportional and integral gain of its input signal. The op-amp having input at inverting terminal gives the output which proportional to its input. The same input signal is also given to integrator circuit of the op-amp for its integral output. The obtained two outputs are then given to summer circuit with unity gain (R_1=R_f) for the required PI output as shown in Fig. 6.13.1. It can be noted that input passes through two inverting stages, thus the phase shift of 180° is nullified.

The design of op-amp circuit with given proportional gain 'k' is same as discussed above. For the design of op-amp circuit with given integral gain 'k_i' is given by,

$$k_i = \left| \frac{1}{C_f R_1} \right| \quad (6.13.1)$$

The output of the PI controller circuit is given by,

$$Vo = kV_1 + K_i \int V_1 dt \quad (6.13.2)$$

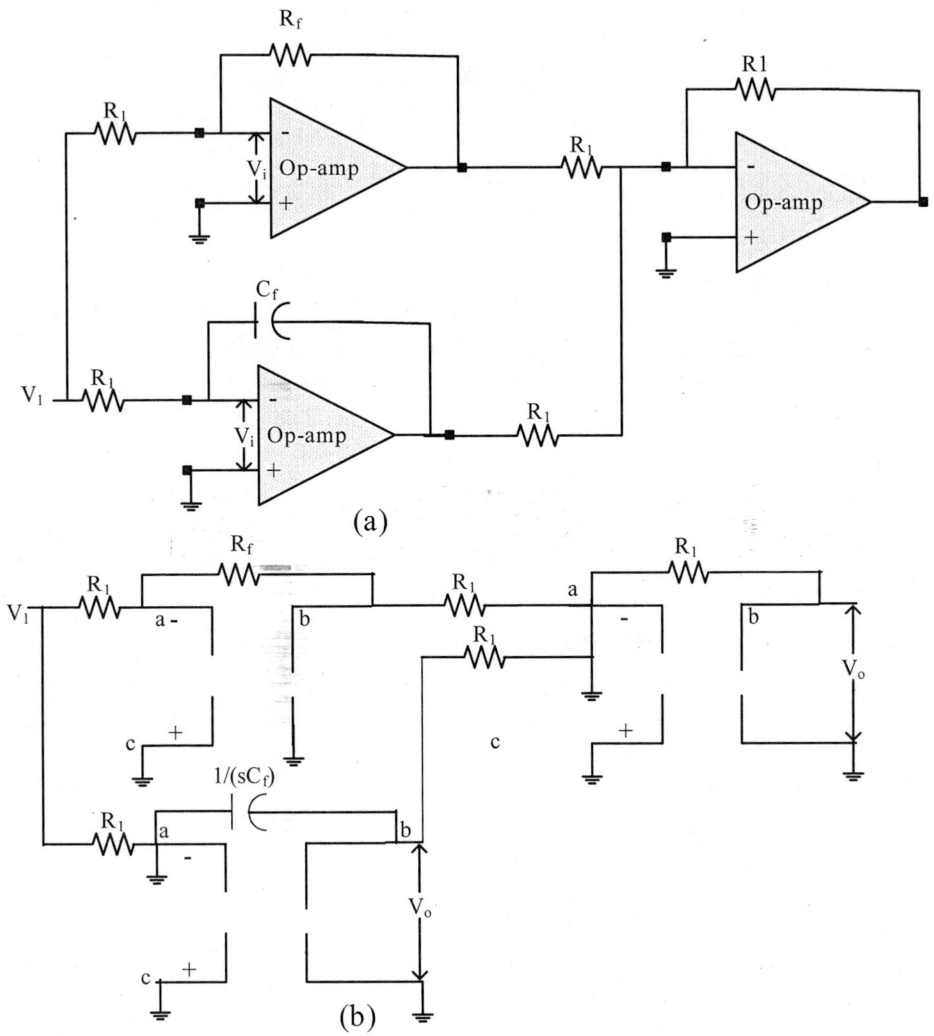

Fig. 6.13.1 Proportional integral controller circuit using op-amp (a) circuit schematic, (b) ideal equivalent circuit schematic.

6.14 PROPORTIONAL INTEGRAL DERIVATIVE CONTROLLER

Proportional Integral Derivative (PID) controller is the extension of proportional integral (PI) controller where a derivative block is added. The PID controller gives output which is sum of proportional, integral and derivative of its input signal. The op-amp having input at inverting terminal gives the output which proportional to its input. The same input signal is also given to integrator and derivative circuit of the op-amp for its integral and derivative output. The obtained three outputs are then given to summer circuit with unity gain (R_1=R_f) for the required PID output as shown in Fig. 6.14.1. It can be noted that input passes through two inverting stages, thus the phase shift of 180° is nullified. The design of op-amp circuit with given proportional and integral gain 'k' and 'k_i' respectively is same as discussed above. For the design of op-amp circuit with given derivative gain 'k_D' is given by,

$$k_D = |C_1 R_f| \tag{6.14.1}$$

The output of the PI controller circuit is given by,

$$Vo = kV_1 + k_i \int V_1 dt + k_D \frac{\partial V_1}{dt} \tag{6.14.2}$$

6.15 ANALOG TO DIGITAL CONVERTER

Before starting the analog to digital circuit using op-amp let us understand the ladder network and its usage in digital to analog conversion. Ladder network is the arrangement of R and 2R resistors in a specific manner as shown in Fig. 6.15.1. The given arrangement discretizes the given voltage 'V_{ref}' into different levels based on number of digits in the binary output. For example in case of 4 bit binary number the voltage level is divided into 2^4 (16) discrete levels. Thus, each discrete voltage level corresponds to,

$$V_{dis\_n} = \frac{V_{ref}}{2^n}$$

$$V_{dis\_4} = \frac{V_{ref}}{2^4} \tag{6.15.1}$$

Where, V_{dis_n} is minimum resolution voltage corresponding to a bit value. The voltage corresponding to the 1101 is given by,

$$V_o = \frac{D_o \times 2^0 + D_1 \times 2^1 + D_2 \times 2^2 + D_3 \times 2^3}{V_{ref}} \times V_{ref}$$

$$V_o = \frac{1\times 2^0 + 0\times 2^1 + 1\times 2^2 + 1\times 2^3}{V_{ref}} \times V_{ref} = 13V \tag{6.15.2}$$

The ladder network converts the binary input to its analogue output equivalent voltage. A simple 4-bit binary number was considered. However, same is true for any n-bit binary number. As the number of bits increases the voltage resolution of binary bit decreases or accuracy increases. Thus, a ladder network is typically used for digital to analogue conversion. The same ladder network is also employed in analogue to digital converter (ADC) for converting input analogue voltage to its equivalent digital value. The ADC IC typically consists ladder network, op-amp as a comparator, clock, control logic and

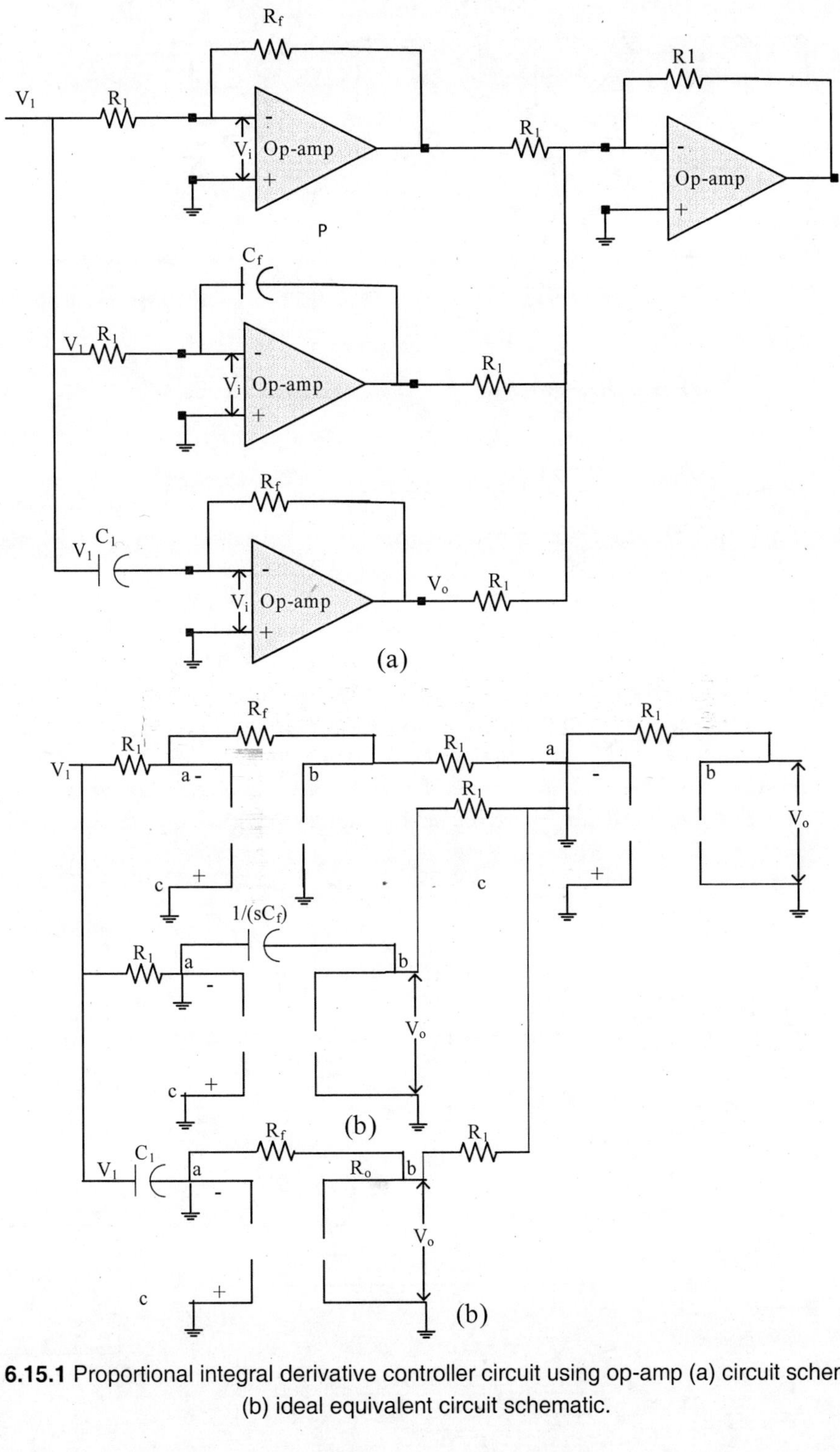

Fig. 6.15.1 Proportional integral derivative controller circuit using op-amp (a) circuit schematic, (b) ideal equivalent circuit schematic.

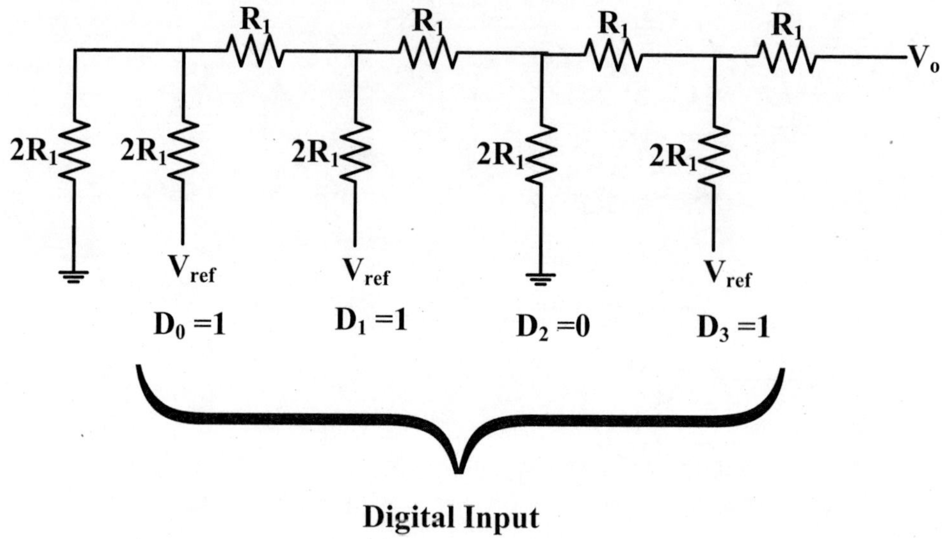

Fig. 6.15.2 Ladder network using resistor network for converting digital value to its analog equivalent.

a digital counter as shown in Fig. 6.15.3. The comparator output is given to the control logic which controls the digital counter output. The digital counter output is given to the ladder circuit which gives the analog voltage output V_L corresponding to the digital input from the counter. The output voltage of the ladder circuit is given to the positive input of comparator op-amp. The output voltage of the ladder circuit increases in steps as shown in Fig. 6.15.4 with the increment in the counter value. When the ladder voltage becomes equal to or greater than the analog input voltage, the comparator output becomes high and control logic stops the digital counter. The obtained digital value from the digital counter corresponds to the digital equivalent of the given analog input.

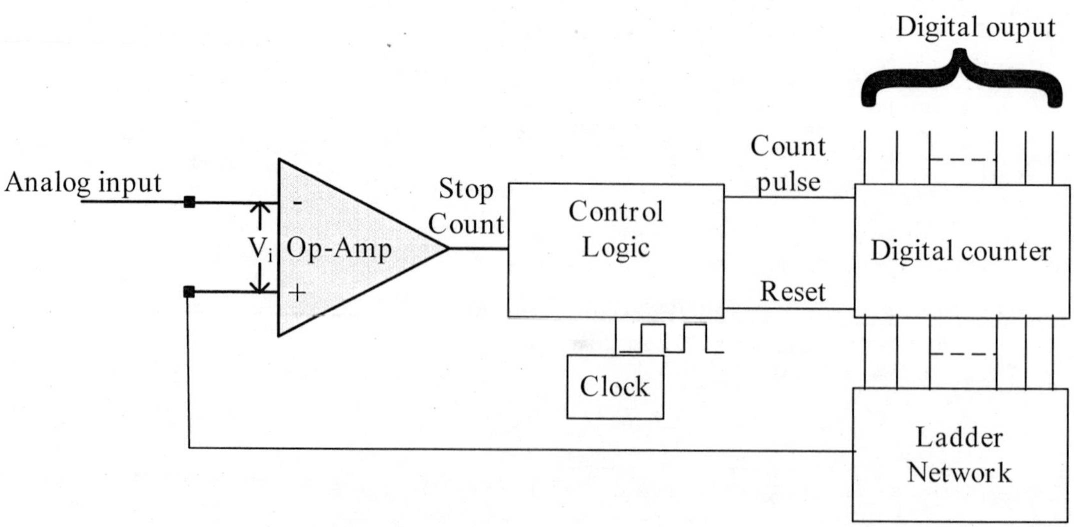

Fig. 6.15.3 Typical ADC IC using op-amp.

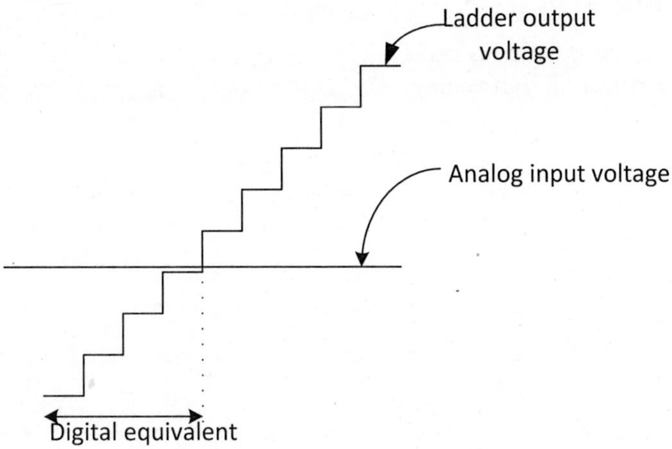

Fig. 6.15.4 Ladder output voltage corresponding to digital input from counter.

6.16 VOLTAGE REGULATOR

The voltage regulator circuit is employed for the regulation of the voltage across load. It consists of transistor which operates in active region apart from zener diode and op-amp. The output regulated with the help of BJT operating in active region maintaining the required voltage drop across its emitter collector terminal. The employed zener diode with break down voltage of V_Z. regulates the output voltage to the value as derived. The circuit diagram is shown in Fig. 6.16.1. Applying virtual ground concept,

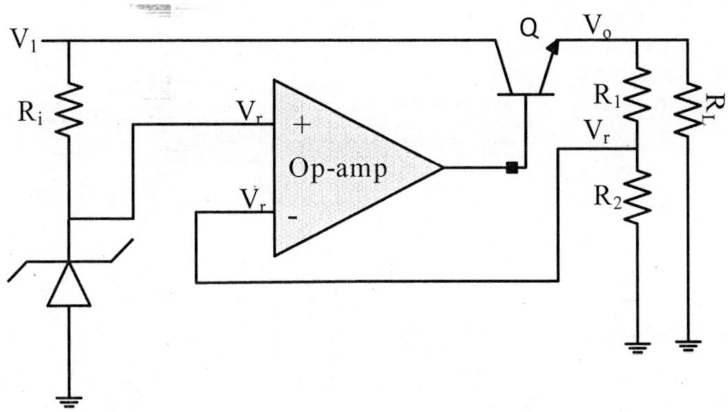

Fig. 6.16.1 Voltage regulator circuit using op-amp.

$$V_z = V_r$$

$$V_r = \frac{V_o R_2}{R_2 + R_1}$$

$$V_z = \frac{V_o R_2}{R_2 + R_1}$$

$$V_o = \frac{V_z (R_2 + R_1)}{R_2} \qquad (6.16.1)$$

6.17 Phase Shift Oscillator

The positive feedback amplifier circuits satisfying phase conditions and having the gain greater than unity forms an oscillator. If the output voltage varies in sinusoidal manner forms a sinusoidal oscillator. If the output voltage has step change in magnitude from positive to negative or vice vérsa forming square wave, it represents square wave oscillator. A simple circuit of phase shift oscillator using op-amp is shown in Fig. 6.17.1. The inverting input of the op-amp gets its input from the three stage resistor capacitor RC network as shown in Fig. 6.17.1. The RC network is so chosen that the phase shift of 180°

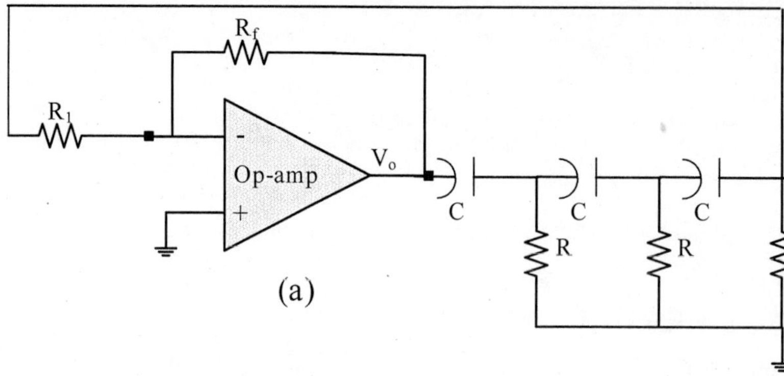

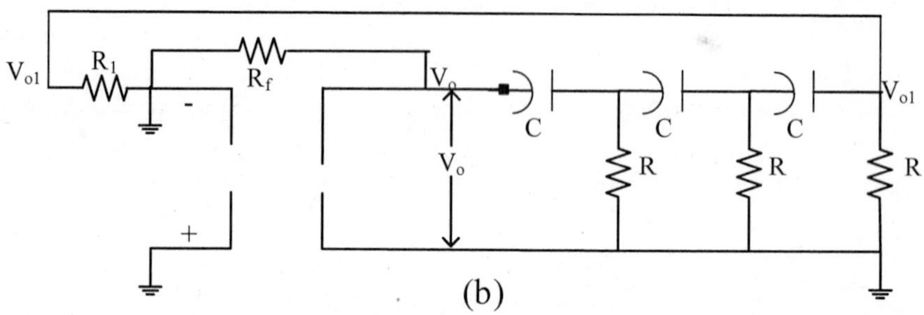

Fig. 6.17.1 Phase shift oscillator circuit using op-amp (a) circuit schematic, (b) ideal equivalent circuit schematic.

is added to op-amp output voltage, which is again to the inverting input as shown in Fig. 6.17.1. The attenuation of the op-amp output voltage due three stage RC network by 1/29 is taken care by having the ration of R_f/R_1 greater than 29. Thus, the gain of the feedback is maintained greater than unity forming forms an phase shift oscillator. The frequency and the attenuation of the op-amp output voltage in the oscillatory circuit is computed as,

$$f = \frac{1}{2\pi RC\sqrt{6}}$$

$$A = \frac{1}{29} \tag{6.17.1}$$

The op-amp gain is given by,

$$A_v = \frac{R_f}{R_1} \tag{6.17.2}$$

Where R_f and R_1 are so chosen that the op-amp gain A_v is greater than 29 to maintain over-all gain greater than unity.

6.18 WIEN BRIDGE OSCILLATOR

Wien bridge oscillator is most practically employed op-amp based oscillator circuit. The circuit uses resistor and capacitor components in series and parallel combination to form oscillator network. The oscillator network is connected to op-amp output via bridge node 'a'. The other nodes of the bridge circuit 'b' and 'd' are connected to inverting and non-inverting terminal of the op-amp respectively. The left node 'c' of the wien bridge circuit is grounded as shown in Fig. 5.18.1. The resistors R_1 and R_f defines the gain of the op-amp circuit which is given by,

$$A_v = \frac{R_f}{R_1}$$

$$f = \frac{1}{2\pi\sqrt{R_2C_2R_1C_1}} \tag{6.18.1}$$

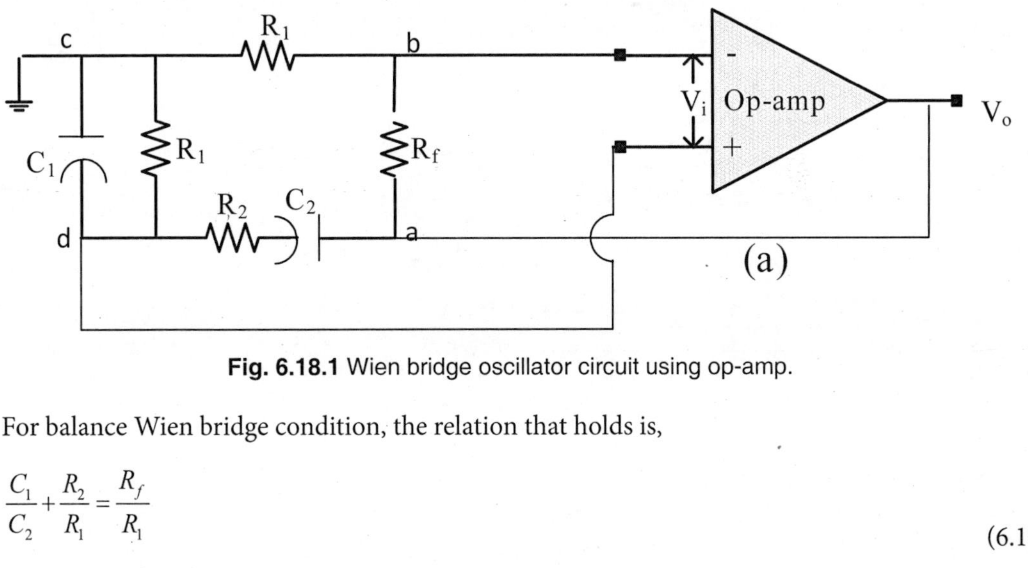

Fig. 6.18.1 Wien bridge oscillator circuit using op-amp.

For balance Wien bridge condition, the relation that holds is,

$$\frac{C_1}{C_2} + \frac{R_2}{R_1} = \frac{R_f}{R_1} \tag{6.18.2}$$

6.19 LAG LEAD COMPENSATOR

Lag lead compensators are typically employed for the adjustment of the phase delay or lead in the voltage waveform. The circuit for the lead–lag compensator is shown in Fig. 6.19.1. It consists of parallel combination of resistor and capacitor components at input inverting terminal and feedback loop as shown in Fig. 6.19.1. The values of resistor and capacitor components are so chosen to obtain the

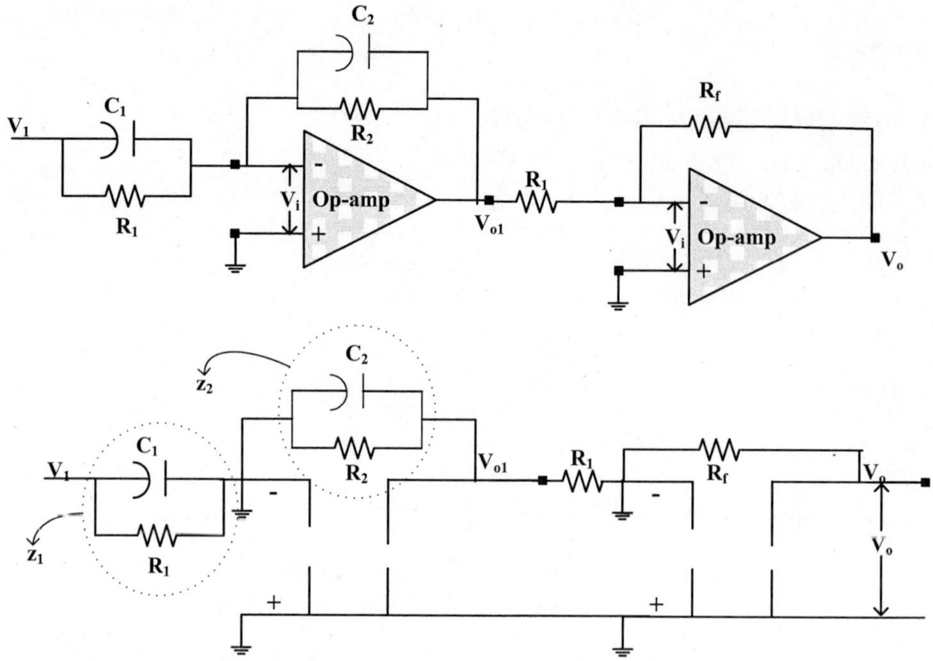

Fig. 6.19.1 Lag lead compensator circuit using op-amp (a) circuit schematic, (b) ideal equivalent circuit schematic.

required lead or lag in the input voltage waveform. The equivalent impedance for resistor and capacitor components at inverting terminal and feedback loop is given by,

$$\frac{1}{z_1} = y_1 = \frac{1}{R_1} + sC_1 = \frac{1+sC_1R_1}{R_1} = \frac{C_1(\frac{1}{C_1R_1}+s)}{1}$$

$$\frac{1}{z_2} = y_2 = \frac{1}{R_2} + sC_2 = \frac{1+sC_2R_2}{R_2} = \frac{C_2(\frac{1}{C_2R_2}+s)}{1} \quad (6.19.1)$$

The voltage gain of the op-amp is given by,

$$\frac{0-V_1}{z_1} + \frac{0-V_{o1}}{z_2} = 0$$

$$\frac{V_{o1}}{V_1} = \frac{-z_1}{z_2} = \frac{-C_2(\frac{1}{C_2R_2}+s)}{C_1(\frac{1}{C_1R_1}+s)}$$

$$\frac{V_{o1}}{V_1} = \frac{-k_a(a+s)}{(b+s)}$$

$$k_a = \frac{C_2}{C_1}; a = \frac{1}{C_2R_2}; b = \frac{1}{C_1R_1} \quad (6.19.2)$$

Thus, by adjusting the values of resistor and capacitor components the phase voltage of the output can be adjusted with respect to input voltage. In other words, by adjusting the values of the pole ('b') and zeros ('a') the phase of the output voltage can be made to lag or lead the input voltage. For lag compensator the value of 'a' should be greater than 'b' and for lead the value of 'a' should be lesser than 'b'. The gain or attenuation due value of k_a can be adjusted with the R_1 and R_f in the cascading op-amp apart from 180° phase displacement.

$$k_a = (\frac{R_f}{R_1})^{-1} \tag{6.19.3}$$

UNIT SUMMARY

The chapter presents the application of the operational amplifier for various signals conditioning circuits. The chapter gives the details for analysis of the basic inverting and the non-inverting operational amplifiers which represents the fundamental circuits for the op-amp. Then, various important signal conditioning circuits like adder, integrator, differentiator, filter circuits, analogue to digital converter, instrumentation amplifier circuit etc are discussed in detail. Using the given concept, the student will not only learn how to analyze the op-amps circuits but can also design the new op-amps circuit for given applications. Further, using the given approaches for analysis of op-amps circuit the student will be able investigate and derive the output expressions for new op-amps circuits. Thus, the chapter gives complete details for different linear applications of the op-amps circuits.

EXERCISES

Multiple Choice Questions

6.1 Ideal op-amps has

a. Infinite open loop gain of the op-amp circuit in open loop is.

b. Infinite input resistance Ri

c. Zero output resistance Ro

d. All the above

6.2 The value of resistance between input terminals of op-amps is

a. Zero

b. 1kΩForward bias diode

c. 10kΩ Breakdown diode

d. 100MΩ

6.3 Give the output of the below op-amps circuit. Assume

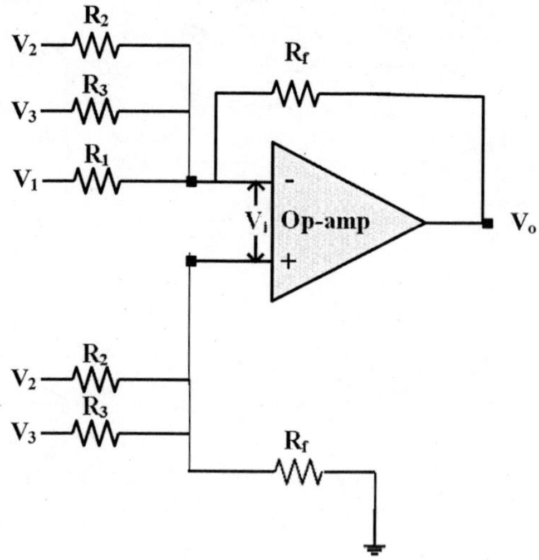

a. $V_1+V_2+V_3$

b. $V_1+2V_2+2V_3$

c. V_1

d. None of above

6.4 Given below circuit of op-amps circuit is for

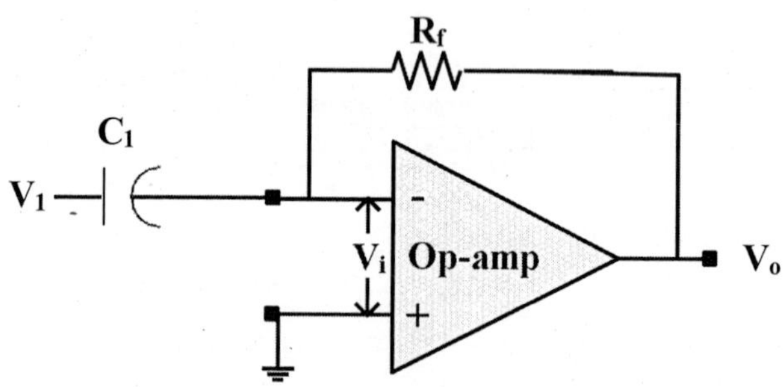

a. differentiator, high pass filter

b. high pass filter

c. differentiator

d. none of the above

Answers for Multiple Choice Questions			
6.1 **d**	6.2 **d**	6.3 **c**	6.4 **a**

Short Answer Type Questions

6.1 Compare properties of ideal and practical op-amps?

6.2 Give the situation for the usage of voltage follower op-amps circuit in a system?

6.3 Explain the working of phase shift oscillator?

6.4 Design an circuit using op-amps which has input of voltages V_1,V_2 and V_3. Gives the output $V_1+V_2-V_3$?

Long Answer Type Questions

6.1 Design a low pass filter with cut-off frequency of 10kHz and gain of 11?

6.2 Design a 8 bit ADC. If input voltage is of 5V, find voltage resolution of the designed ADC?

6.3 Design band pass filter which passes the signals between the frequency range of 30Hz to 100Hz?

6.4 Design an Op-Amp circuit having triangular wave output using input of square wave of 5kHz?

KNOW MORE

For Additional Info Scan Me

REFERENCES

1. *Ramakanth A. Gayakwad, "OP-amps and Linear Integrated Circuits", PHI, Fourth Edition, 2010.*
2. *Adel S. Sedra , Kenneth C. Smith , Arun N. Chandorkar , "Microelectronic Circuits: Theory And Applications", Oxford University Press, Seventh Edition, (1 June 2017).*

07 NONLINEAR APPLICATIONS OF OPERATIONAL AMPLIFIERS

UNIT SPECIFICS

Through this unit we have discussed the following aspects:

- *Introduction to nonlinear applications of operational amplifiers;*
- *Hysteresis comparators;*
- *Zero crossing detector;*
- *Square-wave and triangular-wave generators;*
- *Precision rectifier;*
- *Peak detector;*
- *Monostable multivibrator.*

RATIONALE

This fundamental unit discusses the nonlinear applications of the op-amp. Firstly, the operation of hysteresis comparators is discussed. This helps the students in designing the electronic circuits whose input signal is signal high noise or fluctuations. Next, the operation of the zero crossing detector is discussed. This helps students in using the zero crossing detector in applications where synchronization of AC signal is necessary. Next, the operation of square wave and triangular wave generators is discussed. The precision rectifiers, peak detectors and monostable multivibrators are also discussed in the unit. The laboratory experiments related execution of astable and monostable multivibrators, execution of precision rectifier and peak detector circuits, square waveform, and triangular waveform generators are presented in the appendix of the book.

PRE-REQUISITES

Nil

UNIT OUTCOMES

List of outcomes of this unit is as follows:

U7-O1: Understand the operation of the hysteresis comparators and zero crossing detectors

U7-O2: Understand the working of square wave and triangular wave generators

U7-O3: Understand the working of precision rectifiers

U7-O4: Understand the working of peak detector

U7-O5: Understand the working of monostable multivibrator

Unit-7 Outcomes	EXPECTED MAPPING WITH COURSE OUTCOMES (1- Weak Correlation; 2- Medium correlation; 3- Strong Correlation)				
	CO-1	CO-2	CO-3	CO-4	CO-5
U7-O1	3	3	1	-	3
U7-O2	3	3	1	-	3
U7-O3	3	3	1	-	3
U7-O4	3	2	1	-	3
U7-O5	3	2	1	-	3

7.1 INTRODUCTION

The linear behaviour of an op-amp is typically achieved by using negative feedback, where a portion of the output signal is fed back to the input in a way that stabilizes and controls the amplifier's overall gain and response. When op-amps are used with positive feedback or operated in an open-loop configuration (without feedback), their behaviour becomes nonlinear. In this mode, the op-amp amplifies the input signal without any stabilizing feedback, and the output can exhibit characteristics such as saturation, distortion, and oscillation. However, it is also possible to achieve non-linear behaviour while using negative feedback by incorporating additional components, such as diodes or bipolar junction transistors (BJTs), into the feedback network. These elements introduce non-linearities into the system, allowing for the creation of various specialized circuits and functions. Some examples of applications that leverage the non-linear behaviour of op-amps include hysteresis comparators, zero crossing detectors, Square-wave and triangular-wave generators, precision rectifiers, Peak detectors and monostable multivibrators. In summary, while negative feedback is primarily used to achieve linear behaviour in op-amp circuits, incorporating positive feedback or non-linear components like diodes and BJTs can enable the creation of specialized circuits with non-linear characteristics for various applications.

7.2 HYSTERESIS COMPARATORS

These circuits use positive feedback and exploit the op-amp's non-linear response to create a two-level output that toggles between high and low states based on the input voltage crossing predetermined thresholds (upper threshold limit and lower threshold limit). Hysteresis comparators, such as Schmitt triggers, are commonly used in applications requiring noise rejection. The hysteresis comparators can be of two types, as described below:

7.2.1 Noninverting Hysteresis Comparators

The circuit of the noninverting hysteresis comparator is shown in Fig. 7.2.1.1. The input signal V_S is applied to the noninverting terminal of the op-amp through a resistance R_1. The feedback resistance R_F is connected between the noninverting terminal and the output terminal of the op-amp. The inverting terminal of the op-amp is grounded.

The expression for the current I_S is given by

$$I_S = \frac{V_O}{R_F} \tag{7.2.1.1}$$

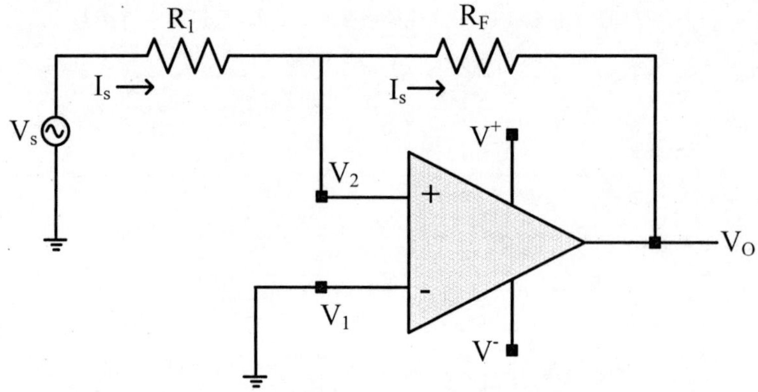

Fig. 7.2.1.1 Circuit of the noninverting hysteresis comparator.

The output voltage V_O can saturate at two values: the positive supply voltage V^+ and the negative supply voltage V^- based on the input voltage V_S crossing predetermined thresholds (upper threshold limit and lower threshold limit). Let us assume the positive supply voltage V^+ is equal $+V_{sat}$ and the negative supply voltage V^- is equal to $-V_{sat}$. The thresholds are computed with Vs=0. Now to calculate the upper threshold limit, the output voltage saturates at the positive supply voltage V^+. During this case, the upper threshold limit V_{UT} is given by the voltage across the resistor R_1 i.e.,

$$V_{UT} = I_S R_1 = \frac{R_1 V_{sat}}{R_F} \tag{7.2.1.2}$$

Now to calculate the lower threshold limit, the output voltage saturates at the negative supply voltage V^-. During this case, the lower threshold limit V_{LT} is given by the voltage across the resistor R_1 i.e.,

$$V_{LT} = I_S R_1 = -\frac{R_1 V_{sat}}{R_F} \tag{7.2.1.3}$$

The hysteresis range V_T is obtained by subtracting the upper and lower threshold limits. Therefore,

$$V_H = V_{UT} - V_{LT} = 2\frac{R_1 V_{sat}}{R_F} \tag{7.2.1.4}$$

The input and output waveforms of the non-inverting hysteresis comparator are shown in Fig. 7.2.1.2. Let the input voltage V_s be a sinusoidal waveform. The voltages V_{UT} and V_{LT} are the upper and lower threshold limits of the non-inverting hysteresis comparator. Whenever the magnitude of the input voltage V_s is greater than the upper threshold limit V_{UT}, the output voltage V_O saturates around the value $+V_{sat}$. The magnitude of the output voltage V_O will saturate around $+V_{sat}$ even if the magnitude of the input voltage becomes less than V_{UT}. This will continue until the magnitude of the input voltage becomes less than V_{LT}. Once the magnitude of the input voltage becomes less than V_{LT}, the magnitude of the output voltage V_O will change to $-V_{sat}$. The magnitude of the output voltage V_O is maintained around $-V_{sat}$ until the magnitude of the input voltage becomes greater than V_{UT}. Let T be the period of the output voltage.

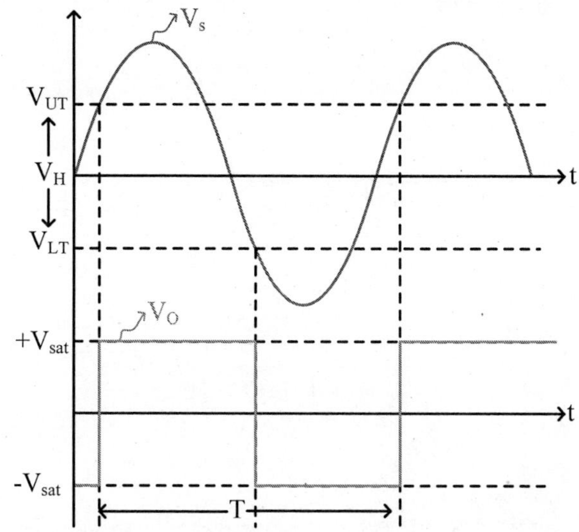

Fig. 7.2.1.2 Input and output waveforms of noninverting hysteresis comparator.

The transfer characteristics of the noninverting hysteresis comparator are shown in Fig. 7.2.1.3. The transfer characteristics are plotted between the input voltage V_s and the output voltage V_O.

7.2.2 Inverting Hysteresis Comparators

The circuit of the inverting hysteresis comparator is shown in Fig. 7.2.2.1. The input signal V_S is applied to the inverting terminal of the op-amp. The feedback resistance R_F is connected between the

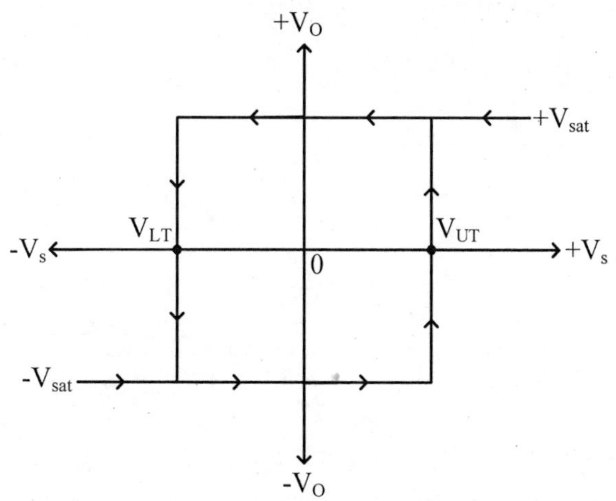

Fig. 7.2.1.3 Transfer characteristics of noninverting hysteresis comparator.

noninverting terminal and the output terminal of the op-amp and a resistance R_1 is connected between the noninverting terminal and ground.

The voltage V_1 can be obtained by

$$V_1 = \frac{V_O R_1}{R_1 + R_F} \tag{7.2.2.1}$$

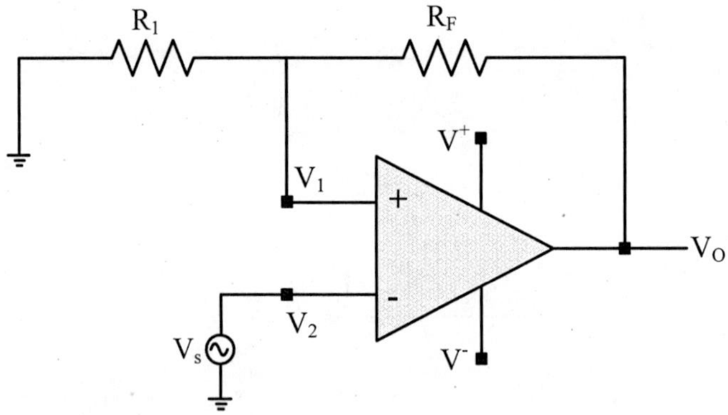

Fig. 7.2.2.1 Circuit of the inverting hysteresis comparator.

The output voltage V_O can saturate at two values: the positive supply voltage V^+ and the negative supply voltage V^- based on the input voltage V_S crossing predetermined thresholds (upper threshold limit and lower threshold limit). Let us assume the positive supply voltage V^+ is equal $+V_{sat}$ and the negative supply voltage V^- is equal $-V_{sat}$. During this case, the upper threshold limit V_{UT} is given by

$$V_{UT} = \frac{R_1 V_{sat}}{R_1 + R_F} \tag{7.2.2.2}$$

Now to calculate the lower threshold limit, the output voltage saturates at the positive supply voltage V. During this case, the lower threshold limit V_{LT} is given by

$$V_{LT} = -\frac{R_1 V_{sat}}{R_1 + R_F} \tag{7.2.2.3}$$

The hysteresis range V_T is obtained by subtracting the upper and lower threshold limits. Therefore,

$$V_H = V_{UT} - V_{LT} = 2\frac{R_1 V_{sat}}{R_1 + R_F} \tag{7.2.2.4}$$

The input and output waveforms of the inverting hysteresis comparator are shown in Fig. 7.2.2.2. Let the input voltage V_s be a sinusoidal waveform. The voltages V_{UT} and V_{LT} are the upper and lower threshold limits of the non-inverting hysteresis comparator. Whenever the magnitude of the input voltage V_s is greater than the upper threshold limit V_{UT}, the output voltage V_O saturates around the value $-V_{sat}$. The magnitude of the output voltage V_O will saturate around $-V_{sat}$ even if the magnitude of the input voltage becomes less than V_{UT}. This will continue until the magnitude of the input voltage becomes less than V_{LT}. Once the magnitude of the input voltage becomes less than V_{LT}, the magnitude of the output voltage V_O will change to $+V_{sat}$. The magnitude of the output voltage V_O is maintained around $+V_{sat}$ until the magnitude of the input voltage becomes greater than V_{UT}. Let T be the period of the output voltage. The transfer characteristics of the inverting hysteresis comparator are shown in Fig. 7.2.2.3. The transfer characteristics are plotted between the input voltage V_s and the output voltage V_O.

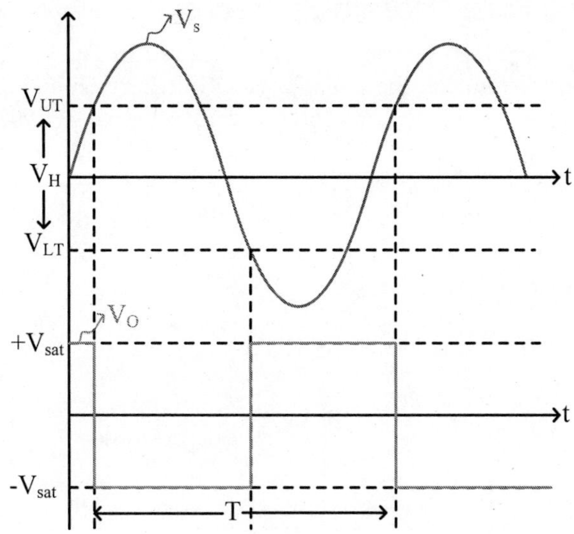

Fig. 7.2.2.2 Input and output waveforms of inverting hysteresis comparator.

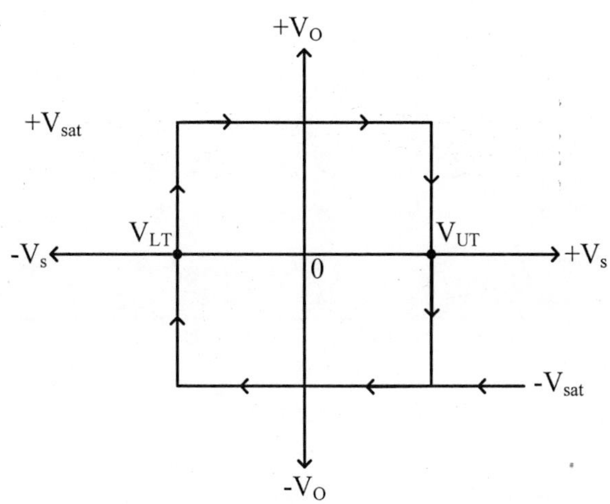

Fig. 7.2.1.3 Transfer characteristics of inverting hysteresis comparator.

7.3 ZERO CROSSING DETECTOR

By utilizing positive feedback, op-amp circuits can detect when an input signal crosses zero. These detectors are frequently used in applications like phase control, frequency measurement, and AC power control. The zero crossing detectors can also be called zero crossing comparators. The zero crossing detector can be of two types:

7.3.1 Noninverting Zero Crossing Detector

The circuit of the noninverting zero crossing detector is shown in Fig. 7:3.1.1. The input signal V_S is applied to the noninverting terminal of the op-amp. The inverting terminal of the op-amp is grounded due to which the inverting terminal of the op-amp is at zero potential. For the positive half-cycle of the input voltage, the potential of V_1 is greater than or equal to zero. This results in the clamping of the output voltage of the op-amp to the positive supply voltage V^+. Similarly, during the negative half-cycle of the input voltage, the potential of V_1 is less than zero. This results in the clamping of the output voltage of the op-amp to the negative supply voltage V^-. The input and output waveforms of the noninverting zero crossing detector are shown in Fig. 7.3.1.2.

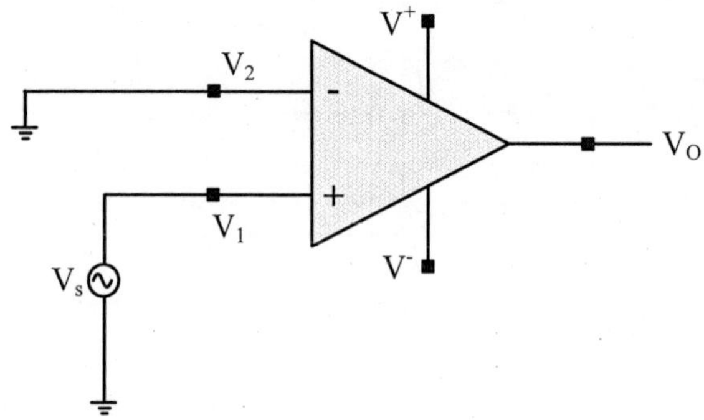

Fig. 7.3.1.1 Circuit of the noninverting zero crossing detector.

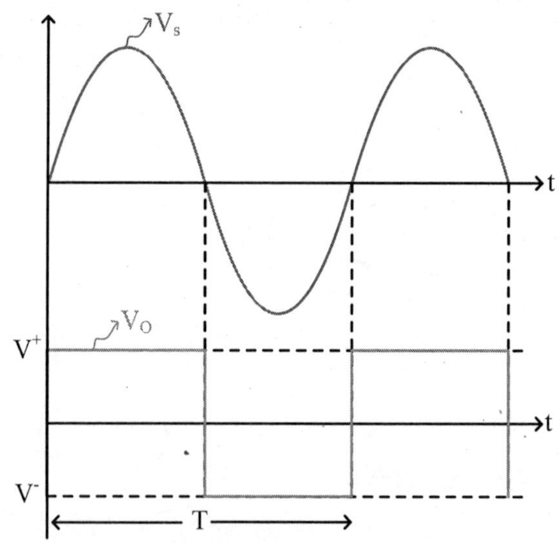

Fig. 7.3.1.2 Input and output waveforms of noninverting zero crossing detector.

7.3.2 Inverting Zero Crossing Detector

The circuit of the inverting zero crossing detector is shown in Fig. 7.3.2.1. The input signal V_S is applied to the inverting terminal of the op-amp. The noninverting terminal of the op-amp is grounded due to which the noninverting terminal of the op-amp is at zero potential. For the positive

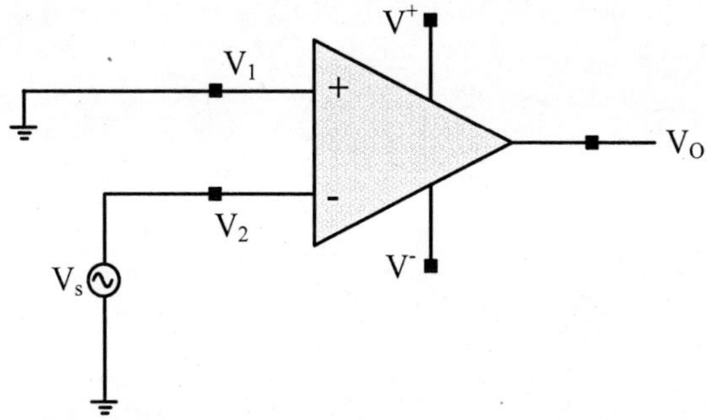

Fig. 7.3.2.1 Circuit of the noninverting zero crossing detector.

half-cycle of the input voltage, the potential of V_2 is greater than or equal to zero. This results in the clamping of the output voltage of the op-amp to the negative supply voltage V^-. Similarly, during the negative half-cycle of the input voltage, the potential of V_2 is less than zero. This results in the clamping of the output voltage of the op-amp to the positive supply voltage V^+. The input and output waveforms of the inverting zero crossing detector are shown in Fig. 7.3.2.2.

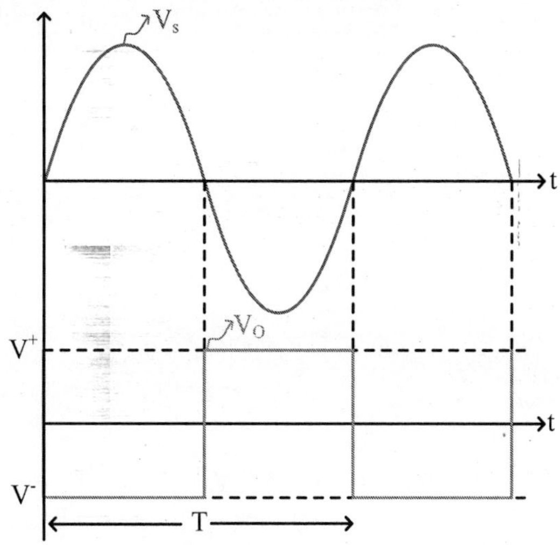

Fig. 7.3.2.2 Input and output waveforms of inverting zero crossing detector.

7.4 SQUARE-WAVE AND TRIANGULAR-WAVE GENERATORS

Non-linear op-amp circuits can generate square or triangular waveforms by adjusting the feedback network appropriately. These waveforms find applications in various fields, including signal processing and communications.

7.4.1 Square-Wave Generator

The circuit diagram of the square-wave generator is shown in Fig. 7.4.1.1. The noninverting terminal of the op-amp is connected to the output terminal of the op-amp using the feedback resistors R_1 and R_2. Similarly, the inverting terminal of the op-amp is connected to the output terminal using the feedback resistance R_F. The capacitor C is connected between the inverting terminal of the op-amp and the ground.

The output voltage V_O can saturate at two values: the positive supply voltage V^+ and the negative supply voltage V^- depending on the upper threshold and lower threshold limits. Let us assume the positive supply voltage V^+ is equal $+V_{sat}$ and the negative supply voltage V^- is equal $-V_{sat}$. The value of the upper threshold limit V_{UT} is obtained by

$$V_{UT} = \frac{R_1 V_{sat}}{R_1 + R_2} \tag{7.4.1.1}$$

Similarly, the value of the lower threshold limit V_{LT} is obtained by

$$V_{LT} = -\frac{R_1 V_{sat}}{R_1 + R_2} \tag{7.4.1.2}$$

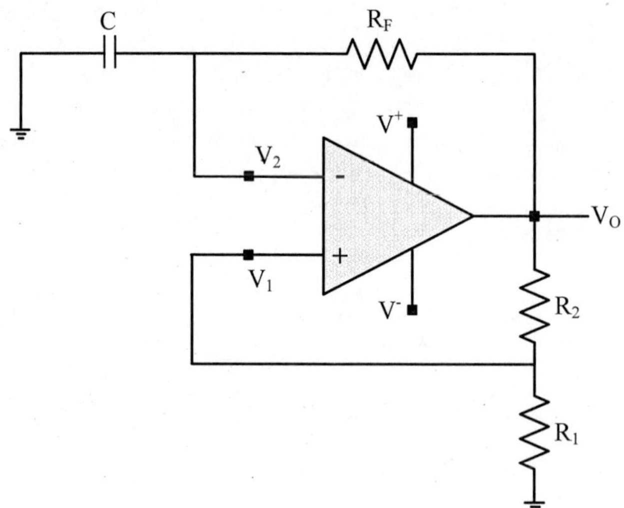

Fig. 7.4.1.1 Circuit of the square-wave generator.

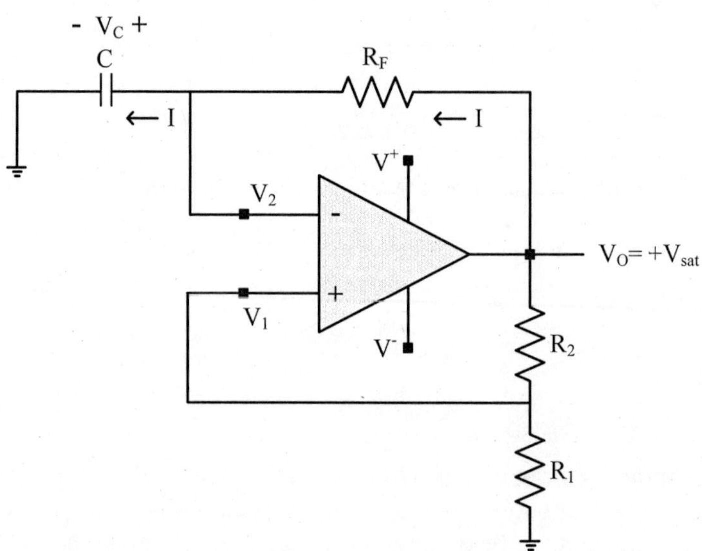

Fig. 7.4.1.2 Operation of square wave generator when $V_O = +V_{sat}$.

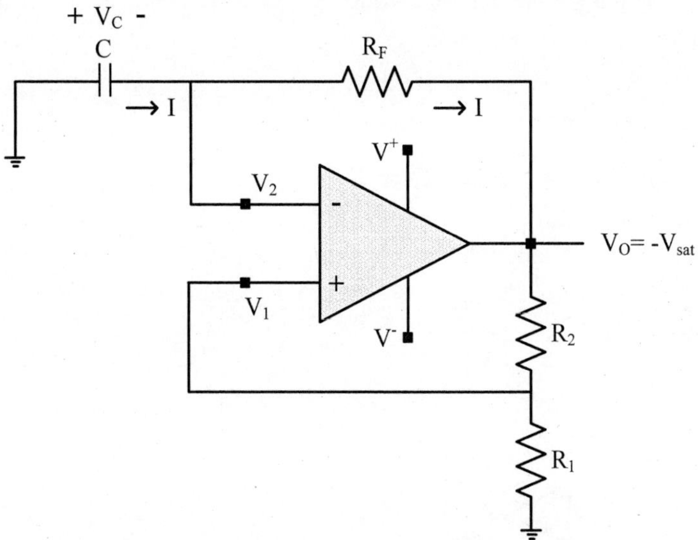

Fig. 7.4.1.3 Operation of square wave generator when V_O=-V_{sat}.

Initially let us assume that the output voltage of the square-wave generator is saturated to +V_{sat}. During this, condition, the capacitor C connected between the inverting terminal and ground, charges through the feedback resistance R_F as shown in Fig. 7.4.1.2. Once the capacitor C is charged to the upper threshold limit V_{UT}, the output voltage of the square-wave generator is changed from +V_{sat} to -V_{sat}. Due to this capacitor C now discharges through the feedback resistor R_F as shown in Fig. 7.4.1.3. The capacitor C will discharge up to the lower threshold limit V_{LT}. Once the capacitor voltage reaches the lower threshold limit V_{LT}, the output voltage of the square-wave generator is changed from -V_{sat} to +V_{sat}. In this way, a periodic square waveform is obtained in the output voltage V_O. Fig. 7.4.1.4 shows the waveforms of the output voltage V_O along with the waveform of the voltage across capacitor C.

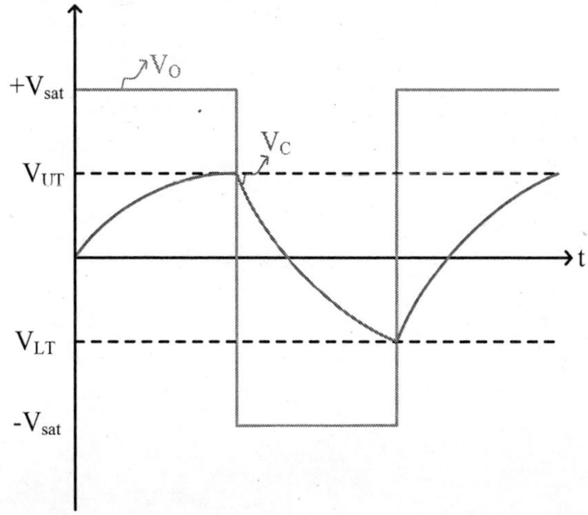

Fig. 7.4.1.4 Waveforms of the output voltage V_O and waveform of the voltage across the capacitor V_C in the square-wave generator.

The expression for the frequency of the output voltage V_O is given by

$$f = \frac{1}{2R_F C \ln\left(\frac{+V_{sat} - V_{LT}}{+V_{sat} - V_{UT}}\right)} \tag{7.4.1.3}$$

7.4.2 Triangular-Wave Generator

For obtaining the triangular-wave, the output of the square-wave is connected to the input of the integrator. The circuit of the triangular-wave is shown in Fig. 7.4.2.1. Fig. 7.4.2.2 shows the output voltage waveforms of the triangular-wave generator. The triangular waveform will oscillate between the upper limit and lower limit.

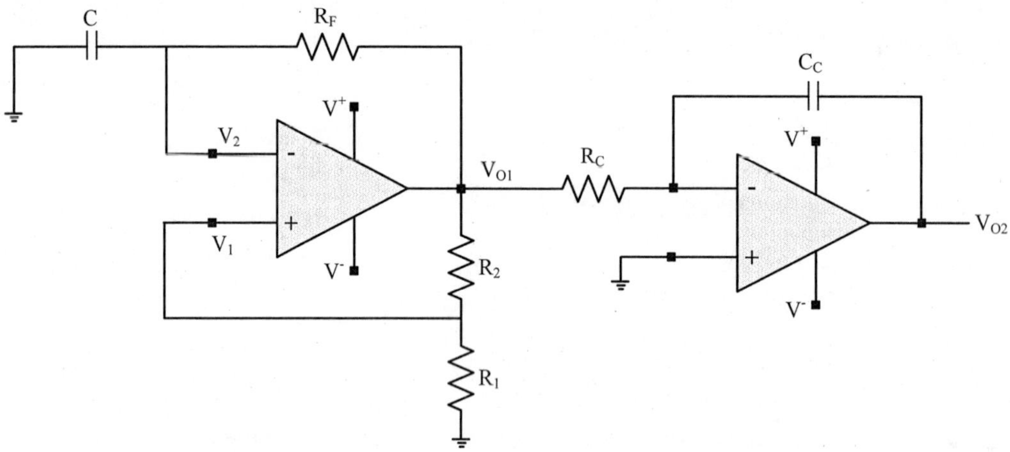

Fig. 7.4.2.1 Circuit of the triangular-wave generator.

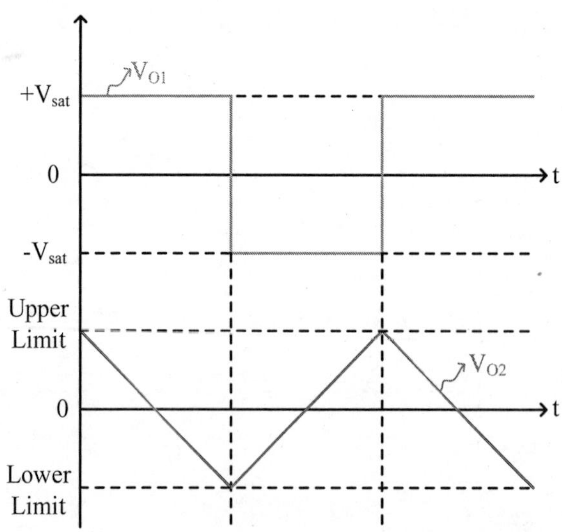

Fig. 7.4.2.2 Waveforms of triangular-wave generator.

7.5 PRECISION RECTIFIER

Op-amp circuits with diodes in the feedback path can rectify AC signals with minimal distortion, allowing for precise signal processing and measurement. Conventional diode rectifier has the drawback of distortion in the output rectified signal due to diode drop. The presence of the diode drop voltage results negative -0.6 or -0.7V in the positive rectified output voltage instead of the zero voltage. The distortion in rectifier signal due to diode drop is taken care in the precision rectifier. Precision rectifier takes care for the distortion by including the diode in the feedback part or keeping diode voltage drop a part of open loop circuit. Rectifiers are basically of two types:

i. Half wave rectifier

ii. Full wave rectifier

7.5.1 Half wave rectifier

In this type of circuit, only one half either positive or negative half cycle of the ac input, is given as output. The circuit diagram for the half wave precision rectifier is given in Fig. 7.5.1.1. It consists of two diodes D_1 and D_2 with input and feedback resistor connected, as shown in Fig. 7.5.1.1. The diode D_1 is kept in feedback loop diode D_2 prevents saturation of output voltage during the positive cycle. The feedback and input resistors take care of the gain in the circuit. During a positive half-cycle, when the input voltage at the non-inverting terminal is positive, then the output voltage of Op-Amps is negative, making the diode D_1 reverse biased and D_2 forward biased as shown in Fig. 7.5.1.2(a). Further, due to virtual ground, the inverting terminal of the Op-Amps is grounded. Thus, there is no current in the feed-back resistor leading to zero voltage at node 'a' or output. Now for the negative half cycle or when the input voltage is less than zero, the diode D_1 gets forward biased while the diode D_2 gets reverse biased as shown in Fig. 7.5.1.2(b). Thus, the output of is positive voltage with a gain defined by the ratio of the feedback resistor to the input resistor (R_f/R_1). The circuit has the benefit of avoiding the saturation condition due to the presence of diode D_2. The output impedance of the Op-Amps is low when diode D_1 is conducting or in the negative half cycle of the input voltage. In a positive half cycle, the output impedance is infinite. To solve this issue the output of the rectifier is typically given to the voltage follower or unity gain buffer circuit, as shown in Fig. 7.5.1.3.

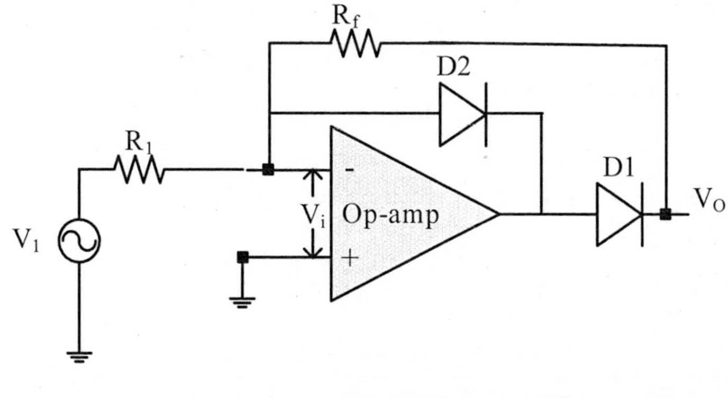

Fig. 7.5.1.1 Circuit schematic of the half wave precision rectifier.

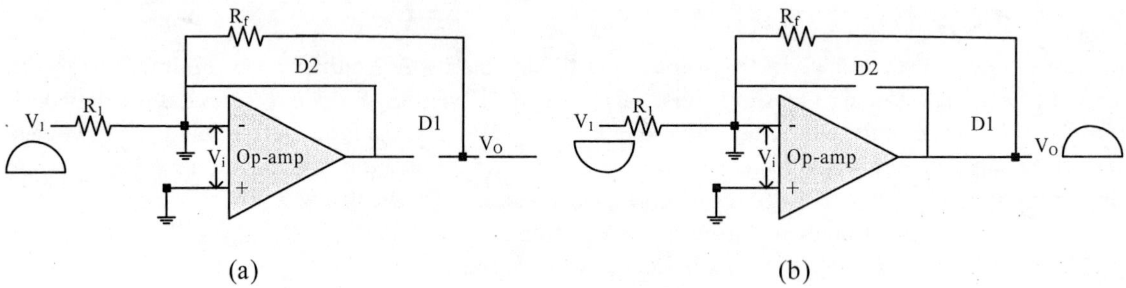

Fig. 7.5.1.2 Circuit schematic of the half wave precision rectifier during (a) positive half cycle of ac input voltage; (b) negative half cycle of ac input voltage.

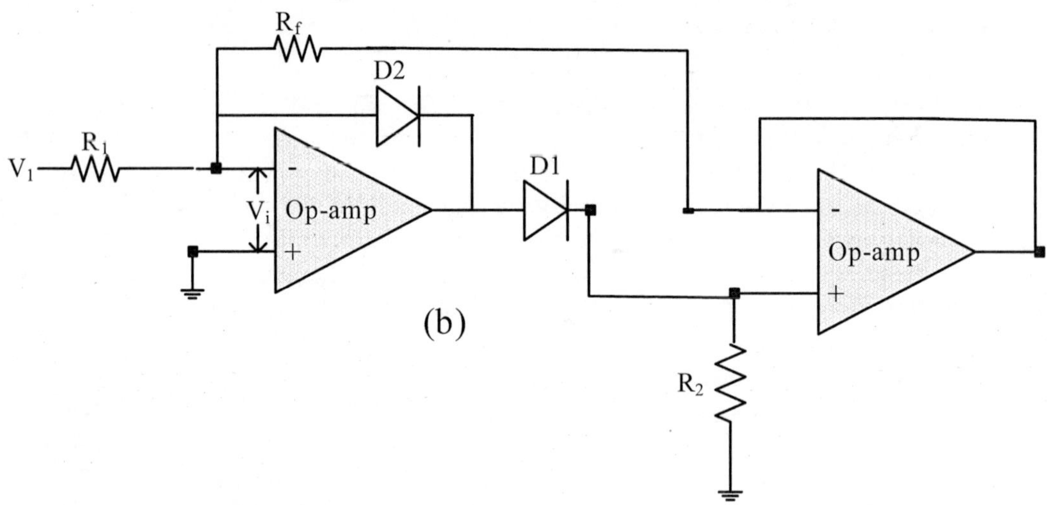

Fig. 7.5.1.3 Circuit schematic of the half wave precision rectifier with zero output impedance.

7.5.2 Full wave rectifier

The circuit diagram for the full wave rectifier is given in Fig. 7.5.2.1. It is the same circuit as the half wave rectifier where both cycles of the input voltage are utilized for output voltage. In the given circuit, the assumption is there that load or output resistance R_L is much higher than resistor 2R employed in full wave circuitry, as shown in Fig. 7.5.2.1. The resistor R_{1a} is introduced, which prevents the grounding of the feedback resistor the positive half cycle of the input voltage and results in the output voltage with an attenuation of half of its input value. During a positive half-cycle, when the input voltage at the non-inverting terminal is positive, as shown in Fig. 7.5.2.2(a), then the output voltage of Op-Amps is negative, making the diode D_1 reverse biased and D_2 forward biased. Further, due to the virtual ground concept, the inverting terminal of the Op-Amps is grounded.

Applying KCL at node 'b' we have,

$$\frac{\frac{V_1}{2} - V_1}{2R} + \frac{V_1}{2R} + \frac{(\frac{V_1}{2} - V_o)}{2R} = 0$$

$$V_o = \frac{V_1}{2} \qquad (7.5.2.1)$$

Now for the negative half cycle or when the input voltage is less than zero, the diode D_1 gets forward biased while the diode D_2 gets reverse biased as seen from Fig. 7.5.2.2(b). Thus, the output of the Op-Amps is positive voltage with a gain defined by the ratio of feedback to input resistor (R_f/R_1) and is given by,

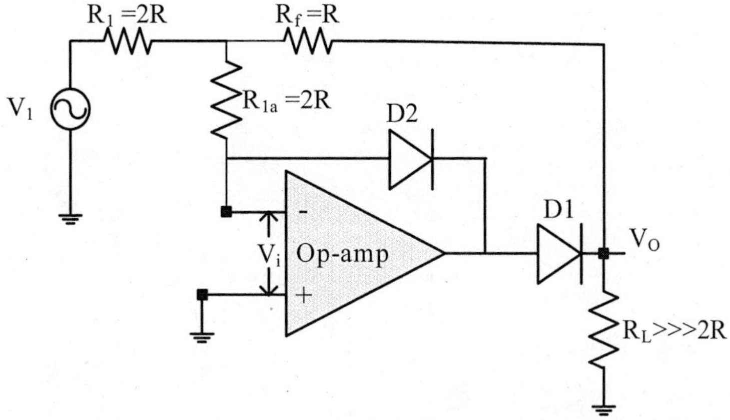

Fig. 7.5.2.1 Circuit schematic of the full wave precision rectifier.

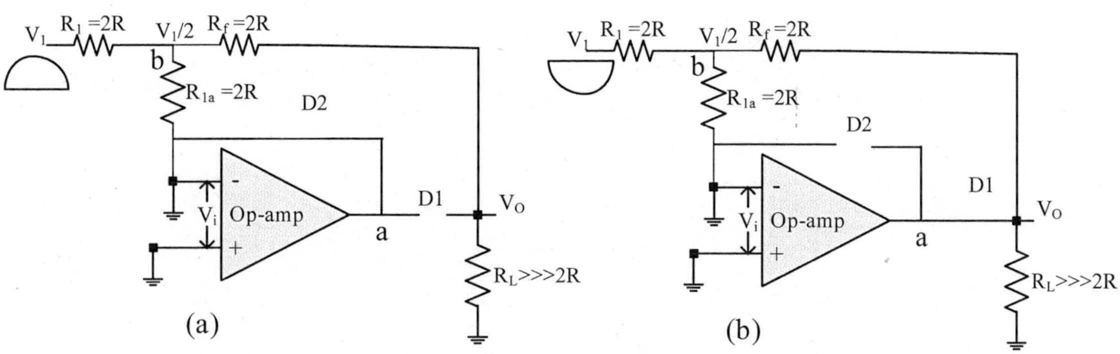

Fig. 7.5.2.2 Circuit schematic of the full wave precision rectifier during (a) positive half cycle of ac input voltage; (b) negative half cycle of ac input voltage.

$$V_o = -(\frac{2R}{2R})\frac{V_1}{2} = -\frac{V_1}{2} \tag{7.5.2.2}$$

Thus, one gets the rectified output voltage from the employed full wave precision circuit.

7.6 PEAK DETECTOR

By employing diodes and capacitors in the feedback network, op-amp peak detectors capture and hold the maximum amplitude of an input signal until reset. These circuits are useful for applications like signal analysis and amplitude measurement. The circuit diagram of the basic peak detector is shown in Fig. 7.6.1. The circuit shown in Fig. 7.6.1 is a positive peak detector. The circuit consists of an op-amp, diode D, capacitor C and a MOSFET reset switch. The input signal V_{in} for which the peak needs to be detected is connected to the non-inverting input terminal via diode D. The inverting input terminal is directly connected to the output of the op-amp. The diode D has a small internal resistance R_F. The time constant of the R_F and C should be as low as possible. Such a low time constant is required for the immediate charging of the capacitor to the peak value. Whenever the input signal reaches to the peak value, the capacitor C charges to the peak value with a small delay. Thereafter with the decrease in the magnitude of the input signal, the diode D will be reverse biased and the capacitor C maintained at the peak value of the input voltage V_{in}. Because of the virtual ground concept, the output voltage V_O now will remain at the peak value of the input voltage V_S. In the next cycle, if the peak of input voltage V_{in} is increased. Then again, the capacitor C is charged to the increased peak value and the output voltage is changed to the increased peak value. Fig. 7.6.2 shows the waveform of the input voltage and the

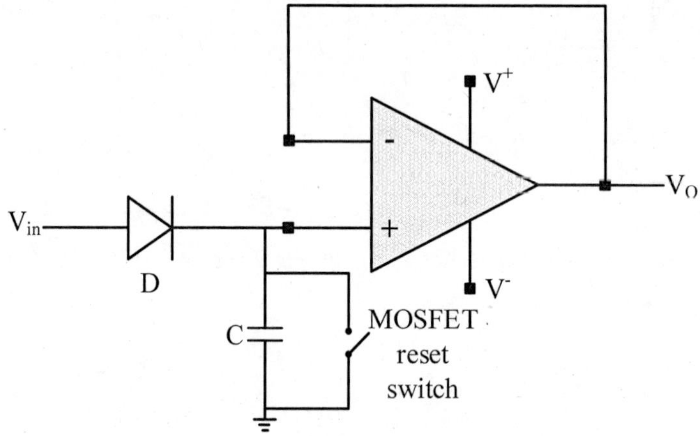

Fig. 7.6.1 Basic positive peak detector.

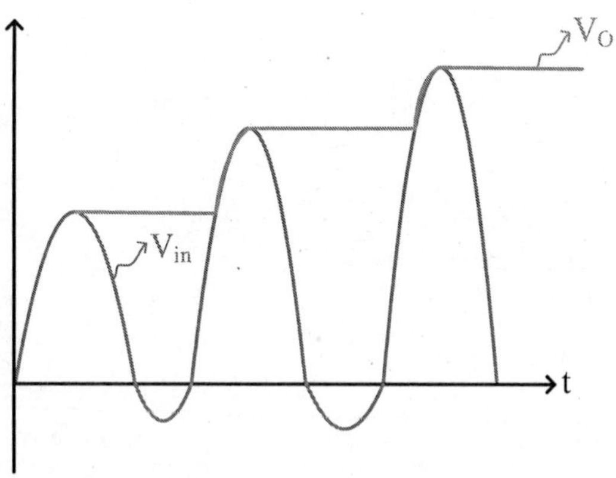

Fig. 7.6.2 Waveforms of basic positive peak detector.

output voltage. For detecting the peak value in the negative half cycle, diode D needs to be reverse-biased.

7.7 MONOSTABLE MULTIVIBRATOR

Non-linear op-amp circuits can generate a pulse of a specific duration in response to a triggering event. Monostable circuits find applications in timing, control, and pulse generation. The circuit diagram of the monostable multivibrator is shown in Fig. 7.7.1. The input trigger V_t is applied to the non-inverting terminal of op-amp using the reverse biased diode D_2. A capacitor C is connected between the inverting terminal of the op-amp and the ground. A clamping diode D_1 is connected across the capacitor C. The resistor R_3 is connected at the output terminal of the op-amp as shown in Fig. 7.7.1. A feedback connection is provided between the non-inverting terminal of the op-amp and the output terminal using the resistors R_1 and R_2 as shown in Fig. 7.7.1. A resistor R is connected between the inverting terminal and the output. An anti-series connected zener diode are connected between the output and ground. Initially let us assume that the output voltage is saturated to the positive supply voltage $+V_{sat}$. Due to this, the diode D_1 will get forward biased and a voltage of 0.7V will appear across the capacitor C and even at the inverting terminal of the op-amp. Similarly, the voltage at the non-inverting terminal (V_1) of the op-amp will be equal to

$$V_1 = \frac{R_2}{R_1 + R_2} V_{sat} = \beta V_{sat} \tag{7.7.1}$$

Whenever a negative trigger is applied at the non-inverting terminal of the op-amp, the effective voltage i.e., (βV_{sat}-V_t) will be less than 0.7V. due to this, the output of the op-amp will change its state to the negative supply voltage -V_{sat}. Due to this negative supply voltage -V_{sat} at the output terminal, the diode D_1 will get reverse biased. Due to this negative supply voltage -V_{sat}, the capacitor C now will charge in the negative direction. In addition to this, the voltage at the non-inverting terminal of the op-amp will be equal to -βV_{sat}. Once the capacitor charges to a negative voltage -βV_{sat}, then the output voltage of the op-amp will change to the positive supply voltage $+V_{sat}$. Due to this transition, the diode D_1 will get forward-biased and the capacitor C will be charged to a voltage of 0.7V. This process will continue until the next negative trigger happens. Fig. 7.7.2 shows the waveforms of the input trigger V_t, output voltage V_O and the voltage across the capacitor V_C.

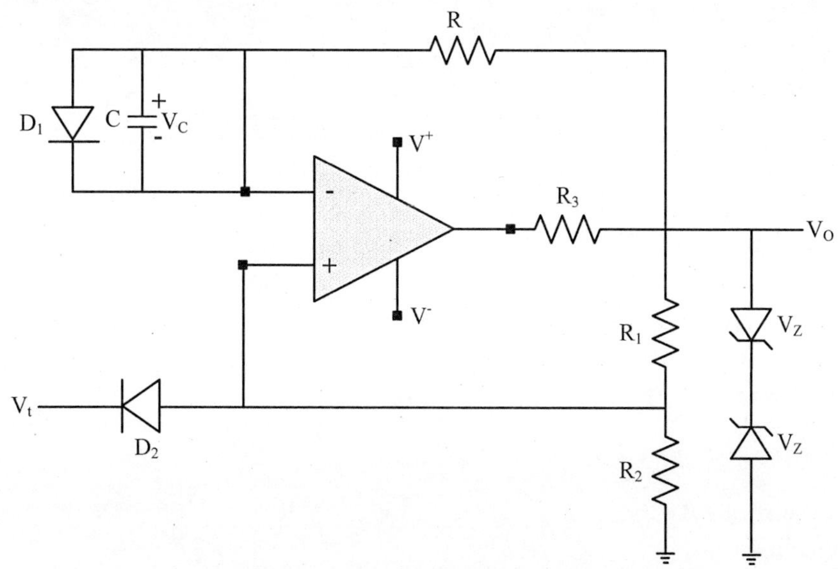

Fig. 7.7.1 Monostable multivibrator.

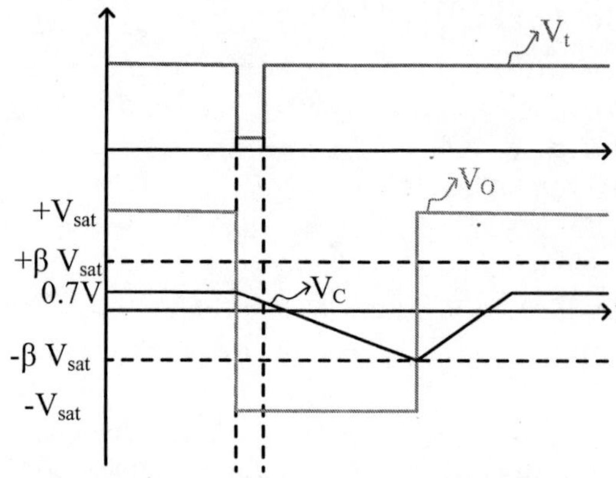

Fig. 7.7.2 Waveforms of monostable multivibrator.

UNIT SUMMARY

The nonlinear uses of the op-amp are covered in this foundational unit. The first topic covered is how hysteresis comparators work. The zero crossing detector's operation is then covered. This facilitates the use of the zero crossing detector by students in situations when AC signal synchronization is required. The functioning of triangle and square wave generators is then covered. The unit also covers monostable multivibrators, peak detectors, and precision rectifiers. The book's appendix contains laboratory experiments pertaining to the operation of astable and monostable multivibrators, precision rectifier and peak detector circuits, square waveform, and triangular waveform generators.

EXERCISES

Multiple Choice Questions

7.1 In hysteresis width, the hysteresis voltage is equal to ___upper & lower threshold voltages (V_{UT} & V_{LT})

a. sum of
b. difference between
c. product of
d. division of

7.2 Which among the following circuits is also regarded/known as ' Threshold Detector '?

a. window detector
b. over voltage indicator
c. level detector
d. zero crossing detector

7.3 Zero crossing detector circuit plays a crucial role in conversion of input sine wave into a perfect at its output.

a. triangular wave
b. square wave
c. saw-tooth wave
d. pulse wave

7.4 The Schmitt trigger can be used as which of the following?

1. square-wave generator
2. comparator
3. astable multivibrator

Select the correct answer

a. 1 and 3 only
b. 1 and 2 only
c. 2 and 3 only
d. 1, 2 and 3

Answers for Multiple Choice Questions			
7.1 **b**	7.2 **c**	7.3 **b**	7.4 **d**

Short Answer and Long Answer Type Questions

7.5 Explain the working of monostable multivibrator?

7.6 Explain the working of peak detector?

7.7 Explain the working of halfwave and full wave rectifiers?

7.8 Explain the working of square and triangular wave generators?

7.9 Explain the working of zero crossing detectors?

7.10 Explain the working of hysteresis compartors?

KNOW MORE

For Additional Info Scan Me

REFERENCES

1. *Ramakanth A. Gayakwad, "OP-amps and Linear Integrated Circuits", PHI, Fourth Edition, 2010.*
2. *Adel S. Sedra , Kenneth C. Smith , Arun N. Chandorkar , "Microelectronic Circuits: Theory And Applications", Oxford University Press, Seventh Edition, (1 June 2017).*
3. *Dr. Sanjay Sharma, "OP-AMPs & Linear Integrated Circuits".*

APPENDIX

APPENDIX

List of Experiments

1. Study and verification of V-I characteristics of a p-n junction diode.
2. Study and verification of V-I characteristics of zener diode.
3. To study the diode clipper and clamper circuits.
4. To study the design of half wave, full wave and bridge rectifier circuits with and without filters.
5. To study the characteristics of BJT in CE configuration.
6. To study the characteristics of MOSFET.
7. To study differential amplifier using BJT.
8. To determine frequency response of negative feedback amplifier using BJT.
9. Measurement of op-amp electrical parameters using simulation.
10. Implementation of inverting and non-inverting amplifier using op-amp.
11. Implementation of comparator, integrator and differentiator using op-amp.
12. Execution of astable and monostable multivibrators using 555 timers.
13. Execution of precision rectifier and peak detector circuits using op-amp.
14. Realization of A/D and D/A converters using op-amp.
15. To study square waveform, and triangular waveform generator using op-amp.

EXPERIMENT NO. 1

AIM: Study and verification of V-I characteristics of a P-N junction diode.

APPARATUS REQUIRED: P-N junction diode, RPS, rheostat, voltmeter, and ammeter.

THEORY: A P-N junction diode is a two terminal junction device. It conducts only in one direction (only on forward biasing).

Forward Bias: On forward biasing, initially no current flows due to barrier potential. As the applied potential exceeds the barrier potential the charge carriers gain sufficient energy to cross the potential barrier and enter the other region. The holes, which are majority carriers in the P-region, become minority carriers on entering the N-regions, and electrons, which are the majority carriers in the N-region, become minority carriers on entering the P-region. This injection of Minority carriers results in the current flow, opposite to the direction of electron movement.

Reverse Bias: On reverse biasing, the majority of charge carriers are attracted toward the terminals due to the applied potential resulting in the widening of the depletion region. Since the charge carriers are pushed towards the terminals, no current flows in the device due to the majority of charge carriers. The device will have some current due to the thermally generated minority carriers. The generation of such carriers is independent of the applied potential; hence, the current is constant for all increasing reverse potential. This current is referred to as Reverse Saturation Current (IO) and it increases with temperature. When the applied reverse voltage is increased beyond the certain limit, it results in breakdown. During breakdown, the diode current increases tremendously.

CIRCUIT DIAGRAM:

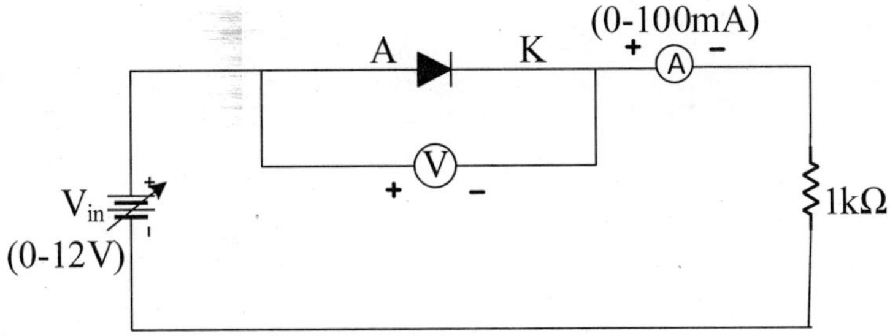

Fig. 1. Forward bias operation

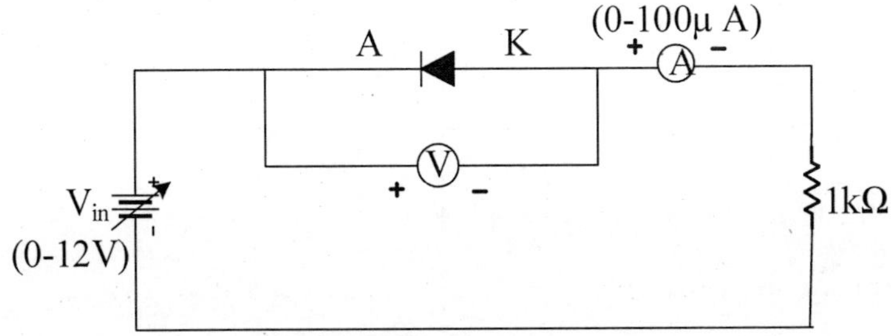

Fig. 2. Reverse bias operation

PROCEDURE:

a. Forward Bias:

1. The circuit connections are made as per the circuit diagram given in Fig.1.
2. For the diode forward bias, the RPS positive terminal is connected to the anode of the diode and RPS negative is connected to the cathode of the diode.
3. Now switch on the power supply (RPS) and increase the input supply voltage in regular voltage interval of 0.1V.
4. Note down the corresponding current flowing through the diode and voltage across the diode for each and every step of the supply input voltage.
5. The recorded reading of voltage and current are tabulated in the tabular form given below.
6. Graph is plotted between voltage (V_f) on X-axis and current (I_f) on Y-axis.

b. Reverse Bias:

1. The circuit connections are made as per the circuit diagram given in Fig. 2.
2. For the diode reverse bias, the RPS negative is connected to the anode of the diode and RPS positive is connected the cathode of the diode.
3. Now switch on the power supply (RPS) and increase the supply input voltage in regular voltage intervals of 1V.
4. Note down the corresponding current flowing through the diode and voltage across the diode for each and every step of the supply input voltage.
5. The recorded reading of voltage and current are tabulated in the tabular form given below.
6. Graph is plotted between voltage (V_R) on X-axis and current (I_R) on Y-axis.

IDEAL WAVEFORMS:

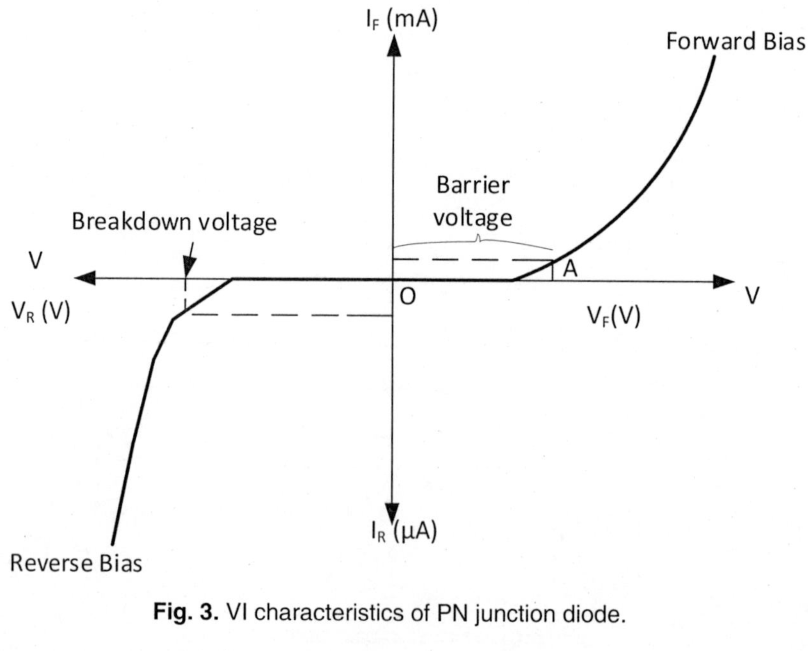

Fig. 3. VI characteristics of PN junction diode.

OBSERVATION TABLE:

S.No	Ammeter(A)	Voltmeter(V)
1		
2		
3		

RESULT:

EXPERIMENT NO. 2

AIM: Study and verification of V-I characteristics of Zener diode.

APPARATUS REQUIRED: Power supply, voltmeter, ammeter, Zener diode and resistor.

THEORY: A Zener diode is a heavily doped P-N junction diode specifically designed to operate in the breakdown region. In a normal P-N junction diode, there is minimal conduction when it is reverse biased. However, when the reverse bias voltage reaches a certain threshold, the Zener diode begins to conduct heavily. This threshold voltage is known as the breakdown voltage. It's important to note that when a Zener diode is conducting heavily, there is a risk of high current flowing through it, potentially causing permanent damage. To prevent this, a resistor is often connected in series with the Zener diode. This resistor limits the current flowing through the diode, protecting it from damage. One significant characteristic of a Zener diode is its ability to maintain a nearly constant voltage across its terminals, regardless of the current passing through it. This is due to its low dynamic resistance. Consequently, Zener diodes are commonly used in voltage regulation circuits, where a stable voltage output is required.

CIRCUIT DIAGRAM:

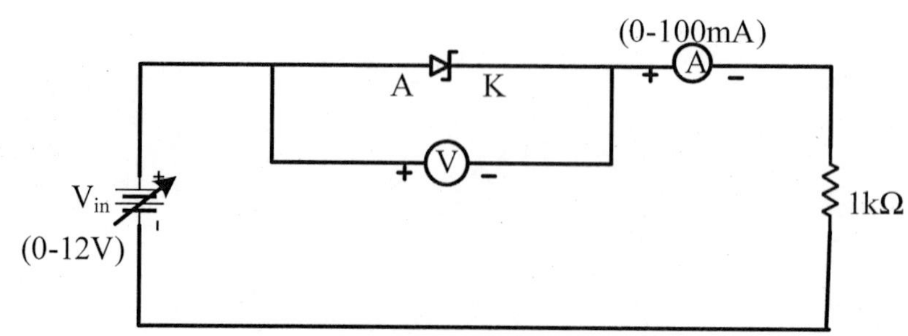

Fig. 1. Static characteristics

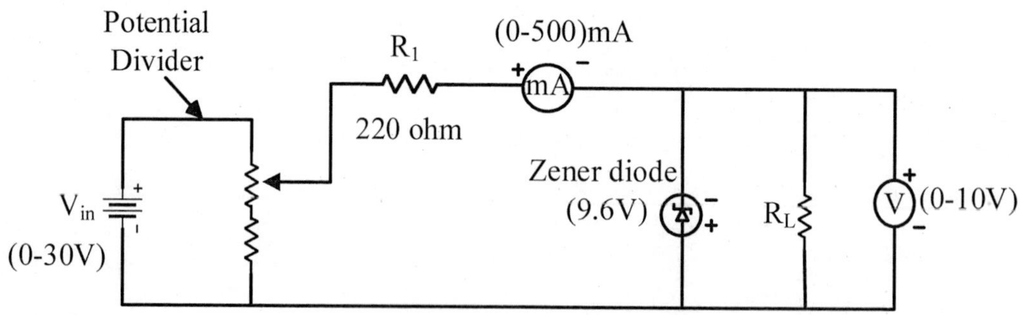

Fig. 2. Regulation characteristics

PROCEDURE:

a. Static characteristics:
   1. Follow the circuit diagram provided in Fig. 1.
   2. Gradually increase the regulated power supply voltage in steps.
   3. Observe and note down the forward current (I_F) and forward voltage (V_F) in a tabular form.
   4. Plot a graph with the forward current (I_F) on the X-axis and the forward voltage (V_F) on the Y-axis.

b. Regulation characteristics:
   1. Set up the circuit as shown in Fig. 2.
   2. Place the load in full load condition and measure the Zener voltage (V_Z), Zener current (I_Z), and load current (I_L).
   3. Repeat the above step by decreasing the value of the load in steps.
   4. Tabulate all the readings obtained.
   5. Calculate the percentage regulation using the following formula:
   6. Percentage Regulation = $((V_Z - V_{load}) / V_{load}) \times 100$.
   7. Voltage regulation is typically expressed as a percentage. Where V_Z is the Zener voltage, and V_{load} is the load voltage.

IDEAL WAVEFORMS:

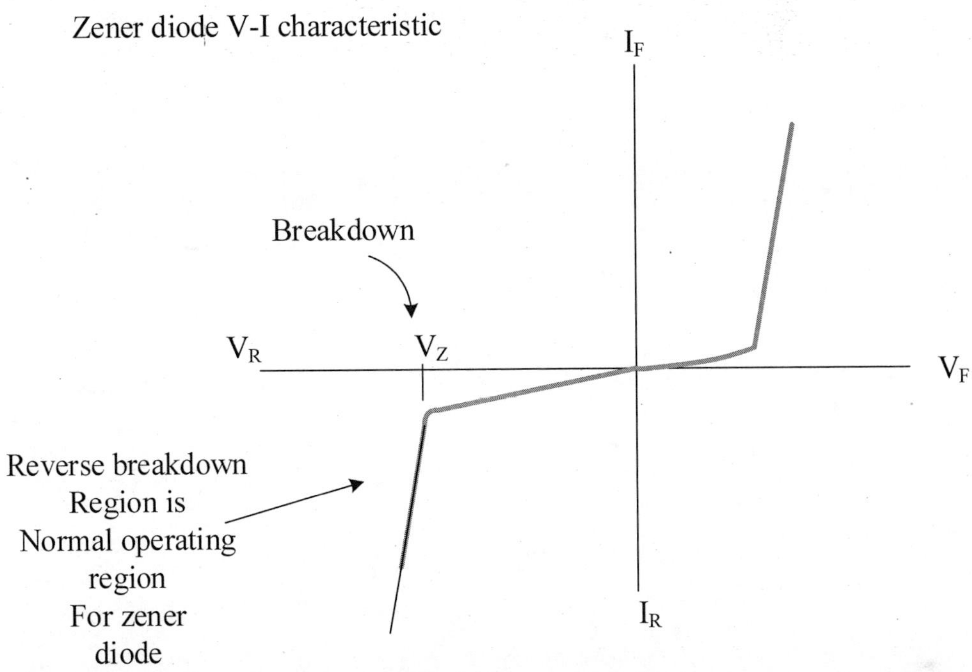

Fig. 3. VI characteristics of Zener diode.

OBSERVATION TABLE:

S.No	Ammeter(A)	Voltmeter(V)
1		
2		
3		

RESULT:

EXPERIMENT NO. 3

AIM: To design and construct Clipper and Clamper circuits.

APPARATUS REQUIRED: Regulated power supply, Voltmeter, Ammeters, Resistor, Capacitor, DSO.

THEORY: Clipping circuits are utilized to remove portions of a signal that exceed or fall below a specific reference level. We previously examined an example of a clipper in the context of a half-wave rectifier circuit. In that circuit, the signal was cut off at the reference level of zero, allowing only the positive portion of the input waveform to pass through. On the other hand, clamping circuits, also known as dc restorers or damped capacitors, shift an input signal by a predetermined amount defined by an independent voltage source. While clippers restrict the portion of the input signal that reaches the output based on a reference level, a clamping circuit allows the entire input signal to reach the output. However, the signal is shifted so that the maximum or minimum value of the input is "clamped" to the voltage level set by the independent source. In essence, clamping circuits modify the DC level of the input signal without altering its shape or waveform. This can be useful in various applications, such as in television and video systems, where clamping circuits help maintain a consistent reference voltage level for synchronization purposes. By employing clamping circuits, it becomes possible to shift and anchor the input signal to a desired reference level, ensuring accurate signal processing and maintaining compatibility with subsequent stages of the circuit.

CIRCUIT DIAGRAM:

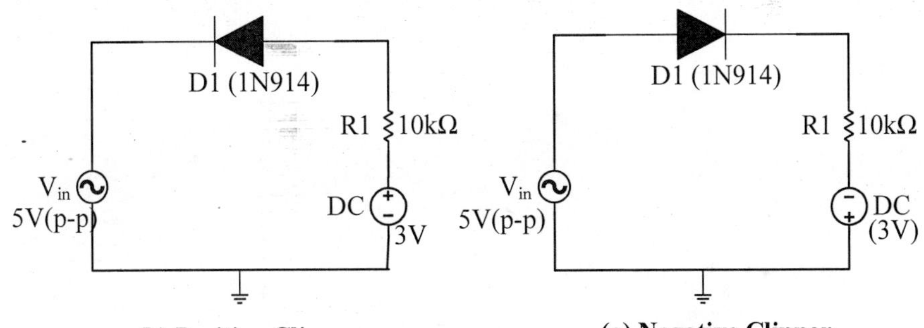

(b) Positive Clipper **(a) Negative Clipper**

Fig.1. Connection diagram for clipper circuit

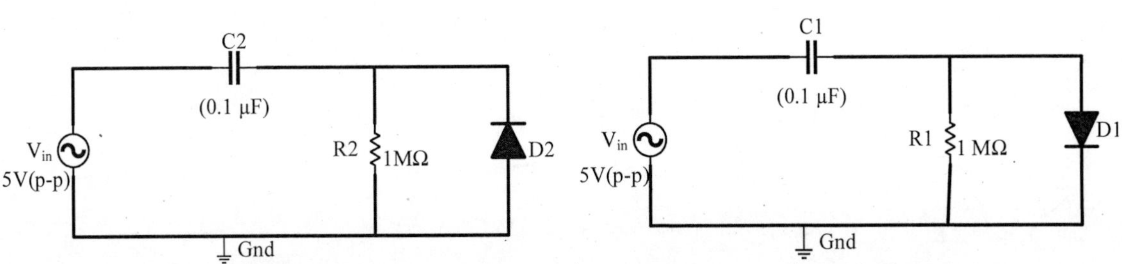

Positive Clamper **Negative Clamper**

Fig.2. Connection diagram for clamper circuit

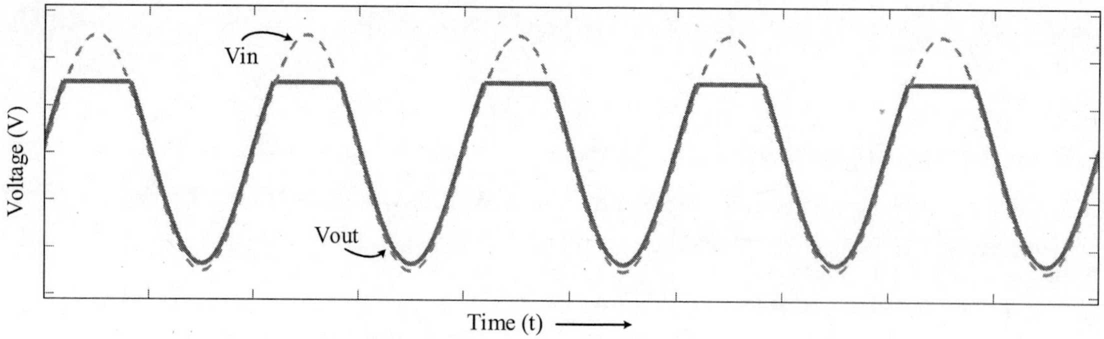

Fig. 3. Waveform of positive clipper circuit

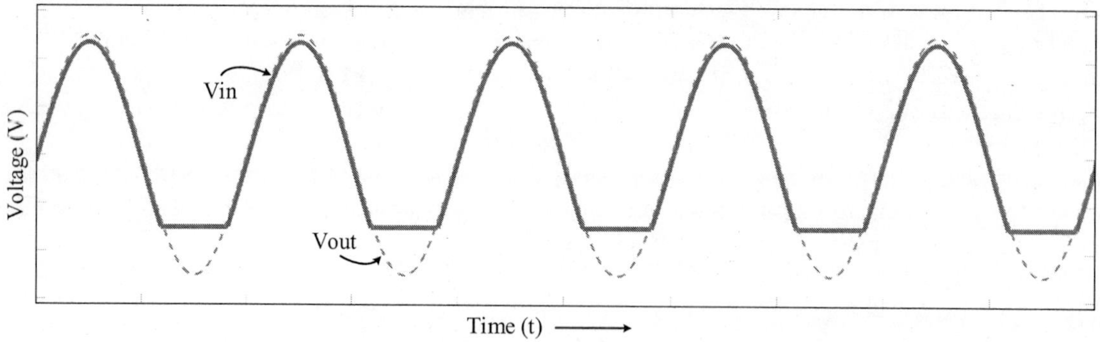

Fig. 4. Waveform of negative clipper circuit

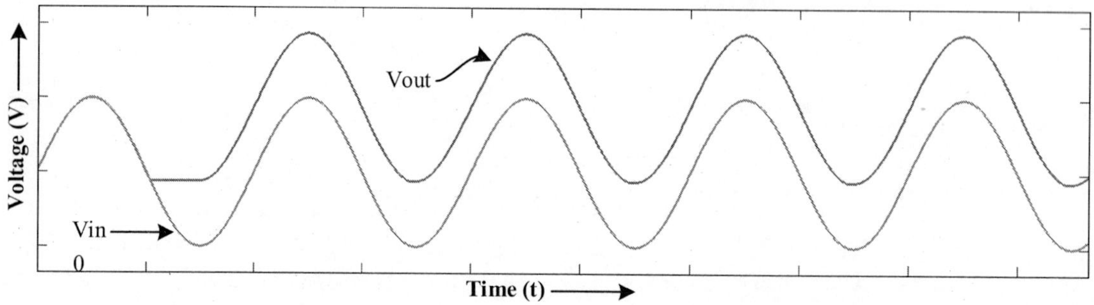

Fig. 5. Waveform of positive clamper circuit

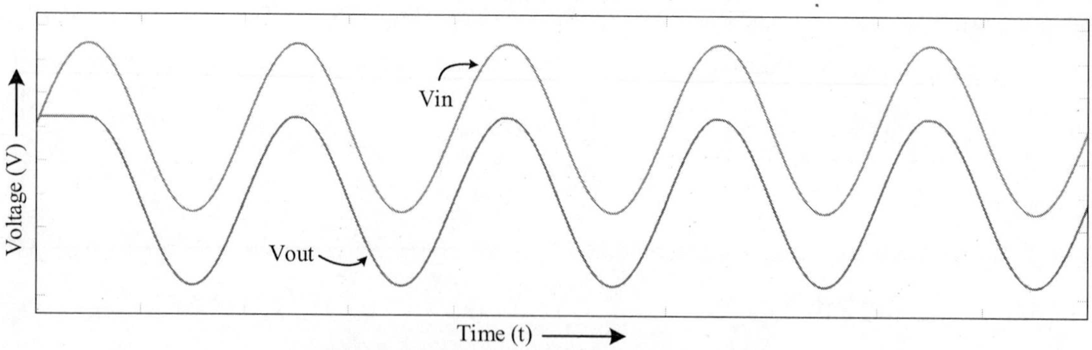

Fig. 6. Waveform of negative clamper circuit

PROCEDURE:

a. For Clipper Circuit

To perform the experiment as described:

1. Set up the circuit according to the circuit diagram given in Fig. 1.
2. Connect the diode with the reference voltage for the positive clipper, as shown in Fig. 1(a).
3. For the negative clipper, reverse the direction of the diode and the reference voltage, as shown in Fig. 1(b).
4. Apply a sinusoidal input of 5V peak-to-peak (p-p) at a frequency of 1kHz to the positive clipper and the negative clipper.
5. Observe the corresponding output waveforms on a Cathode Ray Oscilloscope (CRO) connected to the output of each clipper. By following these steps, you will be able to analyze the effect of the positive and negative clippers on the input signal and observe the resulting waveforms on the CRO.
6. Regenerate response Fig. 3 and Fig. 4 shows the model waveform for the positive and negative clipper circuit.

b. For Clamper Circuit

1. Set up the circuit according to the circuit diagram given in Figure 2.
2. Use a function generator to generate an input signal voltage of 5V peak-to-peak (p-p) at a frequency of 1kHz. Connect the function generator's output to the input of the circuit.
3. Connect an oscilloscope (CRO) to the output of the circuit to observe the waveform.
4. Adjust the settings of the oscilloscope to properly view the waveform, such as adjusting the time base and voltage scales.
5. Observe the output waveform on the oscilloscope and note its shape and characteristics.
6. Sketch the observed waveform on a graph with time (horizontal axis) and voltage (vertical axis). Care to accurately represent the waveform's shape, amplitude, and other relevant details.
7. By following these steps, you will be able to generate and observe the output waveform of the circuit on the oscilloscope and then sketch it on a graph to visually represent its characteristics. Figs. 5 and 6 show the model waveform for the positive and negative clamper circuit.

OBSERVATION TABLE:

a. Tabular column for clipper:

Vm = Vmax= f = 1000Hz Vmin=

POSTIVE CLIPPER	DC Voltage(V)	V_O	
		Vmax	Vmin
	2V 4V 6V		

NEGATIVE CLIPPER	DC Voltage(V)	V_O	
		Vmax	Vmin
	2V 4V 6V		

b. Tabular column of clamper:

Vm= f = 1000Hz

	INPUT		OUTPUT	
	V_{max}	V_{min}	V_{max}	V_{min}
POSITIVE CLAMPER				
NEGATIVE CLAMPER				

RESULT :

EXPERIMENT NO. 4

AIM: To study the characteristics of half wave, full wave and bridge rectifier with and without filter.

APPARATUS REQUIRED: Diodes, Resistor, Transformer, Voltmeter, Ammeter, Breadboard and CRO

THEORY: The rectifier changes ac supply to dc supply and it is an essential part of power supply. The unique property of a diode, permitting the current to flow in one direction, is utilized in rectifiers.

Half Wave Rectifier: The mains power supply is applied at the primary of the step-down transformer. All the positive half cycles of the stepped down ac supply pass through the diode and all the negative half cycles get eliminated. The peak value of the output voltage is less than the peak value of the input voltage by 0.6V because of the voltage drop across the diode. For a half wave rectifier, $V_{rms} = V_m/2$ and $V_{dc} = V_m/\pi$: where V_{rms} = RMS value of input, V_{dc} = Average value of input and V_m = peak value of output.

Full Wave Rectifier: During the positive half cycle of the transformer's secondary voltage, diode D1 is forward biased and diode D2 gets reverse biased (mark D1 and D2 in Fig.). So, a current flow through the diode, load resistor and upper half of the transformer winding. During the negative half cycle, diode D2 is forward biased and diode D1 gets reverse biased (mark D1 and D2 in Fig.).. The current then flows through the diode, load resistor, and lower half of the transformer winding. Current flows through the load resistor in the same direction during both half cycles. The peak value of the output voltage is less than the peak value of the input voltage by 0.6V because of the voltage drop across the diode.

$$\text{For a full wave rectifier, } Vrm\ s = \frac{Vm}{\sqrt{2}},\ Vdc = \frac{2Vm}{pi}$$

Bridge Rectifier: During the positive half cycle of the secondary voltage, diodes D_1 and $D2$ are forward biased, and diodes D_3 and D_4 are reverse biased. Therefore, current flows through the secondary winding, diode D_1, load resistor R_L, and diode $D2$. During the negative half cycle, D_3 and D_4 are forward biased and diodes D_1 and D_2 are reverse biased. Therefore, current flows through the secondary winding, diode D_3. Load resistor R_L and D_4. During both half cycles, the current flows through the load resistor in the same direction. The peak value of the output voltage is less than the peak value of the input voltage by 1.2V due to the voltage drop across two diodes. The ripple factor of the bridge rectifier is the same as that of the full wave rectifier.

Rectifiers with Filter: Rectifiers convert AC supply signals into DC supply, but they often produce an undesirable amount of ripple along with the desired DC component. To remove this ripple and eliminate AC components, a filter is connected to the output of the rectifier. There are several types of filters commonly used, including capacitor input filters, choke input filters, RC filters, CRC filters, LC filters, and CLC filters. Among them, the capacitor input filter is the simplest and most cost-effective option. It involves connecting a high-value capacitor (C) in parallel with the load resistor (RL).

During the half cycle of the input waveform, the capacitor charges to the peak voltage (Vm) at the output. As the peak value is exceeded, the capacitor gradually discharges through the load resistor. This slow discharge occurs because the diode becomes reverse biased due to the voltage across the capacitor. Before the capacitor voltage drops significantly, the next output cycle arrives, causing the capacitor to recharge to the peak voltage.To calculate the RMS value of the filtered output, an approximation is made by assuming the output waveform resembles a triangular wave. The formula used is based on this assumption, where the RMS value is calculated as:

$$V_r, rms = \text{Vrpp}/2\sqrt{3}$$

where V_{rpp} is the peak-to-peak value of the ripple voltage.

$$V_{dc} = V_m - (V_{rpp}/2), \text{ Ripple factor r} = Vr, rms\ /\ V_{dc}$$

For a half wave rectifier, the ripple factor is also expressed as a function of capacitance and load resistance, r = $1/2\sqrt{3}$ f – RLC. For a full wave rectifier, it is given by the expression, r = $1/4\sqrt{3}$ f – R_LC, where f is the mains supply frequency 50 Hz.

PROCEDURE: To perform the experiment as described:

1. Set up the circuit according to the circuit diagram provided in Fig. 1 for the half-wave rectifier.
2. Connect all the components of the half-wave rectifier circuit without the capacitor. Ensure all the connections are secure.
3. Switch on the main power supply. Using a Cathode Ray Oscilloscope (CRO), simultaneously observe the waveform of the transformer secondary voltage and the output voltage across the load resistor. Take note of the peak voltage (V_m) from the transformer secondary waveform. Calculate the peak rectified output voltage (V_{rms}) and the DC component voltage (V_{dc}).
4. Calculate the ripple factor, rectifier efficiency, and percentage regulation using the following formulas:
5. Ripple factor = (V_{rms}–V_{dc}) / V_{dc}
6. Rectifier efficiency = (V_{dc} / V_m) * 100
7. Percentage regulation = ((V_m–V_{dc}) / V_{dc}) * 100
8. V_{rms} is the root mean square value of the rectified output voltage, V_{dc} is the DC component voltage, and V_m is the peak voltage from the transformer secondary waveform.
9. Connect the capacitor filter to the circuit and observe the waveforms using the CRO. Note down the peak voltage (V_m) and the peak-to-peak ripple voltage (V_{rpp}). Calculate the ripple factor, rectifier efficiency, and percentage regulation using the same formulas as in step 4. Repeat these calculations for different capacitor values.
10. Repeat the above steps for full-wave and bridge rectifiers, following the respective circuit diagrams. Make the necessary connections, observe the waveforms, calculate the required parameters (V_m, V_{rpp}, V_{rms}, Vdc), and determine the ripple factor, rectifier efficiency, and percentage regulation for each rectifier type.

By following these steps, you will be able to perform the experiments and calculations to analyze the characteristics of half-wave, full-wave, and bridge rectifiers with and without capacitor filters.

CIRCUIT DIAGRAMS

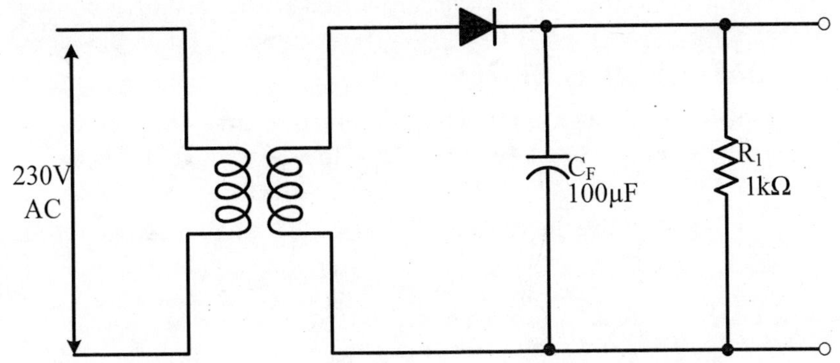

Fig. 1 Half wave rectifier with filter

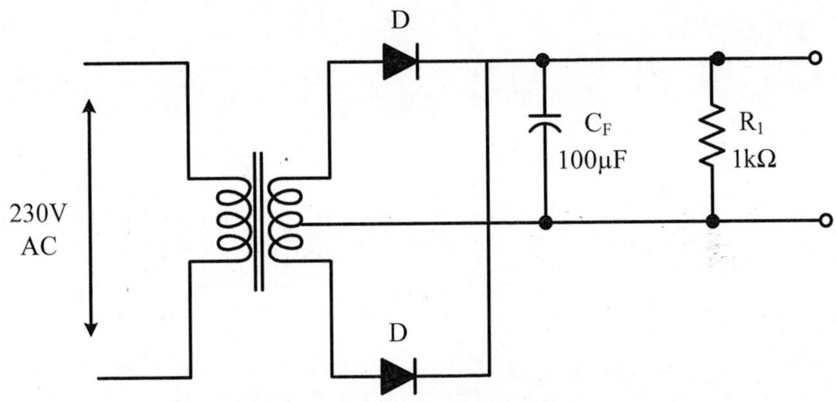

Fig. 2 Full wave rectifier with filter

Bridge rectifier with filter:

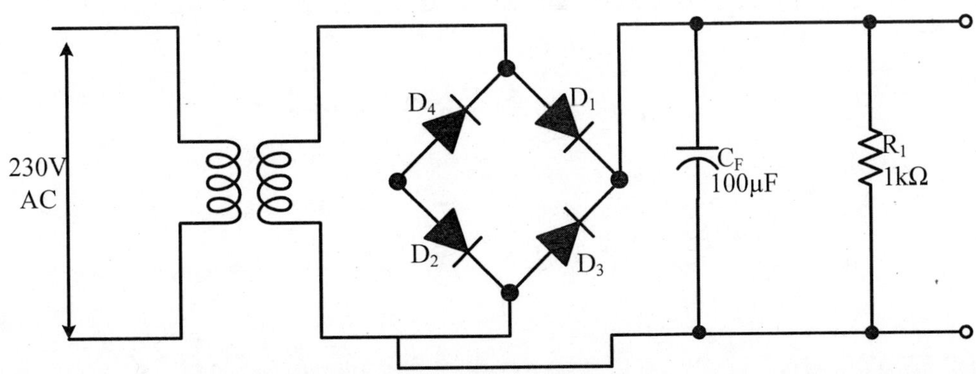

Fig. 3 Bridge rectifier with filter

DESIGN

Select 230V/6V-0-6V, 100 mA center-tapped transformer and diodes IN4001.

DESIGN OF LOAD RESISTOR *RL*:

Load resistor R_L should be high enough to make the capacitor discharge slowly. Same time it should limit the current through the diodes. Assume a current of 5mA to flow through the diodes.

Then $R_L = \dfrac{6\sqrt{2}-1.4}{0.005}\Omega$ Because the voltage drop across the two diodes together are 1.4V. Select $R_L = 1k\Omega$.

DESIGN OF CAPACITANCE C:

The required ripple factor of capacitor input filter is 3%. Theoretical value of r = ¼ $\sqrt{3}f\, R_L X$. Power supply frequency f = 50 Hz. Assume R_L =1 kΩ. Then C ≈ 100μF.

TABULAR COLUMN

Table 1: Rectifier without filter

HWR	V_m	$V_{rms} = V_m / 2$	$V_{dc} = V_m / \pi$	$r = \sqrt{(V_{rms} / V_{dc})^2 - 1}$
FWR	V_m	$V_{rms} = V_m / \sqrt{2}$	$V_{dc} = 2V_m / \pi$	$r = \sqrt{(V_{rms} / V_{dc})^2 - 1}$
BR	V_m	$V_{rms} = V_m / \sqrt{2}$	$V_{dc} = 2V_m / \pi$	$r = \sqrt{(V_{rms} / V_{dc})^2 - 1}$

Table 2: Rectifiers with capacitor filter

Type	V_m	V_{rpp}	$V_{r,rms} = V_{rpp}\ 2\sqrt{3}$	$V_{dc} = V_m - V_{rpp} / 2$	$r = V_{r,rms} / V_{dc}$
HWR					
FWR					
BR					

Waveforms: Typical waveforms of half wave rectifier without filter and with filter are shown in the figure below.

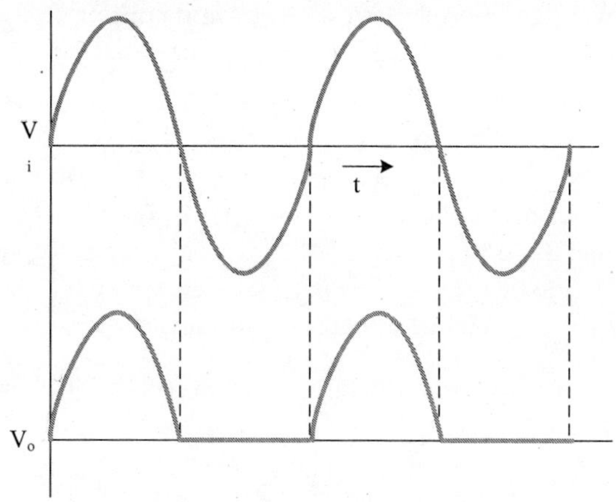

Fig. 4 Half wave rectifier without filter

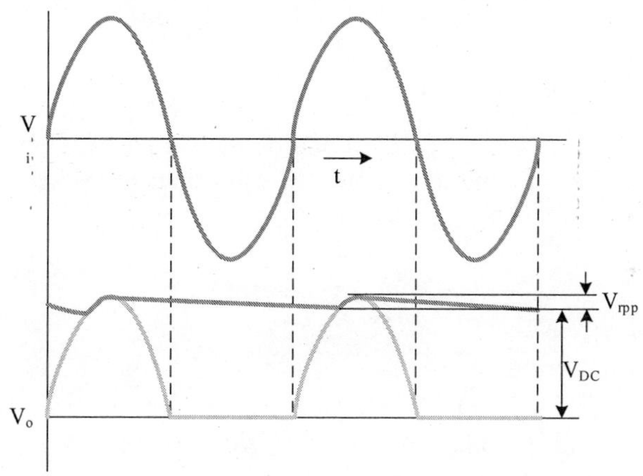

Fig. 5 Half wave rectifier with filter

RESULT:

EXPERIMENT NO. 5

AIM: To draw the input and output characteristics of common emitter configuration of BJT.

APPARATUS REQUIRED: Two voltmeters, two ammeters, multimeter, connecting wires and transistor (n-p-n)

THEORY: When a transistor is connected in a circuit, there are three possible configurations: common base, common emitter, and common collector. In this experiment, our focus will be on the common emitter configuration. To understand the behavior of a transistor, we can analyze its DC current-voltage (I-V) curves, which are known as the static characteristic curves of the device. These curves provide valuable insights into the transistor's operation and performance.

There are two important characteristics of a transistor that we will be studying:

i. i. Input Characteristics:

ii. ii. Output Characteristics:

Input Characteristics: The input characteristics of a transistor depict the relationship between the input current and the input voltage. In the common emitter configuration, the input characteristics show the variation of the base current (I_B) with respect to the base-emitter voltage (V_{BE}), while keeping the collector-emitter voltage (VCE) fixed

Input Dynamic Resistance (ri): This is defined as the ratio of change in base emitter voltage (ΔV_{BE}) to the resulting change in base current (ΔI_B) at constant collector- emitter voltage (V_{CE}). This is dynamic and it can be seen from the input characteristic, its value varies with the operating current in the transistor:

$$r_o = \frac{\Delta V_{BE}}{\Delta I_B} \text{ at constant VCE}$$

The value of r_i can be anything from a few hundred to a few thousand ohms.

Output Characteristics: The output characteristics of a transistor illustrate the relationship between the output current and the output voltage. In the common emitter configuration, the output characteristics show the variation of the collector current (I_C) with respect to the collector-emitter voltage (V_{CE}), while keeping the base current (I_B) constant. These characteristics give us insights into the transistor's output behavior, such as output resistance, saturation region, and cutoff region. By studying the input and output characteristics of a transistor in the common emitter configuration, we can gain a deeper understanding of its performance and use that knowledge in various electronic applications.

Output Dynamic Resistance (ro): This is defined as the ratio of change in collector- emitter voltage (ΔV_{CE}) to the change in collector current (ΔI_C) at a constant base current IB.

$$r_o = \frac{\Delta V_{CE}}{\Delta I_C} \text{ at constant IB}$$

The high magnitude of the output resistance (of the order of 100 kW) is due to the reverse-biased state of this diode.

CIRCUIT DIAGRAM: -

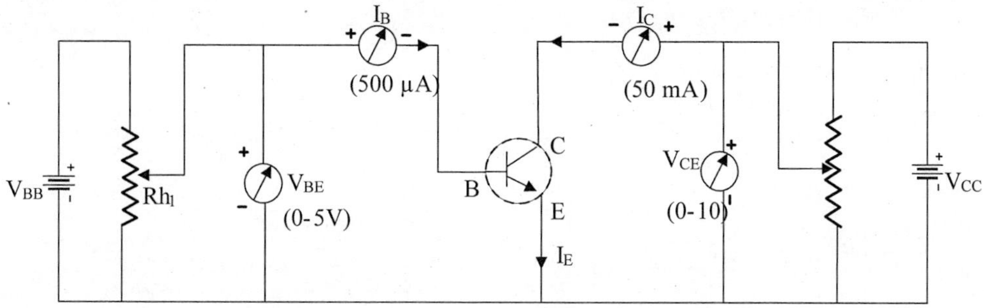

Fig. 1. Common emitter configuration of BJT

PROCEDURE:

a. **Input Characteristics:**

To perform the experiment as described:

1. Set up the circuit according to the circuit diagram provided in Fig. 1 for the common emitter configuration.
2. Set the collector-emitter voltage (V_{CE}) to 1V by adjusting the appropriate voltage source. Keep V_{CE} constant throughout the experiment.
3. Vary the base current (I_B) by adjusting the voltage V_{BB}. Measure the corresponding base-emitter voltage (V_{BE}) using a voltmeter or multimeter.
4. Repeat the above step for different values of I_B, ensuring to record the corresponding V_{BE} measurements each time.
5. Change the V_{CE} to 2V by adjusting the voltage source.
6. Repeat steps 3 and 4, varying I_B with V_{BB} and measuring V_{BE} for the new V_{CE} value.
7. Record all the measured values of V_{BE} for each corresponding I_B and V_{CE} combination.
8. Plot graph of I_B v/s V_{BE}.

b. **Output Characteristics:**

To perform the experiment as described:

1. Set the base current (I_B) to a constant value, such as 15 μA, by adjusting the appropriate current source. Keep I_B constant throughout this step.
2. Vary the collector-emitter voltage (V_{CE}) by adjusting the voltage source. Measure and note down the corresponding collector current (I_C) using a current measuring instrument, such as an ammeter.
3. Repeat the above step for different values of V_{CE}, ensuring to record the corresponding I_C measurements each time.
4. Change the I_B to a new constant value, such as 30 μA, by adjusting the current source.
5. Repeat step 2, varying V_{CE} and measuring I_C for the new I_B value.
6. Record all the measured values of IC for each corresponding V_{CE} and I_B combination.
7. Plot graph of I_B v/s V_{CE}.

OBSERVATION TABLE:

Input Characteristic:

S. No	V_{CE}		V_{CE}		V_{CE}	
	V_{BE}	I_B	V_{BE}	I_B	V_{BE}	I_B
1						
2						
3						

Output Characteristic:

S. No	I_B		I_B		I_B	
	V_{CE}	I_C	V_{CE}	I_C	V_{CE}	I_C
1						
2						
3						

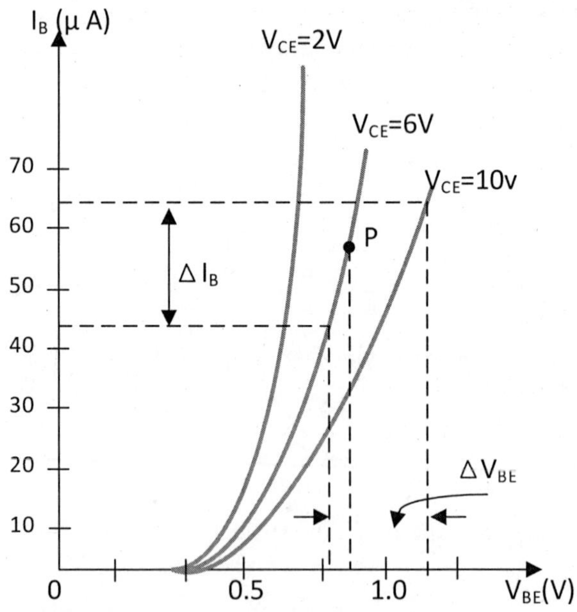

Fig. 2. Input characteristics CE configuration

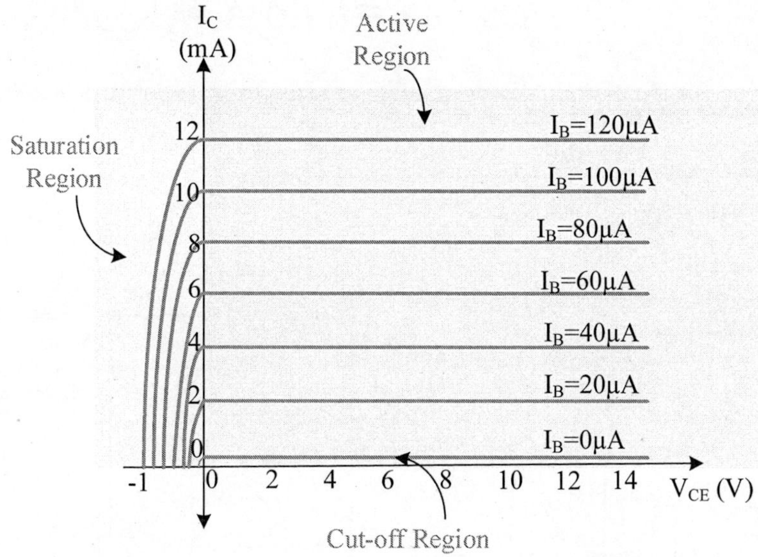

Fig. 3. Output characteristics CE configuration

RESULT:

EXPERIMENT NO. 6

AIM: Study and verification of drain characteristics and transfer characteristics of a MOSFET.

APPARATUS REQUIRED: DC power supply, multimeter, resistors, MOSFET.

CIRCUIT DIAGRAM:

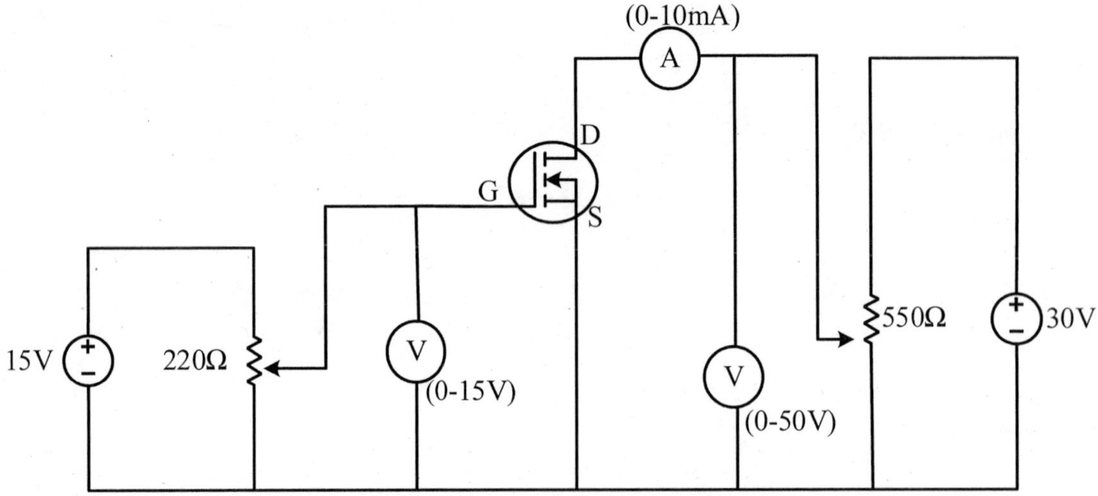

Fig. 1. Circuit Diagram of MOSFET.

THEORY: MOSFET form family of FET's (Field Effect Transistor). MOSFET derives its name from the fact that its metal gate is insulated by a very thin oxide layer from semiconductor channel. It is also known as insulated gate FET or IGFET.

MOSFET's are of two types: depletion type and enhancement type. Each of them may be n- channel MOSFET or p-channel MOSFET.

Depletion Mode:

In a depletion mode n-channel D-MOSFET (Depletion Metal-Oxide-Semiconductor Field-Effect Transistor), the behavior is such that the gate voltage is negative, causing a depletion of electrons in the channel. This depletion occurs because the negative charge on the gate repels the free electrons in the channel, creating a region with positive ions. By depleting the n-channel of some of its electrons, the number of available free electrons for current conduction is reduced. This results in an increased resistance in the channel, as there are fewer carriers available to carry the current. As a result, the current flowing from the source to the drain is affected by the resistance of the n-channel. By changing the negative voltage on the gate, we can control the resistance of the n-channel and, consequently, the current flow from source to drain. This is due to the depletion of the channel and the resulting change in the available carriers for conduction. The operation of a depletion mode n-channel D-MOSFET with a negative gate voltage is referred to as depletion mode, as it involves depleting the channel of free electrons and influencing the resistance and current characteristics of the device.

ENHANCEMENT MODE:

In an enhancement mode n-channel MOSFET (Metal-Oxide-Semiconductor Field-Effect Transistor), the behavior is such that the gate voltage is positive. In this case, the gate acts as a capacitor and induces a negative charge in the n-channel. Due to the positive gate voltage, additional free electrons are attracted to the n-channel. These electrons combine with the existing electrons in the channel, increasing the overall number of electrons available for conduction. This increase in the number of electrons enhances or increases the conductivity of the channel. As the positive gate voltage increases, the induced negative charge and the number of electrons in the channel also increase. Consequently, the conductivity of the channel increases, allowing for a greater flow of current from the source to the drain. Therefore, in enhancement mode operation, the positive gate voltage enhances or increases the conductivity of the n-channel, leading to a higher current flow from the source to the drain. The greater the positive voltage on the gate, the greater the enhancement and the resulting conduction in the channel.

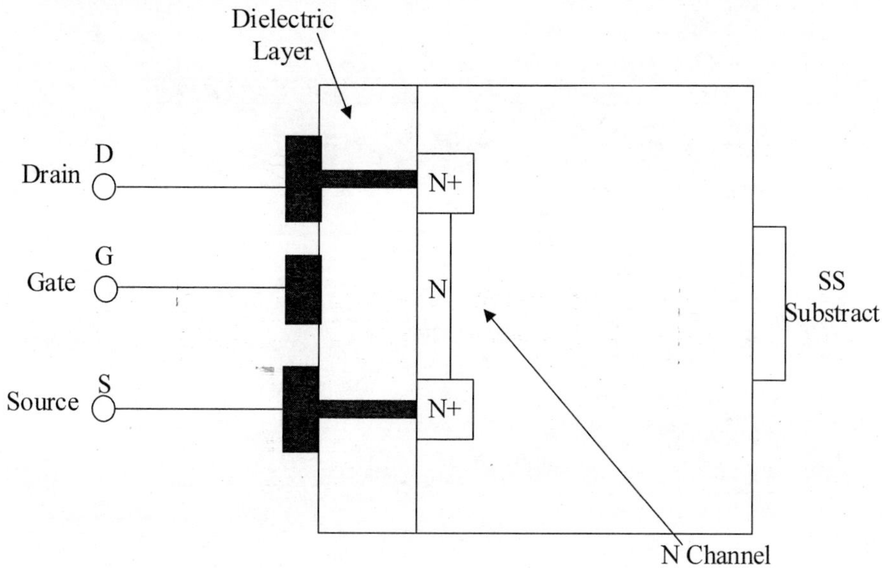

Fig. 2. A Typical N-Channel MOSFET

PROCEDURE:

a. Drain Characteristics:

Drain characteristics is drawn between I_D and V_{DS} keeping V_{GS} as constant.

1. Connect the circuit as shown in figure 1.
2. Let for V_{DS}= (0, 5, 10, 15, 20, 25) V and measure I_D respectively for different values of V_{GS}
3. Plot the drain characteristics in the graph.

b. Transfer Characteristics:

Transfer characteristics is drawn between I_D and V_{GS} keeping V_{DS} as constant.

1. Let for V_{GS} = (-4, -3 ,-2, -1, 0, 1, 2, 3, 4, 5)V and measure I_D respectively for different values of V_{DS} .
2. Plot the transfer characteristics in the graph.

OBSERVATION TABLE:

Drain Characteristic:

S. No	V_{GS}		V_{GS}		V_{GS}	
	V_{DS}	I_D	V_{DS}	I_D	V_{DS}	I_D
1						
2						
3						

Transfer Characteristic:

S. No	V_{DS}		V_{DS}		V_{DS}	
	V_{GS}	I_D	V_{GS}	I_D	V_{GS}	I_D
1						
2						
3						

IDEAL CHARACTERISTICS:

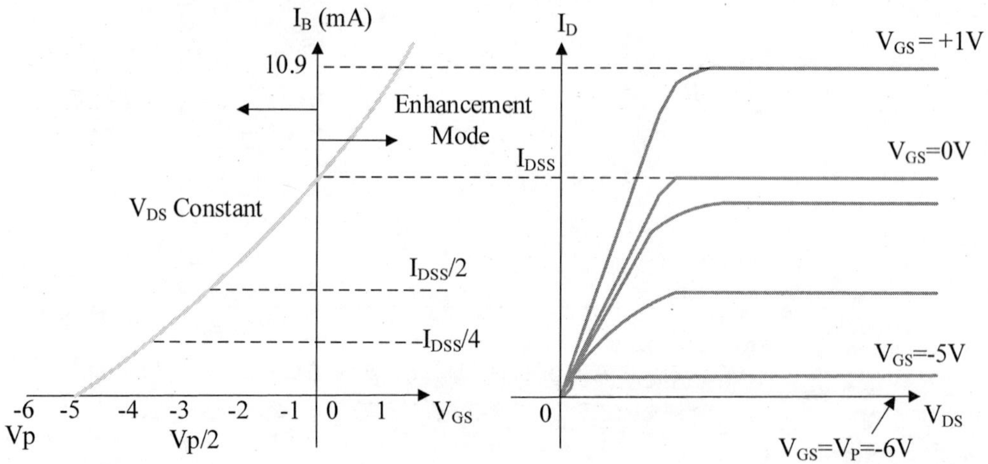

Fig. 3. Drain and transfer characteristics for an n-channel depletion type MOSFET.

RESULT:

EXPERIMENT NO. 7

AIM: To study Differential amplifier using BJT

APPARATUS REQUIRED: CRO, Power Supply, resistance, bread-board, capacitor, connecting wires

THEORY:

Differential Amplifier: A differential amplifier is most widely used circuit building block in analog integrated circuit. For instance, input stage of every op-amp amplifier is a differential amplifier. BJT differential amplifier is the basis of a very high-speed logic circuit family called emitter-coupled logic (ECL). The differential amplifier as the name suggests amplifies the difference between two input signal vin1 and vin2. BJT differential pair configuration consists of two matched transistors Q_1 and Q_2, whose emitters are formed together and biased by a constant current source I. Latter is usually implemented by a transistor circuit. The two collectors may be connected to another transistor rather than to resistive loads. It is essential though that the collector circuits be such that Q1 and Q2 can never enter saturation.

Gain of differential amplifier Ad = V_0/V_d where $V_d = V_{in1} - V_{in2}$

Non-Ideal Characteristics:

1. Finite Gain: In real operational amplifiers, the open-loop gain is not infinite as it is in the ideal differential amplifier. This finite gain becomes significant when the closed-loop gain is very high, resulting in a low feedback gain and low loop gain. Consequently, the differential amplifier deviates from ideal behavior in such cases.
2. Finite Input Impedance: The input impedance of a differential amplifier is the impedance between its two inputs, rather than the impedance from each input to ground. In high-gain applications with negative feedback, the input impedance is typically high.
3. Non-Zero Output Impedance: Real amplifiers have a non-zero output impedance, which becomes important when driving low resistance loads. The voltage drop across the output impedance of the amplifier can be significant and affects the maximum power that can be delivered to the load.
4. Input Offset Voltage: Input offset voltage refers to the voltage required across the differential amplifier's input terminals to drive the output voltage to zero. It arises from mismatches in input bias currents. In an ideal amplifier, there would be no input offset voltage.
5. Common Mode Gain: While an ideal differential amplifier amplifies only the voltage difference between its inputs and rejects any voltage that is common to both inputs, the differential input stage of a real differential amplifier is not perfect. As a result, there is some amplification of these common-mode voltages to a certain degree.

CIRCUIT DIAGRAM:

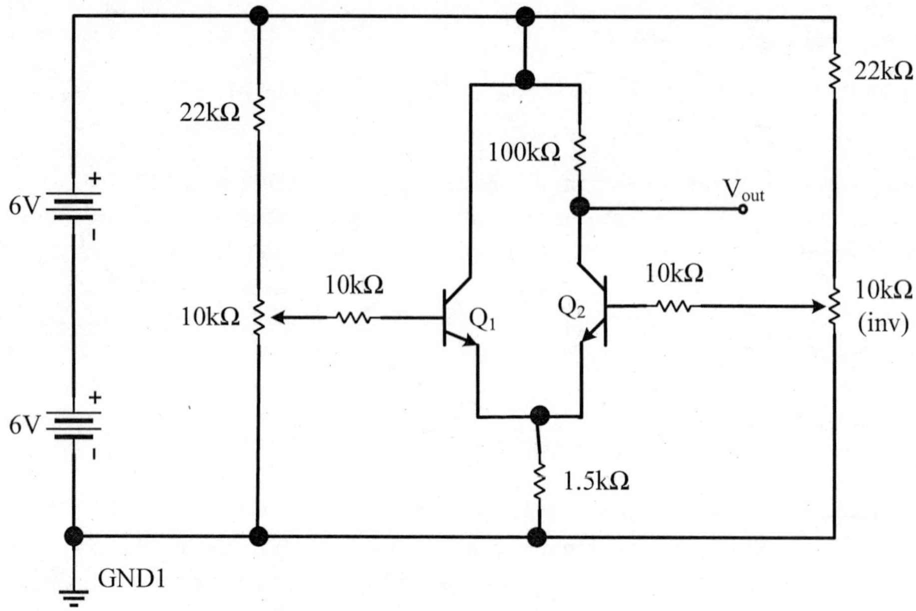

Fig. 1. Differential amplifier

PROCEDURE:

To connect the circuit and measure the output voltage, follow the steps below:

1. Obtain the circuit diagram shown in Fig. 1.
2. Gather the necessary components mentioned in the circuit diagram.
3. Place the components on a breadboard or any suitable circuit prototyping platform.
4. Connect the 12V power supply to the circuit. Ensure the positive terminal of the power supply is connected to the appropriate point in the circuit and the negative terminal is connected to the ground or common reference point.
5. Connect the V_{in1} and V_{in2} voltages to their respective points in the circuit as indicated in the diagram. Use suitable wires or connections to establish the connections securely.

PRECAUTIONS:

1. Once all the connections are made, ensure that there are no loose connections or short circuits in the circuit.
2. Use a voltmeter or multimeter to measure the output voltage. Set the voltmeter to the appropriate voltage range, typically higher than the expected output voltage.
3. Place the voltmeter probes across the output points of the circuit, as indicated in the diagram, to measure the voltage accurately.
4. Turn on the power supply and observe the output voltage reading on the voltmeter.
5. Note down the measured output voltage for further analysis or use.

RESULT:

The gain of a differential amplifier can be calculated using the following formula:

$$\text{Gain} = A_d \times (V_{inp} - V_{inn}) \text{ and is } \ldots\ldots\ldots\ldots\ldots..$$

Gain: The voltage gain of the differential amplifier.

A_d: The differential voltage gain of the amplifier.

V_{inp}: The input voltage at the non-inverting terminal.

V_{inn}: The input voltage at the inverting terminal.

EXPERIMENT NO. 8

AIM: 1. To design a small signal voltage amplifier.

2. To plot its frequency response and to obtain bandwidth.

APPARATUS REQUIRED: CRO, Power Supply, resistance, bread-board, capacitor. BJT

THEORY: Amplifiers are classified as small signal amplifiers and large signal amplifiers depending on the shift in operating point, from the quiescent condition caused by the input signal. If the shift is small, amplifiers are referred to as small signal amplifiers and if the shift is large, they are known as large signal amplifiers. In small signal amplifiers, voltage swing and current swing are small. Large signal amplifiers have large voltage swing and current swing and the signal power handled by such amplifiers remain large. Voltage amplifiers come under small signal amplifiers. Power amplifiers are one in which the output power of the signal is increased. They are called large signal amplifiers. Figure shows the circuit diagram of a common emitter amplifier.

CIRCUIT DIAGRAM:

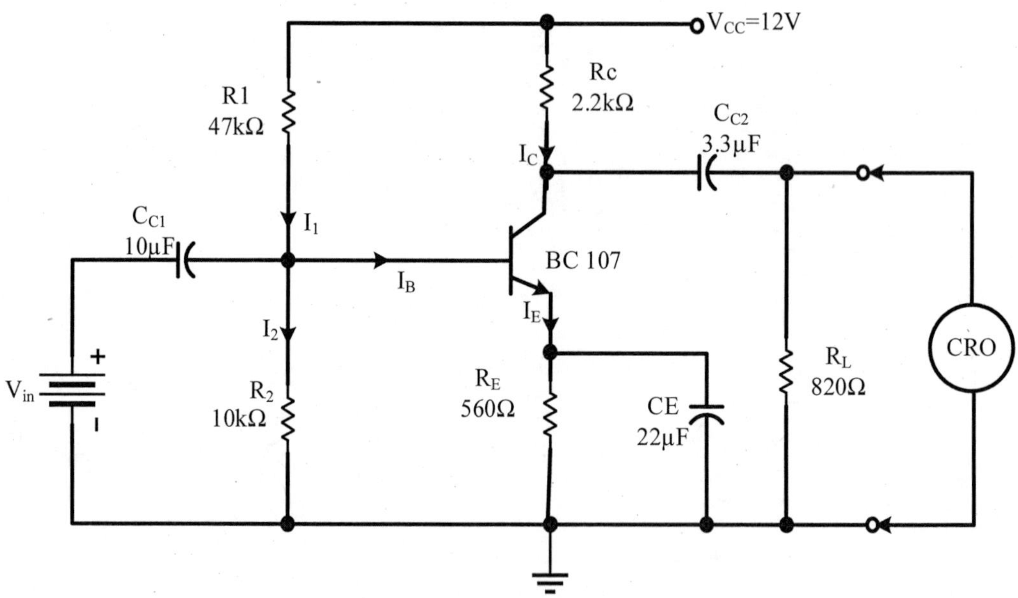

Fig. 1. Circuit diagram

DESIGN: From the transistor datasheet, for BC107,

$$h_{fe} = \beta = 110,\ I_{c\,max} = 100\text{mA},\ V_{CE\,max} = 45\text{V}$$

Let V_{cc} = 12V, I_c = 2mA, Since the quiescent point is in the middle of the laod line for the amplifier, V_{CE} = 50% of V_{cc} = 6V.

$$V_{RE} = 10\% \text{ of } Vcc = 1.2\text{V}$$

Assuming $I_C = I_E$, $V_{RE} = I_C R_E = I_E R_E$

$$1.2 = 2 \times 10^{-3} \times R_E$$

$$R_E = \frac{1.2}{2\times10^{-3}} = 600\Omega$$ Select standard value of resistance 560Ω.

Voltage across collector resistance, $V_{RC}=V_{CC}-V_{CE}-V_{RE}$

$$V_{RC} = 12 - 6 - 1.2 = 4.8V$$

$$R_C = \frac{V_R C}{I_C} = \frac{4.8}{2 \times 10^{-3}} = 2.4k\Omega$$ Select standard value of 2.2kΩ

Base Current, $I_B = \frac{I_c}{\beta} = \frac{2\times10^{-3}}{110} = 18.2\Omega A$

Take $I_2 = 10I_B$ then $I_1 = 10I_B + I_B = 11I_B$

Base voltage, $V_B = V_{RE} + V_{BE} = 1.2 + 0.6 = 1.8V$

$$R_2 = \frac{V_B}{I_2} = \frac{1.8}{10 \times 18.2 \times 10^{-6}} = 9.9k\Omega$$ Select standard value of 10kΩ

$$R_1 = \frac{V_C C - V_B}{I_1} = \frac{12-1.8}{11 \times 18.2 \times 10^{-6}} = 51k\Omega$$ Select standard value of 47kΩ

Design of R_L:

Gain of the common emitter amplifier is given by the expression $Av = -\left(\frac{r_c}{r_c}\right)$

Where $r_c = R_C \,||\, R_L$ and $r_e = \frac{25mV}{2mA} = 12.5\Omega$

For a gain of 50, substitution it in the expression we get, $R_L = 873k\Omega$
Select standard value of 820Ω for R_L.
Design of coupling capacitors C_{C1} and C_{C2}:
X_{C1} should be less than the input impedance of the transistor. Here, R_{in} is the series impedance.

Then $X_{c1} \le \frac{R_{in}}{10}$

Here, $R_{in} = R_1 \,||\, R_2 \,||\, h_{fe}r_e = 47k\Omega \,||\, 10k\Omega \,||\, 110 \times 12.5\Omega = 1.7k\Omega$

We get Rin = 1.17kΩ. Then $X_{C1} \le 117\Omega$

For a lower cut off frequency of 200Hz, $C_{C1} = \frac{1}{2\pi f X_{c1}} = \frac{1}{2\pi \times 200 \times 117} = 6.8\mu F$

Select standard value of $10\mu F$ for C_{C1}

$$\text{Similarly, } X_{C2} \leq \frac{R_{out}}{10} \text{ where } R_{out} = Rc. \text{ Then } X_{CE} \leq 220\Omega$$

$$C_{C2} = \frac{1}{2\pi f\, X_{c2}} = \frac{1}{2\pi \times 200 \times 220} = 3.6\mu F$$

Select standard value of $3.3\mu F$ for C_{C2}

Design of bypass capacitors C_E:

To bypass the lowest frequency (say 200Hz), X_{CE} should be much less than or equal to the resistance R_E.

$$X_{CE} \leq \frac{R_E}{10}$$

$$X_{CE} \leq \frac{560}{10} \quad \text{ie. } X_{CE} \leq 56$$

Apply value of f such that the amplifier has good gain at a lower cutoff frequency of 200Hz

$$C_E \geq \frac{1}{2\pi f\, X_{CE}} = \frac{1}{2\pi \times 200 \times 56} = 14.2¥\text{F}$$

Select standard value of 22μF for C_E

FREQUENCY RESPONSE:

In an actual CE (Common Emitter) amplifier, the frequency response curve will deviate from a straight line due to various factors. These factors include the coupling capacitors, emitter bypass capacitor, internal capacitance of the transistor, and stray capacitance due to wiring. As a result, the gain of the amplifier will vary with different frequencies. Fig. 2 represents the typical frequency response characteristics of a CE amplifier, showing a curve that is flat in the middle range of frequencies. The curve deviates from this flat region at low and high frequencies, indicating a decrease in gain.

Low Frequency (f_L): At low frequencies, the gain of the CE amplifier decreases. This can be attributed to the coupling capacitors and the emitter bypass capacitor acting as high-pass filters, causing a roll-off in gain as the frequency decreases. These capacitors tend to block or attenuate low-frequency signals, resulting in a reduced gain at lower frequencies.

High Frequency (f_H): At high frequencies, the gain of the CE amplifier also decreases. This reduction is mainly due to the internal capacitance of the transistor and stray capacitance in the circuit. These capacitances introduce a low-pass filtering effect, causing a decrease in gain as the frequency increases. The internal and stray capacitances create impedance that becomes significant at higher frequencies, resulting in a loss of signal amplification. Fig. 2 The curve is flat for middle range frequencies. There is only one Low Frequency (f_L) and High Frequency (f_H) beyond which the gains, A_L and A_H are $\frac{1}{\sqrt{2}}$ times the gain A_M (maximum gain) at the middle frequencies. The two frequencies are lower and higher cutoff frequencies. The difference between them is called as the bandwidth.

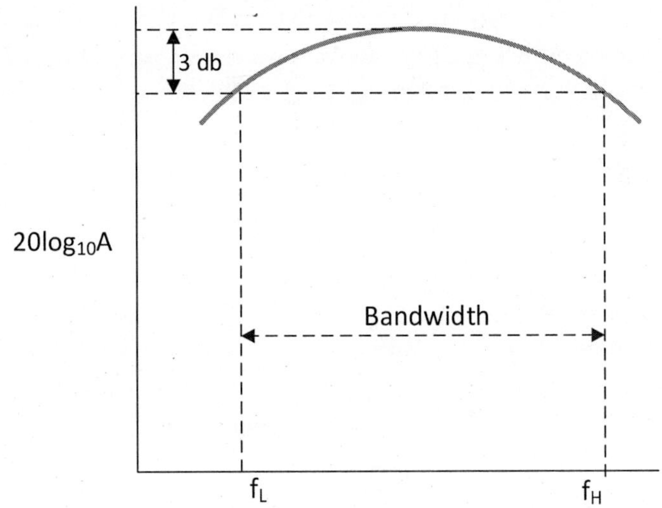

Fig. 2. Frequency response

PROCEDURE:

To perform the experiment the circuit is to be connected as shown in Fig. 1. The frequency response is measured by following the instructions below:

1. Connect the input signal Vs from a signal generator to the appropriate input point of the circuit. Ensure that the signal generator is set to produce a sinusoidal signal.
2. Use a CRO (Cathode Ray Oscilloscope) to measure the magnitude (peak to peak) of the input signal. Connect the probe of the CRO to the input side of the circuit and adjust the settings of the CRO to accurately measure the input signal.
3. Connect the CRO probe to the output side of the circuit to observe the amplified output. Adjust the settings of the CRO to display the output waveform clearly.
4. Start with a specific frequency and increase it in steps. For each frequency step, observe the magnitude of the output voltage (Vo) on the CRO. Note down the corresponding frequency and the magnitude of the output voltage for each step.
5. Continue increasing the frequency and recording the corresponding magnitude of the output voltage until you reach the desired frequency range or until the response deviates significantly from the flat region.
6. Plot the frequency response on a semi-log sheet.

OBSERVATIONS:

Readings are to be taken till Vo decreases appreciably at high frequencies

Vin=………….(p-p)(mV)

Frequency f(Hz)	Vo(p-p) (mV)	Vo/Vin	Gain in db =20log Vo/Vin

RESULT:

The common emitter amplifier is designed, and its frequency response is plotted.

Voltage gain = Vo/Vin =

Lower cutoff frequency =

Upper cutoff frequency =

Bandwidth =

EXPERIMENT NO. 9

AIM:

1. To study the pin configurations, specifications & functioning of IC 741 Op-Amp.
2. To measure the Op-Amp parameters and compare them with ideal characteristics.

APPARATUS REQUIRED:

1. IC μA 741 OP-Amp.
2. CRO.
3. Dual Regulated Power Supply
4. Voltmeter 0-10mV–2 Nos.
5. Ammeters 0-10 μA–2 Nos.
6. Connecting Wires.

PINOUT AND CIRCUIT DIAGRAMS:

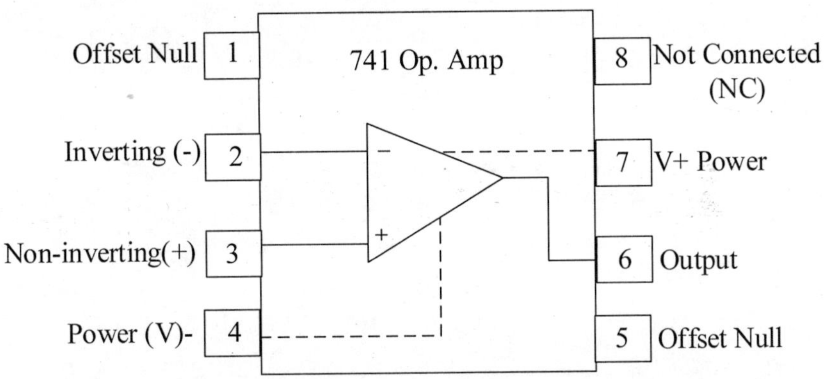

Fig. 1. Pin Configuration of IC 741 Op-amp

CIRCUIT DIAGRAMS:

a. **Input bias current and input offset current**

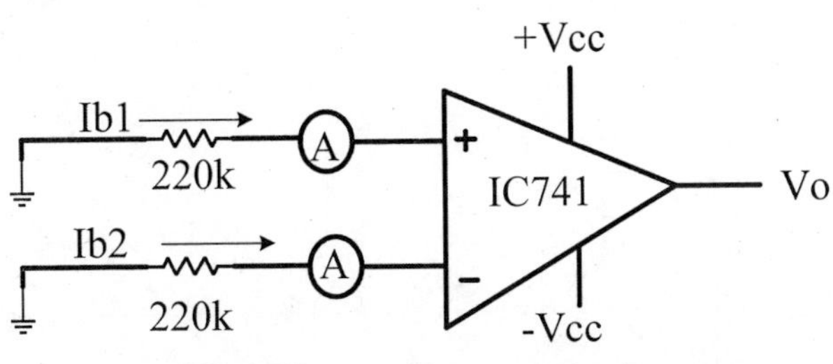

Fig. 2. Input bias current and input offset current

Input Bias current: $(I_{b1} + I_{b2}) / 2$

Input offset current: $|I_{b1} - I_{b2}|$

b. **Input offset voltage**

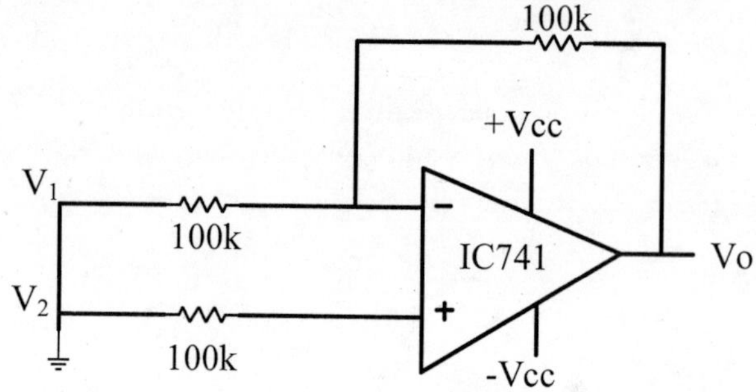

Fig. 3. Input offset voltage

V_{iof} = The input voltage for which the output Vo is zero. Otherwise, it is the output voltage for input zero with unit gain amplification.

c. **CMRR (Common Mode Rejection Ratio)**

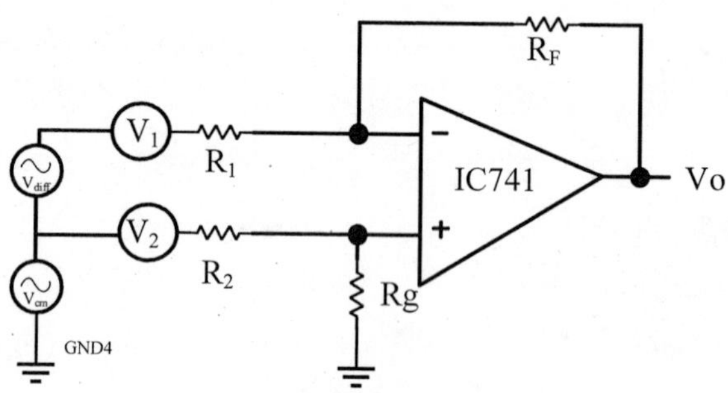

Fig. 4. CMRR

CMRR = A_d / A_c, Differential input voltage is given between V_1 and V_2 and common mode voltage is given as common to both the inputs.

d. **Slew rate and Bandwidth**

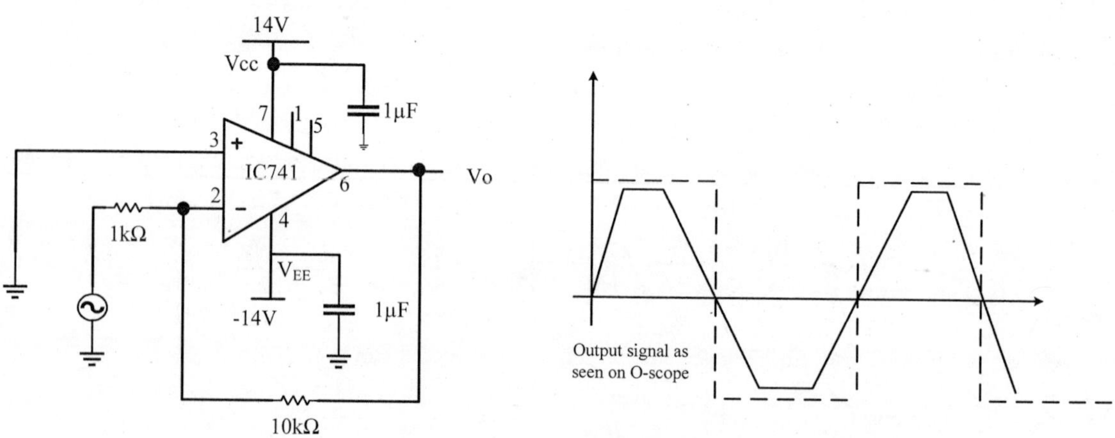

Fig. 5. Slew rate and bandwidth

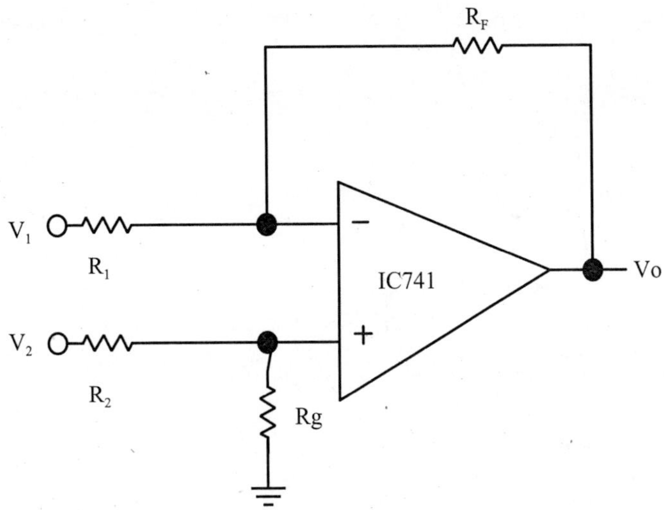

Fig. 6. Bandwidth

THEORY

Operational amplifiers, commonly known as op-amps, are versatile electronic devices widely used in various applications due to their high gain, differential input, and externally controlled response characteristics. Here are some key points related to op-amps:

High Gain: Op-amps have very high voltage gain, typically in the range of tens of thousands to hundreds of thousands. This high gain allows op-amps to amplify small input signals to a much larger output signal.

Differential Amplifier: Op-amps are designed as differential amplifiers, which means they amplify the voltage difference between two input terminals. The output of an op-amp is directly related to the difference between its two inputs.

Negative Feedback: The response characteristics of op-amps are controlled by negative feedback, where a portion of the output signal is fed back to the input in a way that reduces any difference between the input and desired output. Negative feedback stabilizes the op-amp's behavior, making it less sensitive to variations in its internal characteristics.

High Input Impedance: Op-amps have very high input impedance, typically in the megaohm range. This high input impedance ensures that the op-amp draws minimal current from the source connected to its inputs, minimizing any loading effect on the source.

Low Output Impedance: Op-amps have low output impedance, typically less than 100 ohms. This low output impedance allows the op-amp to drive loads without significant signal degradation.

Mathematical Operations: Due to their versatile nature, op-amps can be configured to perform mathematical operations such as summation, integration, differentiation, logarithm, anti-logarithm, etc. These configurations are achieved by connecting external resistors, capacitors, and other components in specific ways.

Application Areas: Op-amps find extensive use in various fields, including communication electronics, instrumentation and control systems, medical electronics, audio and video amplification, oscillators, filters, signal conditioning, and more. Their flexibility and performance characteristics make them suitable for a wide range of applications.

Op-Amp characteristics: An ideal op-amp draws no current from the source and its response is also independent of temperature. However, a real op-amp does not work this way. Current is taken from the source into op-amp inputs. Also, the two inputs respond differently to current and voltage due to mismatch in transistors. A real op-amp also shifts its operation with temperature. These non-ideal characteristics are:

7. Input bias current
8. Input offset current
9. Input offset voltage
10. Thermal drift
11. Slew rate
12. Input and output voltage ranges

PROCEDURE:

1. Connect the IC 741 with supply voltages on a bread board.
2. Connect the CRO as per the requirement for voltage and for current measurement.
3. Measure the parameters of Op-Amp and compare them with ideal one.

1. Input bias current and input offset current:
   1. Connect the circuit as shown in Fig. 2 for Input bias current and input offset current.
   2. Measure the current at inverting and non-inverting terminals of op-amp as IB^+ and IB.
   3. Average these values to find out the input Bias current. Also, find the difference between these two currents to know the input offset current.

IB^+	IB^-	Input Bias Current	Input Offset Current

2. Input offset voltage:
   1. Connect the circuit as shown in Fig. 3. for Input offset voltage measurement.
   2. Give zero volts as input and measure the output.

Output Voltage (V_{out})	Input Voltage at Inverting Terminal V_1(V)	Input Voltage at Non-Inverting Terminal V_2(V)

3. CMRR Measurement:

$$Ad = \frac{V_o}{V_d}$$

Configure the Op-Amp in unit gain configuration, in differential mode operation as shown in Fig. 4. Give 1V as differential input and observe the output as 1V. Keeping the differential input same as

1V, increase the common mode voltage from 0V to 10V insteps of 1V and observe the variation in the output. Calculate A_d and A_c and CMRR.

4. Slew Rate:
   1. Connect the circuit diagram as shown in Fig. 5.
   2. Apply A square wave of low frequency at inverting terminal of op-amp.
   3. The input amplitude of square wave is to be adjusted until the output is 20 volts peak to peak.
   4. The frequency is then to be adjusted until the output becomes triangular.
   5. The rising edge slope of the triangle-wave was taken to be the slew rate that was being sought.

Input Voltage at Inverting Terminal V(V)	Output Slope V/μ Sec	Frequency (Hz)

5. Bandwidth:

Connect the op-amp in unit gain configuration as shown in Fig. 6. Apply 1V sinusoidal input at 10Hz and observe the output. Increase the frequency in steps (semi-logarithmic scale) and note the output voltage values. Identify the -3dB cut-off point, which gives the band width.

RESULT:

The pin configuration, specifications & functioning of different integrated circuits used in the practical applications have been studied.

EXPERIMENT NO. 10

AIM: To design an inverting and non-inverting amplifier using op-amp.

APPARATUS REQUIRED: Signal generator, IC 741, DSO, patch cards, multimeters, R_1 =1K, R_F=0.22K.

PIN DIAGRAM:

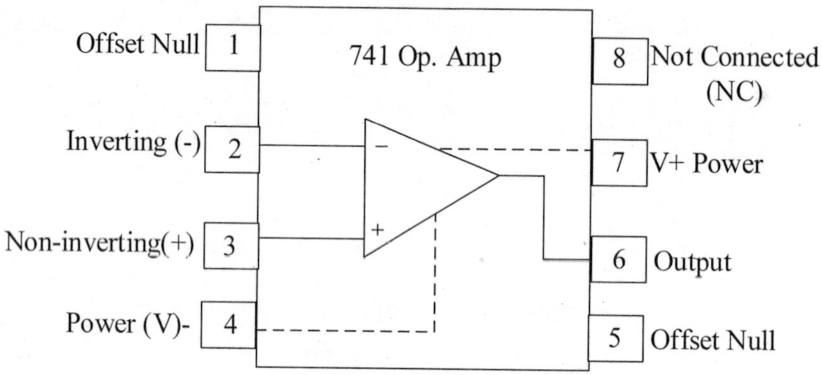

Fig. 1. Pin Configuration of IC 741 Op-amp

CIRCUIT DIAGRAM:

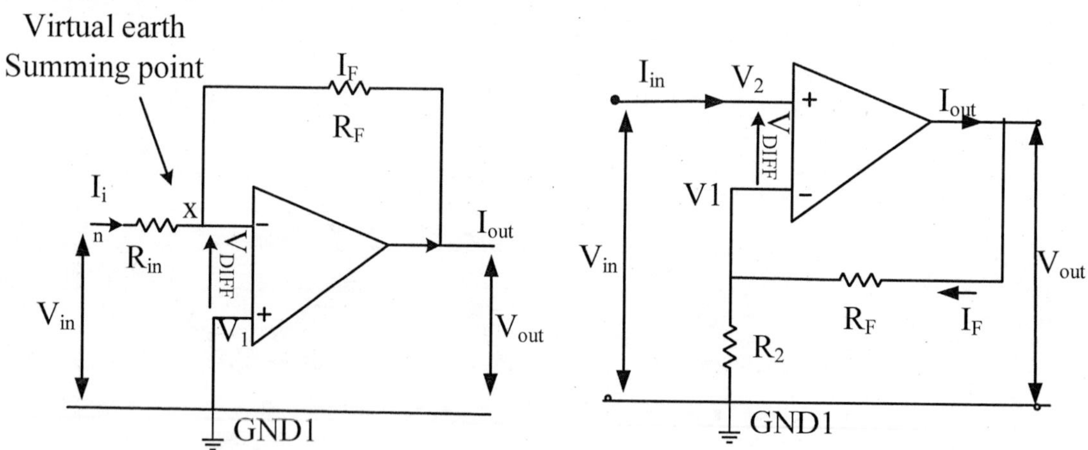

Fig. 2. (a) Inverting **Fig. 2.** (b) Non- Inverting

THEORY: An op-amp can be used for number of applications like amplifier, Subtractor, rectifier, multi vibrators, analogue computer, etc; IC741 is used as inverting and non-inverting amplifier.

When the input is applied to the inverting input terminal (negative terminal) of Op- Amp then Op-Amp is said to be operated in inverting amplifier mode. Whereas when the input is applied to the non-inverting input terminal (positive terminal) of Op-Amp then Op-Amp is said to be operated in non-inverting amplifier mode. The amplified as well as the inverted output signal is obtained from output pin 6 of Op- Amp. This output signal is applied to inverting input via feedback resistor R_F. It forms negative feedback because any increase in the output signal results in a feedback signal into inverting causing a decrease in output signal.

Note that non-inverting terminal is grounded, Feedback circuit used any one resistor R_F and however an extra resistor R_1 is connected in series with the input signal Vin. It is also known as voltage shunt feedback amplifier.

In case of non-inverting amplifier, inverting terminal is connected with O/P V_{out} through Feedback resistor R_F.

CLOSED LOOP VOLTAGE GAIN (For Inverting Amplifier):

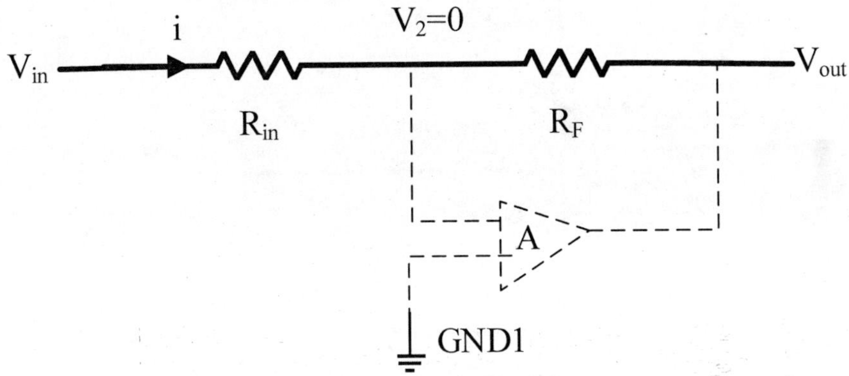

Fig. 3. Diagram of an Inverting Op-Amp Circuit

$$i = \frac{Vin - Vout}{Rin + Rf}$$

$$\text{therefore, } i = \frac{Vin - V2}{Rin} = \frac{V2 - Vout}{RF}$$

$$\text{so, } \frac{Vin}{Rin} = V2\left[\frac{1}{Rin} + \frac{1}{R_F}\right] - \frac{Vout}{R_F}$$

$$\text{and as, } i = \frac{Vin - 0}{Rin} = \frac{0 - Vout}{R_F}$$

$$\frac{R_F}{Rin} = \frac{0 - Vout}{Vin - 0}$$

$$\text{The closed Loop gain (Av) is given as, } \frac{Vout}{Vin} = -\frac{R_F}{Rin}$$

CLOSED LOOP VOLTAGE GAIN (For Non- Inverting Amplifier):

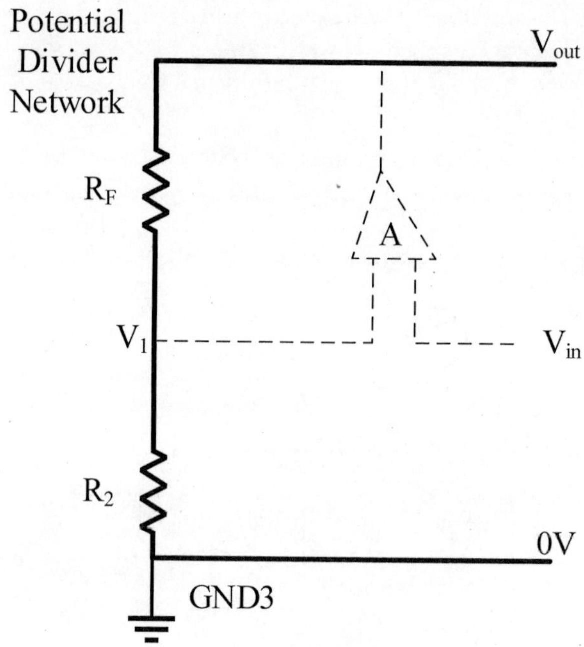

Fig. 4. Diagram of an Non-Inverting Op-Amp Circuit

$$V1 = \frac{R_2}{R_2 + R_F} \times V_{OUT}$$

Ideal Summing Point: $V_1 = V_{IN}$

Voltage Gain, A(V) is equal to $\frac{V_{OUT}}{V_{IN}}$

$$\text{Then, } A(V) = \frac{V_{OUT}}{V_{IN}} = \frac{R_2 + R_F}{R_2}$$

$$\text{Transpose to give: } A(V) = \frac{V_{OUT}}{V_{IN}} = 1 + \frac{R_F}{R_2}$$

PROCEDURE:

For ac input

1. Study the circuit provided on front panel of the kit.
2. Make the circuit as shown above Fig. 2(a)
3. Switch on the power supply.
4. Select desired R_1 and R_F.
5. Connect dual trace CRO at input and output side to observe Vin and Vo respectively.
6. Apply 100Hz, sine wave input Vin from signal generator. Adjust its amplitude so that Op-Amp should not enter in saturation.
7. Observe and note input Vin amplitude and output Vo amplitude on CRO, calculate its gain

using AF = Vo/Vin.

8. Calculate theoretical gain of Op-Amp using selected R_1 and R_F values.
9. Compare the above results.
10. Vary the input frequency, observe change in output.
11. Draw the waveform on graph.
12. Repeat the above procedure for different combinations of R_1 and R_F.

For dc input

1. To observe output at DC input the amplitude of function generator at zero.
2. ON the offset and apply output of function generator i.e. DC output to the DC input of the kit.
3. Adjust the DC offset above and below the zero level (this procedure is applicable if we are giving DC input by function generator).

For applying DC input from power supply

1. Adjust the power supply so that the output of power supply in mV or <1V.
2. In this operating condition apply the power supply to the DC input of the kit and observe the output by connecting DC voltmeter (above no. of procedure is applicable to different input devices to avoid the saturation condition of amplifier).

OBSERVATION TABLE (FOR INVERTING AMPLIFIER):

S. no	Vin	Vo	Theoretical gain AF = $-R_F/R_1$	Practical gain AF = –Vo/Vi	% Error
1					
2					
3					

OBSERVATION TABLE (FOR NON-INVERTING AMPLIFIER):

S. no	Vin	Vo	Theoretical gain AF = $1+R_F/R_1$	Practical gain AF = Vo/Vin	% Error
1					
2					
3					

RESULT:

Thus, IC741 Op-amp can work as a both inverting and non-inverting mode.

EXPERIMENT NO. 11

AIM: To design and set up an integrator and differentiator circuit using op-amp.

APPARATUS REQUIRED: Power supply, CRO, function generator, bread board, op-amp, capacitor and resistors

THEORY:

INTEGRATOR: Refer to the figure 1. This circuit performs the integration of the input waveform. The output voltage Vo can be expressed as $V_O = -\frac{1}{RC}\int Vidt + k$ where k is the constant of integration which depends upon the value of Vo at t=0. The peak of the ouput waveform VT is given by the expression $VT = \frac{VT}{4RC}$, where T is the time period of the input square wave. Integrators are commonly used in analog computers and wave shaping networks.

DIFFERENTIATOR: If the input resistor of the inverting amplifier is replaced by a capacitor, it forms an inverting differentiator. The output of the circuit is the derivative of the input. Gain of the differentiator increases with increase in frequency, which makes the circuit unstable. This is a drawback of the circuit. The output voltage Vo can be expressed as $V_O = -R_F C_i \frac{dVi}{dt}$ Differentiator functions are high pass filter. At high frequency it becomes unstable and breaks into oscillations. Input impedance decreases with increase in frequency which makes the circuit very susceptible to high frequency noise. Both stability and high frequency noise problem can be reduced significantly by additional circuit elements.

DESIGN AND CIRCUIT DIAGRAMS:

DESIGN OF INTEGRATOR: Let the input frequency be 1kHz. The frequency at which the integrator gives unity gain output is given by, $f = \frac{1}{2\pi R_1 C}$

Let C = 0.01μF, then R_1 = 15.9kΩ. Use 15kΩ std.

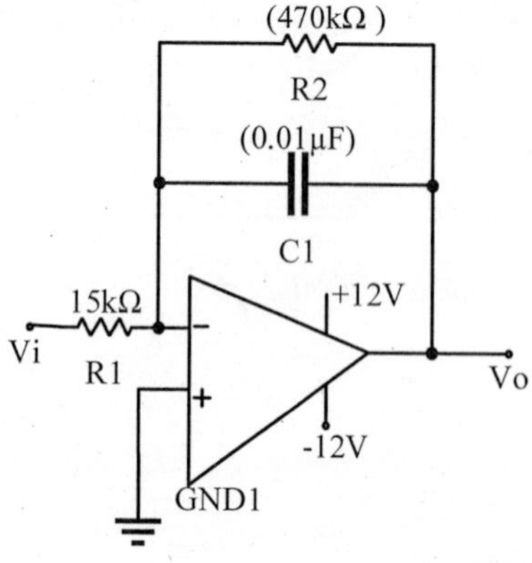

Fig. 1. Circuit diagram of Integrator

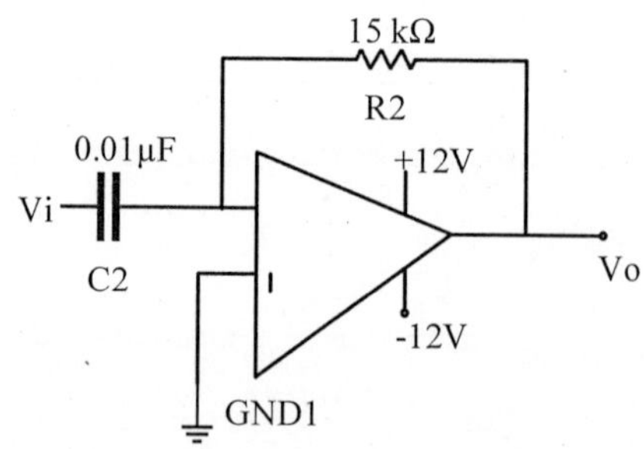

Fig. 2. Circuit diagram of Differentiator

The resistor R_2 in the integrator is provided to attenuate low frequency signals, particularly input dc offset voltage that may be present. Typically, the value of R_2 is selected as 10 times R_1 or more. Select the value of R_2 as 470k.

DESIGN OF DIFFERENTIATOR

$$\text{We have } f = \frac{1}{2\pi RC}$$

Let C = 0.01μF, then R = 15.9kΩ. Use 15kΩ std.

PROCEDURE:

INTEGRATOR

1. Set up the integrator circuit as shown in Fig.1 Give a rectangular wave of ±5V (10Vpp) and 1 kHz frequency at the input and observe the input and output simultaneously on CRO.
2. Vary the dc offset of the square wave input and observe the difference in the output waveform.
3. Repeat the experiment by feeding triangular wave and sine wave at the input and observe the output.

DIFFERENTIATOR

1. Set up the differentiator circuit as shown in Fig. 2. Give a rectangular wave of ±5V (10V pp) and 1 kHz frequency at the input and observe the input and output simultaneously on CRO.
2. Repeat the experiment by feeding triangular wave and sine wave at the input and observe the output.

WAVEFORMS

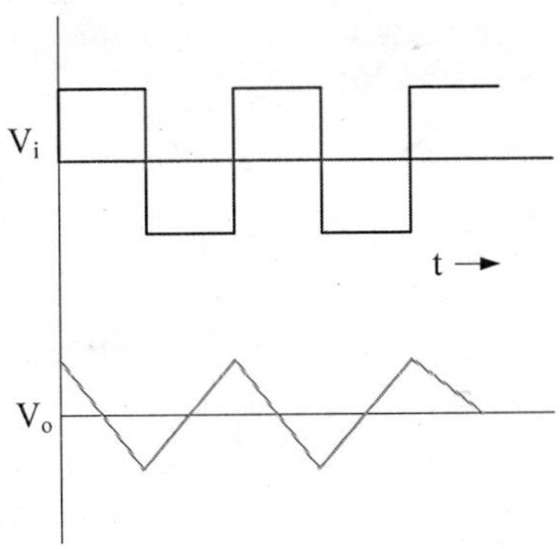

Fig. 3. Integrator output

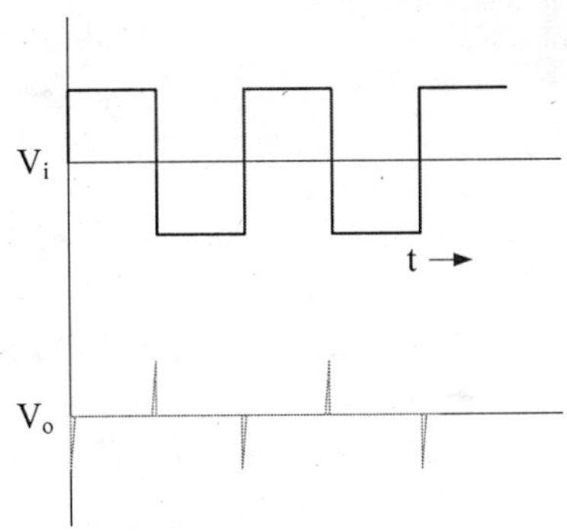

Fig. 4. Differentiator output

RESULT:

Integrator and Differentiator have been studied using op-amp.

EXPERIMENT NO. 12

AIM: Study the operation of Monostable and astable Multivibrator.

APPARATUS REQUIRED: CRO, Resistors, Capacitors, IC 555, Diode

THEORY: The IC 555 is one of the most popular and most widely used IC's. It is a versatile and extremely robust integrated circuit that is used in many applications like timers, wave generators (pulse) and oscillators.

The IC555, popularly known as the 555 Timer, was developed by Hans Camenzind of Signetic Corporation in the year 1971.

Some of the important features of the 555 timer are:

- The 555 timer can be operated at a wide range of power supplies ranging from 5 V to 18 V.
- It is available in 3 different packages: 8-pin Metal Can package, 8-pin DIP and 14-pin DIP.
- Timing can be anywhere from microseconds to hours.
- It can operate in both astable and monostable modes.
- High output current.
- It has an adjustable duty cycle.
- It is TTL compatible due to its high output current.
- The output can source or sink a current of 200mA to the load.
- It has a temperature stability of 0.005% per °C.

Different Modes of Operation: Generally, the 555 timer can be operated in three modes: Astable, Monostable (or one- shot) and Bistable.

Astable Mode: In this mode, the 555 work as a free running mode. The output of astable multivibrator will continuously toggle between low and high, there by generating a train of pulse, which is why it is known as pulse generator. It is a best example for a perfect square wave generator. They are used as an inverter and also used in many of the internal part of the radio. Selecting a Thermistor as a timing resistor allows the use of the 555 in a temperature sensor.

Monostable Mode: In the monostable mode, as the name suggests, it stays in its stable state until and unless an external trigger is applied. In this mode, the 555 functions as a "one-shot" pulse generator. The best application of a monostable is to introduce a time delay in to a system. Applications comprise many things viz., timers, missing pulse detection also included bounce free switches, touch switches as well as frequency divider, capacitance measurement and pulse-width modulation (PWM) and many more.

Bistable Mode: In the bistable mode, the IC 555 acts as a flip-flop as it has two stable states. It can be used to store 1-bit of data. It is not a great choice for implementing a flip-flop.

Pin Configuration of 555 Timer: The 555 Timer is available in 8-pin Metal Can Package, 8-pin Mini Dual in-line Package (DIP) and 14-pin DIP. The 14-pin DIP is IC 556 which consists of two 555 timers. The pin out diagrams of 555 Timer in both the 8-pin packages is shown below. The names and numbers of all the pins along with their descriptions are tabulated below.

CIRCUIT DIAGRAM:

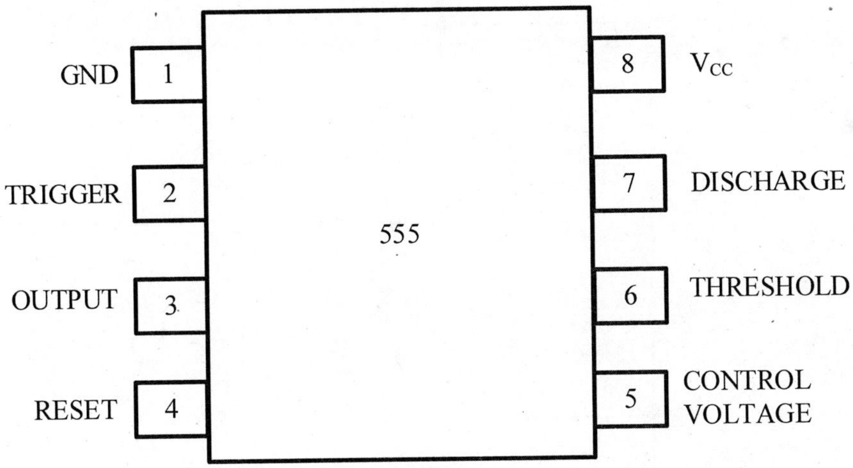

8 – Pin DIP

Pin		I//O	DESCRIPTION
S. NO.	NAME		
1.	GND	O	Ground Reference Voltage
2.	Trigger	I	Responsible for transition of SR flip-flop
3.	Output	O	Output driven waveform
4.	Reset	I	A negative pulse on reset will disable or reset the timer
5.	Control Voltage	I	Controls the width of the output pulse by controlling the threshold and trigger levels
6.	Threshold	I	Compares the voltage applied at the terminal with a reference voltage of 2/3
7.	Discharge	I	Connected to open collector of a transistor which discharges a capacitor between intervals.
8.	Vcc supply	I	Supply voltage

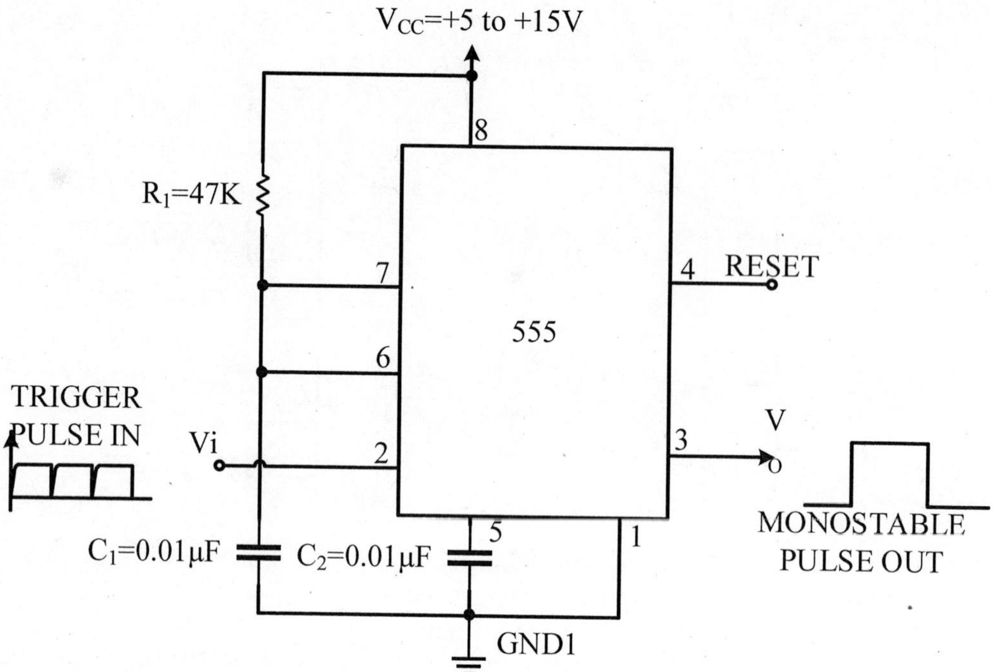

Fig. 1. Monostable multi vibrators

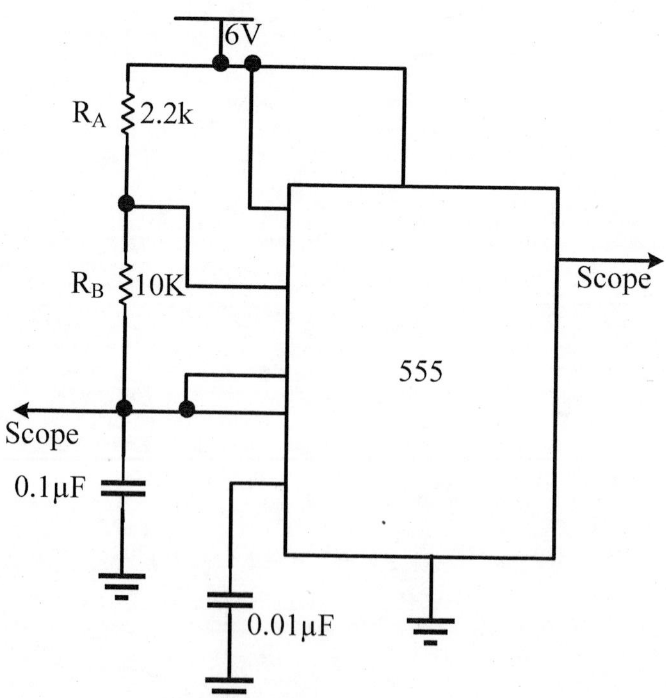

Fig. 2. Astable multi vibrator

PROCEDURE:

Monostable Multivibrator:

1. Connect the circuit as shown in the circuit diagram Fig 1.
2. Apply Negative triggering pulses of frequency 1 KHz at pin 2.
3. Observe the output waveform at pin 3 and measure capacitor voltage across it at pin 6.
4. Theoretically calculate the pulse duration as $T = 1.1R_1C_1$.
5. Compare it with experimental values.
6. Plot the graph for the input and output waveforms.

Astable Multivibrator:

1. Connect the circuit as shown in the circuit diagram Fig 2.
2. Observe the output waveform at pin 3 and measure capacitor voltage across it at pin 6.
3. Theoretically calculate the Time period as $T = 0.69\ R_BC + 0.69\ (R_A + R_B)C$.
4. Compare it with experimental values.
5. Plot the graph for the input and output waveforms.

MODEL WAVE FORMS:

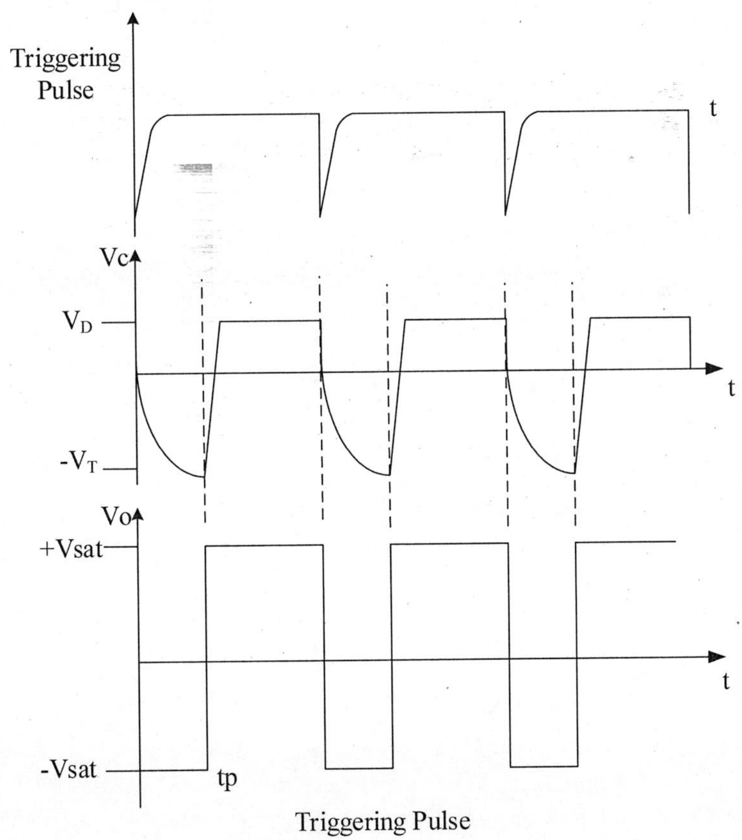

Fig. 3. Monostable multivibrator

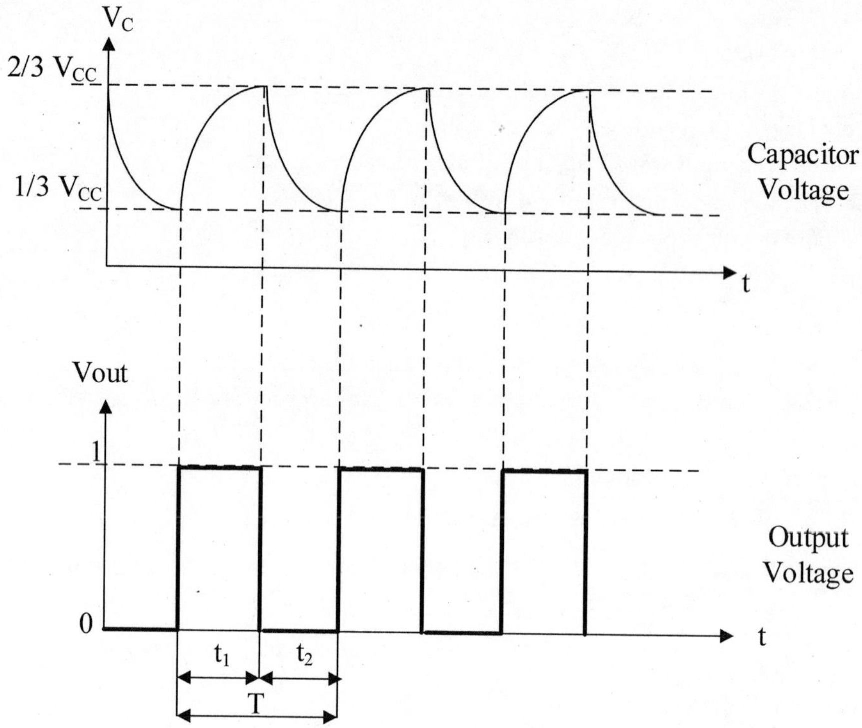

Fig. 4. Astable multivibrator

RESULT:

Designed and verified the waveforms of monostable multivibrator and astable multivibrator using 555 Timer.

EXPERIMENT NO. 13

AIM: To study Precision Rectifier Circuit using Op-amp.

APPARATUS REQUIRED: Dual power supply, CRO, function generator, bread board, op- amp, diodes, and resistors.

THEORY: In a normal diode rectifier, the cut in voltage across the diode will reduce output voltage and inaccuracy of rectification. If an ideal rectifier is needed in an application, a precision rectifier, as shown Fig. 1, may be used. In the circuit, when the input is greater than zero, D_1 will conduct and D_2 is OFF, so the output is zero because the other end of R_2 is connected to the virtual ground and there is no current through R_2. When the input is less than zero, D_2 is ON, D_1 is OFF, and the output is similar to an inverting amplifier with gain = R_2/R_1.

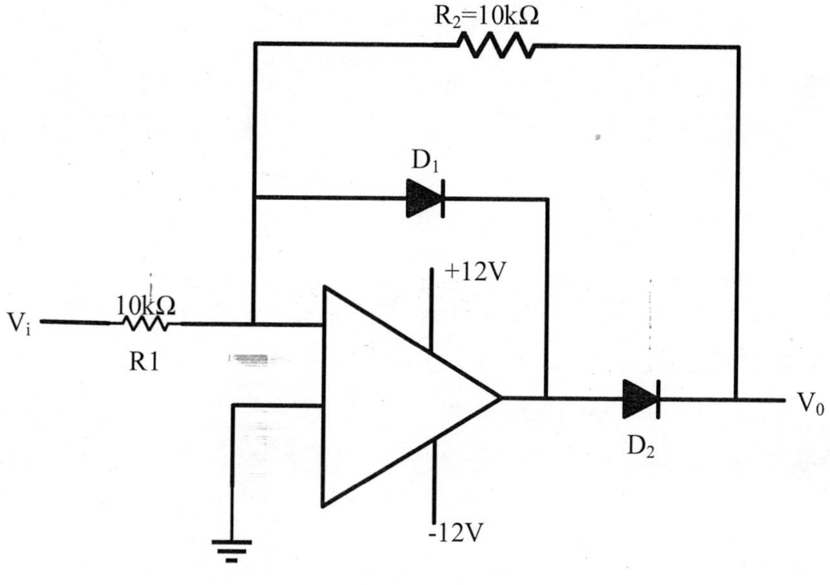

Fig. 1. Circuit Diagram of Precision Rectifier

The value of R_1 and R_2 are selected so that the circuit has reasonable level of input impedance and the gain is unity. Diode D_1 and D_2 are signal diodes.

PROCEDURE:

1. Set up the circuit as shown in Fig.1. Give a sine wave of ±5V peak magnitude and 1 kHz frequency at the input and observe the input and output simultaneously on CRO.
2. Put the CRO into X-Y mode and connect the input signal to X and the output signal to Y. Select a suitable volt per division in both channels and observe the characteristics. The display should look similar to Fig 3.

WAVEFORMS:

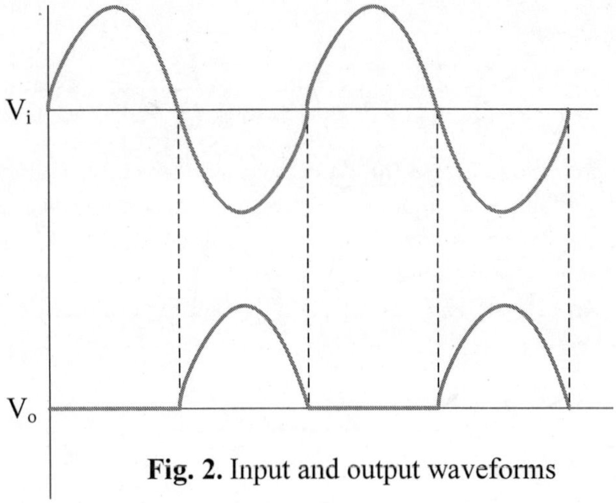

Fig. 2. Input and output waveforms

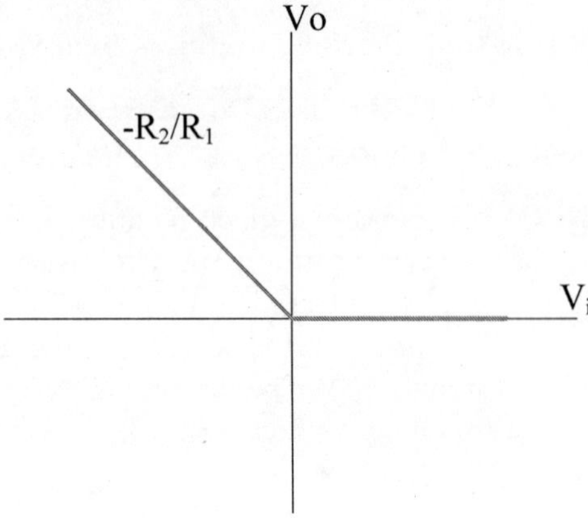

Fig. 3. Transfer characteristics

RESULT:

Designed and verified the waveforms of Precision Rectifier.

EXPERIMENT NO. 14

AIM: To study D/A and A/D converter using Op-Amp

APPARATUS REQUIRED: Patch chord, Digital Multimeter, DAC Trainer kit, ADC Kit,

THEORY:

D/A Converter: A D/A converter, also known as a digital-to-analog converter, transforms a digital code into an analog signal. It takes a binary input and generates an analog output that represents the original digital information. There are two types D/A converter

a. weighted resistor D/A converter
b. D/A converter with ladder N/W.

a. Weighted Resistor D/A Converter:

In a weighted resistor D/A converter, the binary weighted currents are summed using an operational amplifier (op-amp). Each bit of the digital input is associated with a specific resistor that determines the weight of that bit. By combining these weighted currents, the converter produces an output voltage proportional to the digital input value. The weighted resistor D/A converter typically requires a wide range of resistor values for better resolution.

b. D/A Converter with Ladder Network:

In a D/A converter with a ladder network, the reference voltage is applied to one position of a switch, while the other switch positions are connected to ground. The ladder network consists of resistors with values of R and 2R (R-2R ladder network is commonly used). By selectively closing switches based on the binary input, the ladder network generates an analog output voltage. This type of D/A converter can be expanded by adding more ladder sections to increase the number of bits.

A/D Converter:

An A/D converter, also known as an analog-to-digital converter, converts an analog signal into a digital representation. It involves two main steps: quantization and coding.

Quantization: Quantization is the process of dividing the continuous range of analog signal values into a set of discrete levels. The input analog signal is sampled and rounded to the nearest quantization level. The resolution of the A/D converter determines the number of possible quantization levels.

Coding: After quantization, the discrete levels are assigned digital codes. The digital output of the A/D converter represents the quantized levels and is typically in binary form. The number of bits in the digital output determines the resolution of the A/D converter and the range of possible output values. The quantum (Q) represents the smallest step or increment in the digital output. It is calculated as the full-scale range (FSR) divided by 2n, where n is the number of bits in the A/D converter.

$$Q = FSR/N = \text{Full Scale Range}/2n$$

CIRCUIT DIAGRAM:

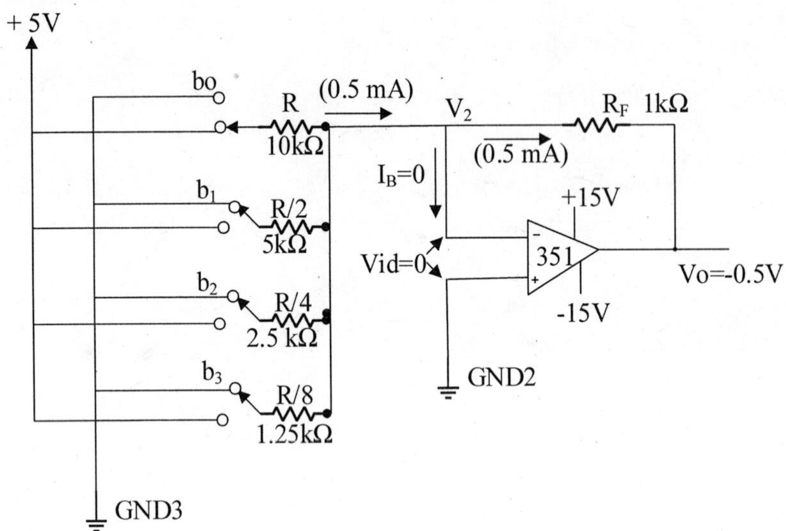

Fig. 1. D/A converter

PROCEDURE:

(D/A converter)

Functional verification of a weighted resistor D/A converter.

1. Make the connection as shown in circuit diagram of Fig. 1.
2. Connect the power supply to the board.
3. Connect the logic switches to the corresponding jacks B0 — B3 of the converter.
4. Set the switches So — S3 to logic level 0.
5. Connect the Vref socket to +5 volts.
6. Connect a multimeter as a voltmeter for DC, to the output V0 of the converter.
7. Switch the logic switches in binary progression & measure & record the output voltage in correspondence of every Combination of the input code.

A/D SIGNAL CONVERTER:

1. Make the connection as shown in Fig.2.
   a. Connect the variable DC supply to the Vi of the converters.
   b. Keep the DC variable in counter-clockwise position.
   c. Place the Reset / Count switch in reset position.
2. Connect the power supply to the board.
3. Connect a multimeter in a DC.voltage measurement mode, to the input Vi of the converter.
4. Set the Reset / Count switch to count position
5. Rotate the pot connected to +5V DC from initial position i.e. ; 0V in clockwise direction to max. voltage i.e. +5V.
6. Read and measure and record the output analog voltage display on.

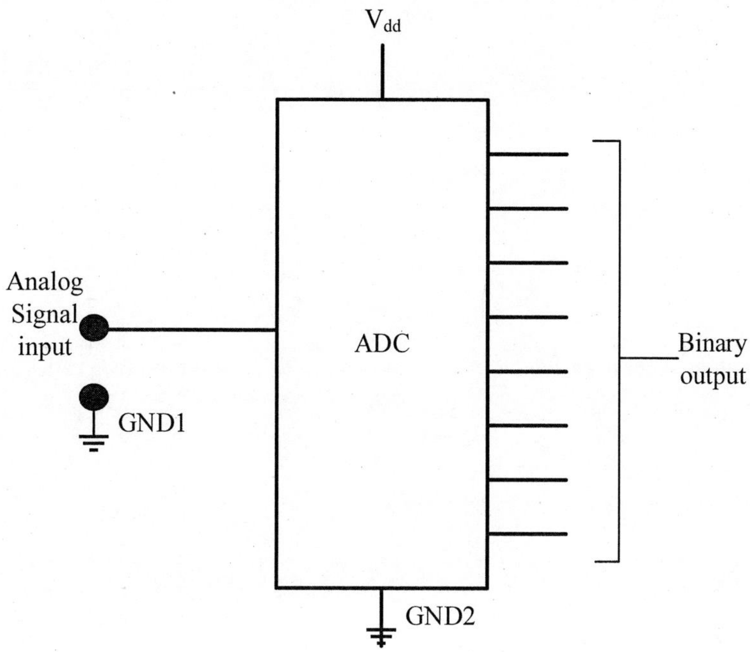

Fig. 2. A/D Converter

OBSERVATION TABLE: [D/A Converter]

S.No.	S3	S2	S1	S0	V_O(volts)
1.	0	0	0	0	
2.	0	0	0	1	
3.	0	0	1	0	
4.	0	0	1	1	
5.	0	1	0	0	
6.	0	1	0	1	

OBSERVATION TABLE: [A/D Converter]

S.No.	Practical Values	S3	S2	S1	S0
1.	0				
2.	0.30				
3.	0.45				
4.	0.88				
5.	1.14				
6.	1.6				

7.	2.				
8.	2.6				
9.	3.0				
10.	3.5				
11.	4.0				
12.	4.5				

RESULT:

The D/A and A/D converter is studied and observations are recorded in observation table.

EXPERIMENT NO. 15

AIM:–To study square waveform, and triangular waveform generator using Op-Amp.

APPARATUS REQUIRED: D.C. supply, IC741, resister, multimeter and connecting wires.

THEORY:

Square wave oscillator: The basic square wave oscillator is based on the charging and discharging of a capacitor. Op-amps inverting input is the capacitor voltage and the non-inverting input is a portion of the output fed back through resistors R_1 and R_2. When the circuit is first turned on, the capacitor is uncharged, and thus the inverting input is at 0V. This makes the output a positive maximum, and the capacitor begins to charge towards voltage at Vo through resistor R. When the capacitor voltage reaches a value equal to the feedback voltage (V_f) on the non-inverting input, the op-amp switches to the maximum negative state. At this point, the capacitor begins to discharge from $+V_f$ towards $-V_f$. When the capacitor voltage reaches $-V_f$, the op-amp switches back to the maximum positive state. This action repeats and a square wave output voltage is obtained.

Expression for period is

$$T = 2RCIn\frac{1+\beta}{1-\beta} \quad \text{where } \beta = \frac{R_2}{R_1 + R_2}$$

If $R_1 = R_2$, the equation for period reduces to T=2RCln3

$$\text{The frequency of oscillation, } f = \frac{1}{2RCIn3}$$

Triangular wave oscillator: This circuit (Fig. 2) uses two operational amplifiers. Op-amp A_1 functions as a comparator, and the op-amp A_2 as an integrator. Comparator compares the voltage at point 'P' continuously with respect to the voltage at the inverting input, which is at ground potential. When the voltage at P goes slightly below zero, the output of A_1 will switch to negative saturation. Suppose the output of A_1 is at positive saturation $+V_{sat}$. Since this voltage is the input of the integrator, the output of A_2 will be a negative going ramp. Thus, one end of the voltage divider R_1-R_2 is at $+V_{sat}$ and the other at the negative going ramp. At time t=t1, when the negative going ramp attains a value of $-V_{ramp}$ the effective voltage at point 'P' becomes slightly less than 0V. This switches the output of A1 from positive saturation to negative saturation level $-V_{sat}$. During the time when the output of A1 is at $-V_{sat}$, the output of A_2 increases in a positive direction. At the instant t=t2, the voltage at point P becomes just above 0V, thereby switching the output of A_1 from $-V_{sat}$ to $+V_{sat}$. The cycle repeats and generates a triangular waveform.

At $t=t_1$

$$\frac{-V_{ramp}}{R_2} = -\frac{V_{sat}}{R_1} \text{ ie. } -V_{ramp} = -\frac{R_2}{R_1}(+V_{sat})$$

Similarly, at $t=t_2$

$$+V_{ramp} = -\frac{R_2}{R_1}(-V_{sat})$$

The peak-to-peak output of the triangular wave is

$$Vo_{(pp)} = +V_{ramp} - (V_{ramp}) = 2\,\frac{R_2}{R_1}\,(V_{sat})$$

During the period 0-t_1, The integrator functions as below.

$$?V_{o(pp)} = \frac{1}{RC}\int_0^{\frac{T}{2}}(-V_{sat})\,dt = \left(\frac{V_{sat}}{RC}\right)\left(\frac{T}{2}\right)f$$

Then

$$T = 2RC\left(\frac{Vo_{(pp)}}{V_{sat}}\right)$$

Substituting for $Vo_{(pp)}$

$$T = \frac{4RCR_2}{R_1}$$

Then, frequency of oscillation,

$$f = \frac{R_1}{4RCR_2}$$

DESIGN AND CIRCUIT DIAGRAMS:

Design of square wave generator

Let the frequency of oscillation be 1kHz

Take $\beta = 0.5$ and R1 = R2 = 10kΩ

Frequency, $f = \dfrac{1}{2RC\,\text{In}3}$ Assume C = 0.1μF

Then,

$$R = \frac{1}{2Cf\,\text{In}3} = \frac{1}{2\times 0.1\times 10^{-6}\times 1000\times \text{In}3} = 4.55\Omega$$

Select standard value of 4.7kΩ for R.

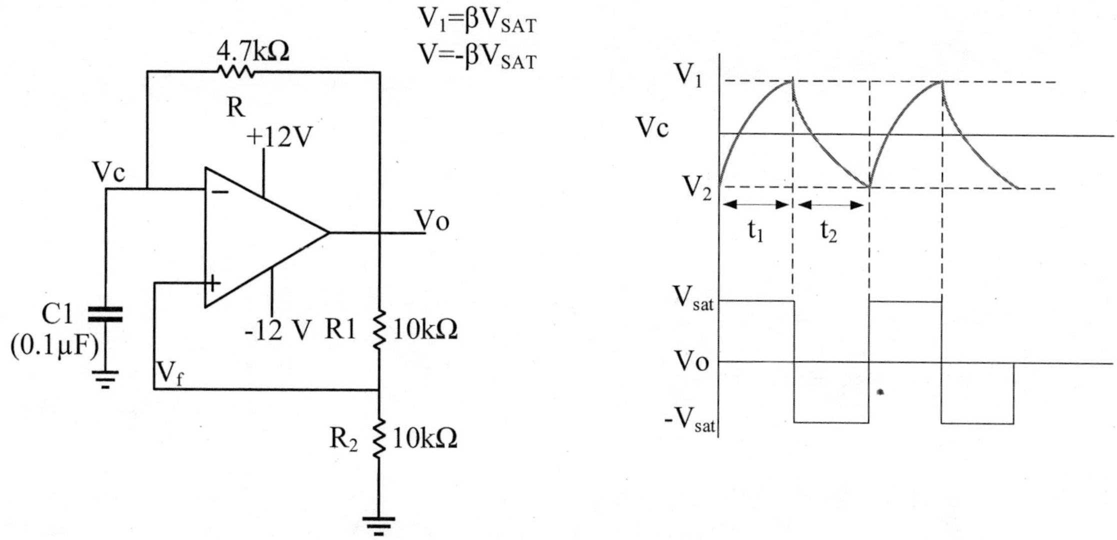

Fig. 1. Square wave generator and waveforms

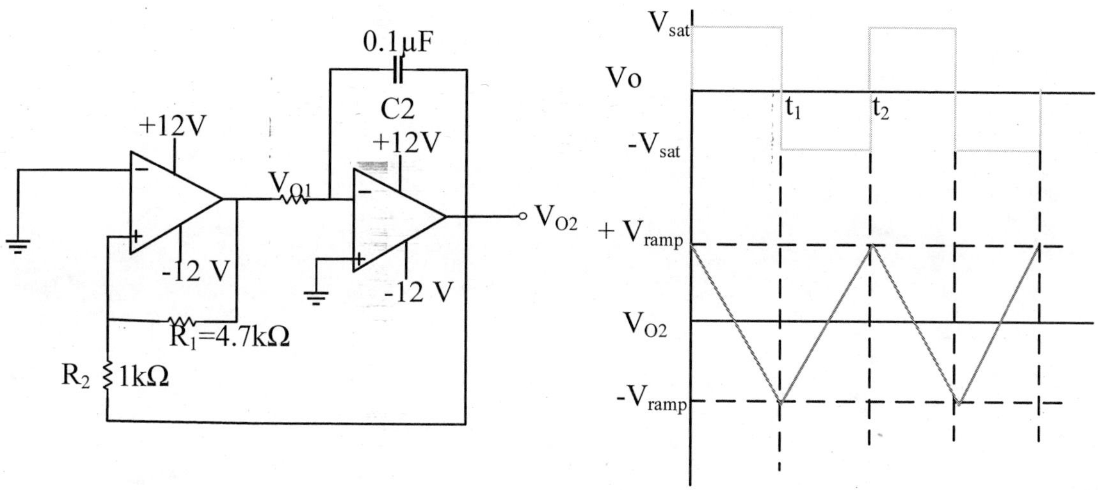

Fig. 2. Triangular wave generator and waveforms

Design of triangular wave generator

Let the frequency of oscillation be 1kHz

We have $f = \dfrac{R_1}{4RC\,R_2}$ $\quad V0_{(pp)} = 2\dfrac{R_2}{R_1}V_{sat}$

Since supply voltage is ±12V, V_{sat} will be approximately 10V

Let $Vo_{(pp)}$ be 5V; Assume $R_2 = 1k\Omega$

Then

$$R_1 = \frac{2V_{sat}}{V_{O(pp)}} R_2 = \frac{2 \times 10}{5} \times 10^3 = 4k\Omega$$

Select standard value, $R_1 = 4.7k\Omega$

Assume $C = 0.1\mu F$

$$R = \frac{R_1}{4fCR_2} = \frac{4.7 \times 10^3}{4 \times 1000 \times 0.1 \times 10^{-6} \times 1 \times 10^3} = 11.7k\Omega$$

Select standard value, $R = 12k\mu$

PROCEDURE:

1. Set up the circuit after testing the components.
2. Set up the square wave generator as shown in Fig. 1, observe the output waveform, and note their amplitudes and frequencies.
3. Set up the triangular wave generator as shown in Fig. 2 and observe the variation in frequencies of output waveform by varying the values of resistances R_1, R_2, and R_3.
4. Move the wiper of the potentiometer in both directions and observe the changes taking place in the waveform.

RESULT:

Circuits of square wave generator and triangular wave generator are designed, setup and waveforms observed.

CO AND PO ATTAINMENT TABLE

Course outcomes (COs) for this course can be mapped with the programme outcomes (POs) after the completion of the course and a correlation can be made for the attainment of POs to analyze the gap. After proper analysis of the gap in the attainment of POs necessary measures can be taken to overcome the gaps.

Table for CO and PO attainment

Course Outcomes	Attainment of Programme Outcomes (1- Weak Correlation; 2- Medium correlation; 3- Strong Correlation)											
	PO-1	PO-2	PO-3	PO-4	PO-5	PO-6	PO-7	PO-8	PO-9	PO-10	PO-11	PO-12
CO-1												
CO-2												
CO-3												
CO-4												
CO-5												

The data filled in the above table can be used for gap analysis.

Index

A

Analog to Digital Converter, 145, 164

B

Biasing of the MOSFET, 97
Bipolar Junction Transistor (BJT), 40
BJT as a Switch, 43, 63

C

Clipper and clampers, 14
Common Base Configuration, 45, 46
Common Collector Configuration, 46, 51
Common Emitter Configuration, 51, 53
Common Gate 926
Current Mirror, 74

D

Differential Amplifiers, 120, 124
Differentiator Circuit, 152

F

Forward-biased PN Junction, 7

H

Hybrid Parameter Model, 66
Hysteresis Comparators 175, 177

I

IC 741, 134, 223
Ideal and Practical Op-amp, 172
Instrumentation Amplifier, 153
Integrator, 151
Internal structure, 129
Inverting and Non-Inverting Amplifiers, 128, 192

L

Lag Lead Compensator, 169, 170

M

Monostable Multivibrators, 174, 190
MOSFET, 92, 212

N

Negative Feedback, 215, 225

O

Op-amp as Band Pass Filters, 160
Op-amp as High Pass Filters, 157
Op-amp as Low Pass Filter, 155
Operating Point, 9
Wien Bridge Oscillator, 169

P

Peak Detector, 187
Piecewise Linearization,11
PN Junction Diode, 1
Precision Rectifiers, 174, 190
Proportional controller, 161

Q

Quiescent Point, 39, 53

R

Rectifiers, 185
Reverse-biased PN Junction, 4

S

Square-wave and Triangular-wave Generators, 174, 181,
Summing, 150, 228

U

Unbiased PN Junction Diode, 3

V

Voltage Divider Bias, 61
Voltage Follower, 149, 150
Voltage Regulator, 167

Z

Zener Diode, 22
Zero Crossing Detectors, 174, 175